Controlling Our Destinies

HISTORICAL, PHILOSOPHICAL, ETHICAL, AND THEOLOGICAL PERSPECTIVES ON THE HUMAN GENOME PROJECT

Edited by

Phillip R. Sloan

University of Notre Dame Press

Notre Dame, Indiana

Library of Congress Cataloging-in-Publication Data

Controlling our destinies : the Human Genome Project from historical,
philosophical, social, and ethical perspectives / edited by
Phillip R. Sloan.
     p.   cm. — (Studies in science and the humanities from the
Reilly Center for Science, Technology, and Values ; v. 5)
   Includes bibliographical references and index.
   ISBN 0-268-00818-3 (alk. paper). — ISBN 0-268-00820-5 (pbk. :
alk. paper)
   1. Human Genome Project.   I. Sloan, Phillip R.   II. Series.
QH445.2.C65   1997
573.2'1—dc20
                                          96-9755
                                          CIP

*For my father, Reid Sloan, who first awakened my
interest in the intellectual life, and for my mother, MaRee Sloan,
who taught me what dedication to ideals means.*

CONTENTS

ILLUSTRATIONS

CONTRIBUTORS

JOHN BEATTY is in the Department of Ecology, Evolution and Behavior, the Program in the History of Science and Technology, and the Minnesota Center for the Philosophy of Science at the University of Minnesota. He studied biology and chemistry at Tulane University and received his Ph.D. in History and Philosophy of Science at Indiana University. Recent publications include "Documenting the Human Genome Project: Challenges and Opportunities" (with Elizabeth Sandager), "Genetics in the Atomic Age: The Atomic Bomb Casualty Commission, 1947-1957," "Scientific Collaboration, Internationalism, and Diplomacy: the Case of the Atomic Bomb Casualty Commission," and "Dobzhansky and the Biology of Democracy: The Moral and Political Significance of Genetic Variation."

ROBERT BUD is Head of Research (Collections) and Head of Life and Communications Technologies, at the Science Museum London and Co-Director of the South Kensington Institute for the History of Technology. He earned his Ph.D. in History and Sociology of Science at the University of Pennsylvania. His publications include *The Uses of Life: A History of Biotechnology, Chemistry in America 1876-1976, Science versus Practice: Chemistry in Victorian Britain,* "Bioengineering in Britain," "100 Years of Biotechnology: Events and Publications that Moulded the Industry's Place in Society," and "Molecular Biology and the Long-term History of Biotechnology." He has recently co-edited with Deborah Warner *Instruments of Science,* and *Manifesting Medicine: Bodies and Machines* with Bernard Finn and Helmuth Trischler.

ARTHUR L. CAPLAN is Trustee Professor of Bioethics and Director of the Center for Bioethics at the University of Pennsylvania. He received his Ph.D. in Philosophy from Columbia University. He has served as Director for the Center for Biomedical Ethics at the University of Minnesota, President of the American Association of Bioethics, and Associate Director of the Hastings Center. His most recent book is *Am I My Brother's Keeper?* He has published extensively and has been featured in numerous press accounts on bioethical issues. He co-edited (with D. Bartels and B. LeRoy) *Prescribing Our Future: Ethical Challenges in Genetic Counseling.* Other contributions to medical genetics include "If Gene Therapy is the Cure, What is the Disease?," "Neutrality is Not Morality: The Ethics of Genetic Counseling," and "Handle with Care: Race, Class, and Genetics," and "Non-Directiveness in Genetic Counseling: A Survey of Practioners."

ALICE DOMURAT DREGER is Assistant Professor of Science and Technology Studies at the Lyman Briggs School, a residential liberal arts program for science undergraduates in the College of Natural Science at Michigan State University. She is also adjunct faculty in the Center for Ethics and Humanities in the Life Sciences at Michigan State University. She received her Ph.D. in 1995 from the Department of History and Philosophy of Science at Indiana University, and her dissertation research, funded by a Charlotte Newcombe fellowship from the Woodrow Wilson Foundation, forms the basis for her first book, *Hermaphrodites and the Medical Invention of Sex.* Her work centers on the role science, medicine, and technology play in mediating the relationships between anatomy and identity.

KEVIN FITZGERALD, S. J. is a Research Associate in Department of Medicine and Medical Humanities Program at the Loyola University Medical Center in Chciago. He received his Ph.D. in molecular genetics from Georgetown University, and is presently completing a doctorate in bioethics, also at Georgetown. The two principal foci of his research efforts are the investigation of abnormal gene regulation in cancer and ethical issues in medical genetics. Recent publications include: "Proposals for Human Cloning: A Review and Ethical Evaluation," and "Human Cloning: Analysis and Evaluation."

JEAN-PAUL GAUDILLIERE is a researcher in history of biology and medicine at the *Institut national de la santé et de la recherches médicale* in Paris. He holds degrees in biochemistry, from the Institute National Agronomique, and in history of science, from the Université Paris VII. His doctoral thesis dealt with the reorganizaion of biomedical research in postwar France focussing on the development of molecular biology. He has held postdoctoral fellowships at Cambridge and MIT, studying the history of molecular biology and cancer research. His is currently working on a comparative history of the relations between genetics, virology and cancer research. Recent publi-cations include "Messenger RNA and the Biochemists: The Birth of a Scientific Network," "NCI and the Spreading Genes: About the Production of Mice, Viruses, and Cancer," and "The Genome Negotiations." His forthcoming book *L'Invention de la* *biomédecine: La Reconstruction des sciences biologiques et de la santé dans la France d'après 1945* will appear this year.

JEAN GAYON is Professor of Epistemology and History of Life and Medical Sciences at the University of Paris 7-Denis Diderot. He is a member of the *Centre de recherches en epistémologie et histoire des sciences exactes et des institutions scientifiques* (REHSEIS) and was elected as senior member of the *Institut universitaire de France* in 1994 for his work in history and philosophy of biology. He has co-edited (with Claude Debru and Jean-François Picard) *Les Sciences biologiques et médicales en France, 1920-1950*. Other publications include "The Singular Fate of Genetics in the History of French Biology, 1900-1940," "Boris Ephrussi and the Synthesis of Genetics and Embryology," "Génétique et recherche médicale en France au XXe siècle," and the chapter on eugenics in the *Précis de génétique humaine*. His most recent book is *Darwinism's Struggle for Survival* .

LILY E. KAY, formally an Associate Professor at MIT, is affiliated with Harvard University. Her undergraduate degree was in physics and mathematics from New York University. She has done postgraduate work in the life sciences, and has done research in biochemistry and in the molecular biology of viruses. She holds a Ph.D. in History of Science from the Johns Hopkins University. Her studies in the history of science have focused primarily on the history of molecuar biology. She is the author of the books *Cells, Molecules, and Life: An Annotated Bibliography of Manuscript Sources on Physiology, Biochemistry, and Biophysics, 1900–1960*, *The Molecular Vision of Life: Caltech, the Rockefeller Foundation, and the Rise of the New Biology*, and *Who Wrote the Book of Life? a History of the Genetic Code* (forthcoming, 2000). Her most recent articles include "A Book of Life? How the Genome Became an Information System and DNA a Language," "Cybernetics, Information, Life: The Emergence of Scriptural Representations of Heredity," "Philanthropy as an Historiographic Problem of Knowledge and Power." She is currently a Guggenheim Fellow working on a book on the history of neural nets.

EVELYN FOX KELLER is a MacArthur Fellow and Professor at MIT, Program in Science, Technology, and Society. She took her Ph.D. in physics from Harvard University and has been granted honorary degrees from Rensselaer Polytechnic Institute, University of Amsterdam, Lulea University, Simmons College, and Mt. Holyoke College. Her work relating to the history and philosophy of biology includes *A Feeling for the Organism: The Life and Work of Barbara McClintock*, *Reflections on Gender and Science*, *Secrets of Life, Secrets of Death: Essays on Language, Gender, and Science*, and *Refiguring Life: Metaphors of Twentieth Century Biology*.

PHILIP KITCHER, Presidential Professor of Philosophy at University of California San Diego, earned his B.A. with first class honours in Mathematics and the History and Philosophy of Science from Christ's College, Cambridge and his Ph.D. in Philosophy at Princeton University. He has worked extensively in the philosophy of biology. He has been a Senior Fellow of the Library of Congress Bio-Ethics Issues in Molecular Genetics, and is currently serving as Editor-in-Chief of *Philosophy of Science*. His books include *Abusing Science: The Case Against Creationism*, *The Nature of Mathematical Knowledge*, *Vaulting Ambition: Sociobiology and the Quest for Human Nature* (for which he received the Imre Lakatos Award), *The Advancement of Science*, and *Lives to Come: The Genetic Revolution and Human Possibilities*. From the fall of 1999 he will be Professor of Philosophy at Columbia University.

TIMOTHY LENOIR is Professor of History and Co-Chair of the Program in History and Philosophy of Science at Stanford University. His doctoral degree was from the Department of the History and Philosophy of Science, Indiana where he worked on Newton's mathematics. His subsequent work has ranged from the history of mathematics to comparisons of the American and German genetics communities to studies on discipline formation. His current work treats the introduction of computing into biomedicine. His publications include *The Strategy of Life: Teleology and Mechanics in Nineteenth Century German Biology*, *Politik im Tempel der Wissenschaft: Forschung und Machtausübung im deutschen Kaiserreich*, *Instituting Science: The Cultural Production of Scientific Disciplines*, "Science for the Clinic: Science Policy and the Formation of Carl Ludwig's Institute in Leipzig," and "The Discipline of Nature and the Nature of Disciplines." He has most recently edited *Inscribing Science: Scientific Texts and the Materiality of Commu-nication*.

EDWARD MANIER is Professor of Philosophy and in the Program in History and Philosophy of Science at Notre Dame. He is a Fellow of the John J. Reilly Center for Science, Technology and Values. He organized the fifth Reilly Center Conference, "Neurobiology and Narrative." Recent publications include "Reductionist

Rhetoric: Expository Strategies and the Development of the Molecular Neurobiology of Behavior," and "Walker Percy: Language, Neuropsychology, and Moral Tradition."

RICHARD A. MCCORMICK, S. J. is emeritus John A. O'Brien professor of Christian Ethics in the Department of Theology at the University of Notre Dame, and formerly served as Rose F. Kennedy Professor of Christian Ethics at the Kennedy Institute of Ethics, Georgetown University. Among other professional acti-vities, he has served as president of the Catholic Theological Society of America, past member of the Ethics Advisory Board, Department of Health, Education and Welfare, as a Fellow of the Institute of Society, Ethics and the Life Sciences of the Hastings Center, and is a member of the Editorial Board of the *Journal of Contemporary Health Law and Policy*. His primary publications include *Readings in Moral Theology I-VIII*, *Health and Medicine in the Catholic Tradition*, *How Brave a New World*, and *The Critical Calling*.

ERNAN MCMULLIN is Professor Emeritus of Philosophy and former Director of the Reilly Center for Science, Technology and Values at Notre Dame. He has also taught at the University of Minnesota, University of Cape Town, UCLA, Princeton, and Yale. He holds degrees in physics, theology, and philosophy. A distinguished philosopher and historian of science, he has received many honors including election as Fellow of the American Academy of Arts and Sciences, the International Academy of the History of Science, and AAAS; the Aquinas Medal of the American Catholic Philosophical Association; and honorary doctorates from Loyola University (Chicago) and the National University of Ireland. He has written and edited numerous works including *The Concept of Matter, Galileo: Man of Science, Newton on Matter and Activity, The Social Dimensions of Science* and *The Inference that Makes Science*.

TIMOTHY MURPHY is Associate Professor and Head of the Medical Humanities Program at the University of Illinois College of Medicine in Chicago. He co-edited *Justice and the Human Genome Project* (with Marc Lappé) and has written extensively on sexual orientation research and ethical issues associated with AIDS. Other recent publications include *Ethics in an Epidemic: AIDS, Morality, and Culture*, "Abortion and the Ethics of Genetic Sexual Orientation Research," and "Genome Mapping and the

Meaning of Difference." His most recent book is *Gay Science: The Ethics of Sexual Orientation Research.*

JOHN M. OPITZ received his M.D. degree from the University of Iowa, and served on the faculty of the University of Wisconsin in pediatrics and medical genetics before becoming the Coordinator of the Shodair Montana Regional Genetics Service Program, and later Chairman of the Department of Medical Genetics at Shodair Children's Hospital in Helena, Montana. Presently he is Professor of Pediatric Medicine at the University of Utah School of Medicine. He was the founder of the *American Journal of Medical Genetics*, which he continues to edit. He is a founding member of the American Board and the American College of Medical Genetics and is board certified in pediatrics and medical genetics. He has specialized as a "syndromologist," but prefers to be known as a pediatric zoologist with strong interests in developmental biology and the relationship between development and evolution. He is the author of 332 papers, and thirteen textbooks chapters, included in such works as *Blastogenesis, Normal and Abnormal, Recent Advances in Ectodermal Dysplasias, Topics in Pediatric Genetic Pathology, The Developmental Field Concept,* and *X-Linked Men-tal Retardation I, II, III.*

DIANE PAUL is Professor of Political Science and Co-Director of the Program in Science, Technology, and Values, University of Massachusetts at Boston. She was a Visiting Professor in the Zoology Department at the University of Otago, Dunedin, New Zealand, a Fellow at the *Wissenschaftskolleg zu Berlin*, a Visiting Scholar and Research Associate at the Museum of Comparative Zoology at Harvard University, and an Exxon Fellow at the Program in Science, Technology, and Society at MIT. She has written on genetics, eugenics, and politics. Her publications include: *Controlling Human Heredity, 1865 to the Present, The Politics of Heredity,* "Eugenic Anxieties, Social Realities, and Political Choices," "Eugenics and the Left," "PKU Screening: Theory and Practice."

ARTHUR PEACOCKE followed an academic scientific career for over twenty-five years in the Universities of Birmingham and Oxford in the field of the physical chemistry of biological macromolecules (particularly DNA) and was awarded a D.Sc. degree by Oxford University. His scientific publications include 125 papers and three books, and his most recent scientific work is *The Physical Chemistry of Biological Organization.* His principal interest for the last twenty-two years has been in the relation of science and theology and associated questions in the philosophy of science and theology. He was awarded a D.D. degree by Oxford and is an ordained priest in the Church of England. His publications include *Intimations of Reality: Critical Realism in Science and Religion, God and the New Biology, Theology for a Scientific Age*, and *God and Science: A Quest for Christian Credibility*

MARTIN S. PERNICK is professor of History and Associate Director of the Program in Society and Medicine at the University of Michigan. He studies the history of value issues in medicine, and the relation between medicine and mass culture. He received his Ph.D. in American history from Columbia University, and has taught at Harvard University and at Penn State University's Hershey Medical Center. He is the author of two books: *A Calculus of Suffering*, on professional and cultural attitudes towards pain in 19th century America; and *The Black Stork*, on eugenic euthanasia in America. His current research projects include the history of uncertainty about the definition of death since the 1700s, and the role of motion pictures in early 20th century public health.

HANS-JÖRG RHEINBERGER is a Director at the Max-Planck-Institut für Wissenschaftsgeschichte in Berlin. He earned his Ph.D. in molecular biology at the Freie Universität Berlin and worked at the Max-Planck-Institute for Molecular Genetics, Berlin-Dahlem. He has held positions at the Free University of Berlin, University of Innsbruck, Austria, the Medical University of Lübeck, the University of Maryland, College Park, the University of Göttingen, the University of Salzburg, and the Institute for Advanced Study in Berlin. His recent work has focused on the history of molecular biology and the epistemology of experimentation in the life sciences. His publications in the field include "The 'Epistemic Thing' and Its Technical Conditions: from Biochemistry to Molecular Biology," "Experiment and Orientation: Early Systems of *in vitro* Protein Synthesis," "Experimental Systems: Historicality, Narration, and Deconstruction," *Experiment, Differenz, Schrift. Anmerkungen zur Geschichte epistemischer Dinge*. His most recent book is *Toward a History of Epistemic Things: Synthesizing Proteins in the Test Tube*.

KENNETH SCHAFFNER is University Professor of Medical Humanities at George Washington University. He earned a Ph.D. in Philosophy from Columbia University and an M.D. from the University of Pittsburgh. While at the University of

Pittsburgh, he served as Chairman for the Department of History and Philosophy of Science, Co-Chairman for the Program for Human Values in Health Care, and Co-Director of the Center for Medical Ethics. His publications include "Approaches to Reduction," "The Watson-Crick Model and Reductionism," "The Peripherality of Reductionism in the Development of Molecular Genetics," "Molecular Genetics, Reductionism, and Disease Concepts in Psychiatry," "Genes, Behavior, and Developmental Emergentism: One Process, Indivisible?" and *Discovery and Explanation in Biology and Medicine.*

PHILLIP R. SLOAN received his training in biology and Marine Biology, specializing in deep-sea biology and plant physiology. His doctoral work in history and philosophy of science was completed at the University of California at San Diego. Since 1974 he has been in the Program of Liberal Studies and the Program in History and Philosophy of Science at the University of Notre Dame, and directed the HPS program from 1994–97. His research work has been in the history of the life sciences in the eighteenth and nineteenth centuries, with publications on Buffon (*From Natural History to the History of Nature* with John Lyon), the human sciences in the Enlightenment ("The Gaze of Natural History"), Darwin ("Darwin, Vital Matter and the Transformism of Species"), and most recently Richard Owen (*Richard Owen's Hunterian Lectures May-June 1837).* He is currently Director of the Reilly Center for Science, Technology and Values at Notre Dame.

JOHN STAUDENMAIER, S. J. is a member of the Society of Jesus, a Professor of History of Technology at the University of Detroit Mercy, and Editor of *Technology and Culture,* the journal of the Society for the History of Technology. He holds degrees from the St. Louis University and the University of Pennsylvania and is an ordained Roman Catholic priest. He has been a Fellow at the Dibner Institute for the History of Science and Technology at MIT and has served as Visiting Faculty for the Program in Science, Technology, and Society at MIT. He currently serves on the board of the Michigan Biotechnology Institute. His publications include *Technology's Storytellers: Reweaving the Human Fabric,* "Rationality vs. Contingency in the History of Technology," "U.S. Technological Style and the Atrophy of Civic Commitment," and "The Politics of Successful Technologies."

ACKNOWLEDGMENTS

Organizing and executing a conference of the range and complexity of the Notre Dame gathering required the help of many people. Primary funding for the conference was obtained from the Department of Energy ELSI Program (Grant# DE-FG02-95ER62049) and the Notre Dame John J. Reilly Center for Science, Technology and Values. Additional assistance was generously provided by the Offices of the Provost, the Dean of Science, the Dean of Arts and Letters, and the Vice President's Office of the University of Notre Dame. I wish to thank each of these supporting agencies. A leave from the University of Notre Dame during 1993–94 was also a considerable aid in the preparation of this conference. Very helpful advice and encouragement of Doris Zallen in the planning of the conference was deeply appreciated.

The Notre Dame steering committee—Richard McCormick, Edward Manier, Christopher Hamlin, David Solomon, David Hyde, and Harvey Bender—gave invaluable advice and assistance in designing this to be the kind of interdisciplinary conference the topic required. I also wish to thank Ms. Laurie Echterling, my administrative assistant during the planning period, for considerable service in the organization of this conference. My graduate research assistant, Amy England-Beery, and my undergraduate assistant, Kelly Puzio, rendered considerable service in the execution of the conference.

I also wish to acknowledge the hospitality extended by Raymond Gesteland, Diane Dunn, Jeffrey Botkin, and Peter Cartwright of the Eccles Center for Human Genetics of the University of Utah. Their receptiveness to my questions and their interest in the project underlined the importance of developing further productive discussions between humanists and scientists on these issues. Mrs. Dorothy Dart of the Eccles Center was also very helpful with audio-visual materials.

I also wish to thank Drs. Leroy Hood, Maynard Olson and Matthew Huang of the Institute of Biotechnology at the University of Washington for their hospitality during a site visit to their mass-sequencing laboratory. Without this exposure to the actual workings of major human genetics centers in action, my understanding of this enterprise would have been considerably less.

My debt must also be acknowledged to Drs. Mortimer Mendelsohn and Frank Church for very helpful responses to questions surrounding the important Alta Utah Conference of 1984.

For assistance in the complex task of manuscript preparation I wish to thank especially my administrative assistant, Mrs. Vicki McMahan for superb assistance during the final preparation stages. My graduate student assistants Amy England-Beery, Patrick McDonald, and undergraduate assistants Sebastian Montufar and Bryan Waldron were all of assistance with manuscript preparation, bibliographical searches, and the preparation of the graphics. I acknowledge a considerable debt to my graduate assistant, Darin Hayton, for his considerable advice and assistance during the final preparation stages of the manuscript. I am also appreciative to Dr. Jeanette Morgenroth and the staff of the Notre Dame Press for their careful editing and assistance in the final publication. Harriet Baldwin and her excellent staff of the Center for Continuing Education were, as always, outstanding assistants in making this conference a success.

Finally I wish to thank my wife Sharon, always my best editor, critic, guide, and friend, for her support and encouragement in my undertaking such a project, considerably distant in many respects from early nineteenth-century biology.

Preface

In October of 1995, the University of Notre Dame hosted a conference exploring humanistic implications of the Human Genome Project (HGP). This was planned as an unusual conference that would bring together philosophers, historians, theologians, ethicists, and working scientists for discussion of the new emerging developments in human genetics. With the sponsorship of the acknowledged agencies, the conference was able to assemble an international group of scholars who spoke from their various disciplinary perspectives on this important scientific enterprise.

The Genome Project (or Initiative) constitutes a massive cooperative biotechnological enterprise whose goal is to provide by the year 2005 the basic sequence information and a locator map of the twenty-three chromosome pairs contained within the nucleus of the cells of the human being. Within four to five years from the publication of this collection of essays, and possibly within three (Wade 1998), it seems probable that the initial goal of this enterprise will have been achieved, although significant technical problems must still be resolved (Marshall 1995; Rowen, Mahairas, and Hood 1997). Conclusion of this initial phase of the HGP will also result in a similar determination of comparative genetic sequence information of at least four additional model organisms— the domestic mouse *Mus domesticus*, the round-worm *Caenorhabditus elegans*, sequenced in December, 1998; the yeast *Saccharomyces cerevisiae*, sequenced 1996; and the bacterium *Escherichia coli*. Completion of this information-gathering phase will expand knowledge of genetic science to unparalleled limits, supplying, in Eric Lander's characterization, a "periodic table" of the human genetic structure (*New York Times* 1996).

There are numerous medically-promising consequences that will flow from this work that have led all major industrial nations to join in on this massive research enterprise. Completion of the HGP in its initial sequencing phase, and the parallel development of the powerful biotechnology needed to accomplish it, promises to open up solutions to some of the great medical problems that face human beings at the end of the century. It will supply fundamental understanding of causes of the classic "genetic" diseases like Huntington's, Cystic Fibrosis, Fragile-X, Sickle Cell and Tay-Sachs syndrome; it promises to unravel many features of the elusive issue of cancer in many of its forms; it will likely assist in the solution to the catastrophic world-health problem of HIV. It can potentially defeat decisively the disease of tuberculosis. Genome research also has more theoretical implications, bearing on the study of evolution and the comparative relations of organisms in the present and in the past. It will considerably advance the understanding of the nature of fundamental cellular mechanisms. Rapid sequencing technology developed by the genome project may result in major conceptual breakthroughs in these areas as an outcome of the comparative aspects of the enterprise.

At the same time, issues presented by the Human Genome Project in specific, and by modern biotechnology in general, pose more ultimate questions about the nature of a human person, about the moral limitations on the technological manipulation of life, and about the definition of individual and social goods. Opposition to the genome project from the point of view of rational resource allocation was expressed by many thoughtful commentators at the outset of the HGP (Rechsteiner 1991; Lewontin 1992a; Tauber and Sarkar 1992). In a historical perspective, the history of the eugenics movement in the early decades of this century presents an ever-present concern about the possible practical social effects of a new form of eugenics (Kevles 1985; Paul 1996; Pernick, Caplan this volume). A tendency to accentuate only the positive benefits of this program in the official justifications of the HGP also concerns many observers, an issue explored by Alice Dreger in this volume. Ethical concerns about the potential applications of new genetic information have also been behind the unusual decision of the United States Congress to require that a significant portion of the genome budget specifically be devoted to an ongoing ethical assessment of the ethical and social implications of the HGP. But what it means to assess such an enterprise "ethically" is not always immediately clear.[1] Such assessment has tended to center on issues of insurance, genetic counseling, and topics more generally delimited by professional bioethics. Unquestionably important issues as they are, nevertheless, these only encompass a portion of the concerns we might have with the new domain of human genetic science.

For this reason, the Notre Dame conference intentionally assembled a broadly interdisciplinary group of scholars, many of whom had not worked specifically on questions of molecular and human genetics, but all of whom had considered some aspect of science and technology from historical, ethical, sociological, legal, or theological perspectives. The primary aim of the Notre Dame conference was to create a new kind of dynamic dialogue across disciplinary and philosophical lines that would generate a sustained academic, and ideally public, discussion of genetic science as it develops over the next decades. It was hoped that this would enable humanistic scholars from several disciplines to keep abreast of this fast-paced field, rather than being placed in the position of belated reaction to its developments. The conference was intended as a contribution to the general educational aims of the Ethical, Legal, and Social Implications (ELSI) component of the Human Genome Project that helped sponsor the conference. The papers in this volume formally continue this dialogue.

It should be obvious from these papers that there were often disagreements between scholars representing many different positions at the conference, and there has been no effort to obliterate these disagreements in this volume. But in this assembly of several different points of view and perspectives represented by this interdisciplinary group, one hopes that the result was mutual learning and continued dialogue. It is difficult to capture in a volume the sense of dynamic interchange and cross-penetration that one often sensed in the discussions at the conference.

Any meaningful dialogue among humanistic scholars about these issues also requires input from the scientific and medical research com-

---

1 The goals of the ELSI program, mandated as a component of the U.S. HGP in the original approval of HGP funding by the United States Congress in 1989, were defined by the original "Working Group" organized by Nancy Wexler to be concerned with four main areas of inquiry: issues of fairness; privacy questions; health-care delivery implications; and public education. See Cook-Deegan 1994, chap. Sixteen; and Cooper 1994, 304 ff.

munity. For this reason, invitations were made to members of the professional genetics establishment, and the contributions of Francis Castellino, David Hyde, Raymond Gesteland, Theodore Dryja, Harvey Bender, John Opitz, Martin Hewlett, and Jessica Davis were invaluable for these discussions, even though their presentations did not uniformly result in written contributions to this volume.

Such a multi-disciplinary approach to pressing issues in the area of contemporary science and technology forms a primary goal of the John J. Reilly Center for Science, Technology and Values at Notre Dame, the generating sponsor of this conference. It also reflects a concern of the Reilly Center to address issues from a combined focus on the interactions of science, technology, ethics, philosophy, history, and theology as they bear directly on human welfare in its several dimensions. The ideal of such a conference was to create a productive dialogue between these components.

It was also an intent of this conference to create a discussion that addressed some pressing questions in science and religion. Official Catholic teaching has in recent decades strongly reaffirmed its commitment to a constructive dialogue with modern science and technology (John Paul II 1990; Russell, Stoeger, and Coyne 1990). In its concern to address pressing issues in the life sciences from a multidisciplinary perspective with attention to these science-theology dimensions, this conference is in several respects a successor to the conference on evolution and creation held at Notre Dame in March of 1983 (McMullin 1985). At the same time, the questions raised by the genome project and modern biotechnology more generally are more complex and more immediately ethically challenging than any raised by the traditional areas of science and religion interaction—cosmology and evolution—and there are significant areas of tension that will increasingly develop as we see our powerful biotechnology increasingly applied to the control of the genetic as well as the reproductive processes of human beings.

Humanistic scholars more generally have many reasons to pay attention to the genome project. Applied ethicists must take account of the issues now confronting professional genetic counselors, lawmakers, insurance actuaries and medical legalists that directly result from the production of such broad genetic information about human beings. The information generated by the HGP will have consequences for medical insurability, the equality of access to health care, and the concrete practices of genetic counselors (Brown and Marshall 1993). The literature on these aspects of the Genome Project is vast and rightly has drawn considerable attention from front-line health delivery personnel and bioethicists (Yesley 1994; Yesley and Ossorio 1995). Most of the funding for the ELSI component of the Human Genome Project has to date been devoted to these issues. Pressing issues in this area often center around the potential for discriminatory uses of genetic information that might threaten persons with handicaps considered to be "genetic," ethnic minorities, and other groups potentially "at risk" from a widespread application of genetic testing.

Another set of problems is more theoretical, involving scholars with theoretical interests that might not ordinarily take part in the current discussions among health science and bioethics professionals. Intellectuals concerned with "science studies" in the broadest sense— history, philosophy, sociology, and ethical dimensions of the sciences— can find in the HGP a dynamic field of scientific development that displays all of the issues involved in the understanding of contemporary science and technology. Genome science is "big" science in the fullest extent of this term, often spoken of in terms of a "Manhattan" or "Apollo" project in biology, as Timothy Lenoir, Marguerite Hays, and John Beatty will explore in this volume. It involves the complex interplay of scientific theory, technological manipulation, governmental politics and industrial capitalism. It also is generating the development of advanced biotechnology with significant commercial dimensions. It is international in scope in a way unmatched by most any other scientific enterprise except high-energy physics. Although the Genome Project was initially conceived with broad internationalist ideals, it has already shown that it can be quickly "balkanized" by competing national and commercial interests (Beatty, Dreger this volume; Wade 1998).

Human genetics is also a science that carries immediate social and ethical consequences. In this dimension, it is easily accompanied by strong ideological components, components that are easily masked by language and hidden background assumptions. As the public receives news of the work of human genetic research in the "gene(s) of the week" reports in the popular media, it seems undeniable that this cascade of information is carrying with it a pervasive sense of reductionism of the human to the biological, a belief in a hard genetic determinism, and in an often crude conception of causation by "genes" of everything from esoteric forms of cancer to homosexuality, nurturing, aggression, obesity and criminal tendency. The inaccuracy and oversimplification represented by such "geneticization" of our popular discourse, to borrow John Opitz' apt term, can be deplored, and the conceptual difficulties surrounding the whole concept of a "gene" are discussed with historical penetration in this volume by Evelyn Fox Keller and the commentary by Jean Gayon. But the sheer complexity of the technical issues surrounding such matters as developmental genetics, incomplete penetrance, gene regulation, epistatic action, and the complexities of the relationship between DNA sequence and eventual phenotypic traits, understandably lends itself to oversimplification. The consequence is a substantial popular misrepresentation of many issues in a form that may frustrate the scientific community, but such misrepresentations have now entered deeply into popular discourse (Turney 1998).

Where this "genetic reductionism" pervading contemporary culture is leading more generally forms a historical unknown. Martin Pernick's important discussion in this volume of the history of the popular eugenics movement and the popular understanding of genetic "defect" in American culture supplies many reasons for thoughtful reflection on this issue, exactly because it illustrates the way in which "popular" understanding of science can influence public discourse and even filter down into the presentations in school curricula.

How we are to deal socially with these issues in the public sphere is a question without easy answers. The scientific community is rightfully concerned to find more effective ways to convey basic genetic and biological understanding within the K-12 school system through educational programs, and the production of educational materials has been one focus of the ELSI dimension of the HGP. But the degree to which this is succeeding is very uncertain, and such presentations must acknowledge their own hidden assumptions (Kay, Dreger this volume).

The role of simplification and misunderstanding in public discussion runs in more than one direction. Humanistic scholars likewise can be distressed by the ignorance of more technical ethical, philosophical, and theological issues in the educational system, in public discussions, and in the deliberations of the scientific community. We need to find some way of creating a more fruitful dialogue across disciplines in ways that will generate further cooperative inquiry into these important issues rather than a perpetuation of two- or three-cultures confrontations. This volume intends to be a contribution to this dialogue.

The papers in this volume are organized around an exploration of historical, philosophical and ethical-theological issues. Following an opening essay in which I have attempted to situate the issues in a broad historical perspective, the volume opens with a historically-situating paper by Timothy Lenoir and Marguerite Hays that began as a commentary on John Beatty's paper. This is followed by historical discussions by French historian and sociologist of science Jean-Paul Gaudillière, historian of molecular biology Lily Kay, and historian and philosopher of biology John Beatty. These situate the genome project against the backdrop of the complex development of technology, physical science, pediatric medicine, and basic biology that followed World-War II. Commentaries by Hans-Jörg Rheinberger and Robert Bud draw out important points in these discussions for further reflection. Alice Dreger's provocative essay then explores the role of "moral metaphors" in the promotion of the Genome project, and their function in the congressional discussion of the American HGP, raising some thought-provoking issues about the ways in which science, politics, language, and moral discourse have interacted in the origins of the genome project.

Turning in the second section to an examination of some of the ethical questions surrounding the HGP, the papers in this section address the troubling question of "eugenics" in its many forms. Particularly when viewed against the events surrounding earlier efforts in this century to apply seemingly valid genetic knowledge to society, the unleashing of a new genetic technology that will potentially permit the development of a new "high technology" eugenics has been at the forefront of public concerns about the HGP. The paper of historian Martin Pernick sets a historical framework for this discussion by examining the popular images of perfection and eugenic health at an earlier period in American history. Bioethicist Arthur Caplan then confronts directly the fears that a new "eugenics" might have the same kind of unforeseen consequences as accompanied the original eugenics movement, and seeks to allay these fears. Philosopher Philip Kitcher then explores the concept of "Utopian" eugenics within a framework that develops upon the concept of rational advantage and social optimality considerations. Commentaries by

Timothy Murphy and Diane Paul raise issues for further reflection on these essays.

The third section of the book raises the more theoretical problems of reductionism and genetic determinism. Evelyn Fox Keller's paper and the commentary upon it by Jean Gayon open up some of the important issues surrounding strong genetic explanations and the more general issue of genetic reductionism. Keller here pursues in specific detail some of her previous critiques of the "master molecule" view of genetics and continue to develop her explorations of a more organismic perspective in molecular biology (Keller 1995; Keller and Longino 1996, chap. Two). Philosopher Kenneth Schaffner, on the other hand, explores the reductionist thematic with a detailed analysis of issues in the genetics of complex traits, dealing particularly with the genetics of human behavior and the degree to which this can be successfully explained by genetics developed on the basis of simpler model systems. Edward Manier's commentary on the Schaffner paper discloses some of the conflict taking place at present in this dynamic area of research.

The issues raised by Keller, Schaffner, Gayon, and Manier flow logically into a set of reflections on theological and philosophical reflections on reductionism that occupy the fourth division of the book. These begin with an essay by Oxford molecular biologist and science-religion scholar, Arthur Peacocke, that develops the implications of emergentist perspectives on the relations of disciplines. This is followed by a paper by philosopher of science Ernan McMullin that originally developed from a commentary on Peacocke's paper, but which has now been expanded to a full contribution. Both of these explore issues surrounding the reduc-tionist program in genetics and biology, and both develop on the arguments for an emergentist perspective that can reconcile contemporary biological knowledge, philosophies of the human person and theological perspectives. This exchange is followed by a contribution by molecular biologist and theological bioethicist Kevin Fitzgerald, who addresses some options in underlying philosophical anthropologies that bear on molecular biology, using some of the perspectives of theologian Karl Rahner. A commentary on Fitzgerald's themes by historian of technology John Staudenmaier places these issues in a broader perspective. This is followed by reflections on the genome project from a

Catholic theological perspective by bioethicist and moral theologian
Richard McCormick.

The final word has been reserved for clinical geneticist and humanistic physician John Opitz, who offers a reflective discussion of contemporary genetics and its role in contemporary society, based on long
years of practical work in developmental biology, pediatrics and human
genetics. It is at this interface where theory and practice must meet in
concrete human decisions that the full implications of the Human
Genome Project will be played out.

1

Introductory Essay:
Completing the Tree of Descartes

Phillip R. Sloan

The growth of the life sciences since the Enlightenment displays how the natural sciences have been able to engage the most fundamental levels of our being, our relations to future generations, and eventually our relations to others. In their philosophical dimensions, these sciences have attempted to give empirical answers to questions that at some point bear intimately on ancient philosophical problems that have exercised the humanistic tradition since Antiquity: "what is human nature?"; "what is life?"; "what are the relations of body and consciousness?"; "why should I behave for good or moral ends?"; "is there transcendence over death?"

The answers the life sciences provide to these questions are often implicit rather than explicit. For the applied medical sciences, it can be argued that the aim of these sciences is not to provide answers to abstract philosophical questions, but is rather, in Claude Bernard's phrase, "to conserve health and cure disease" (Bernard 1957, 1). Nevertheless, we know there is much more at stake in the current scientific explorations of the most fundamental processes of life than the cure of disease. One senses a growing unease in the public sphere about the research in the life sciences, what Jon Turney has recently termed a "fear of Frankenstein" (Turney 1998), a sense that an ethically ambiguous and potentially destructive body of research is taking place behind the scenes, conducted by scientific elites, that promises to alter our lives in some dramatic and even undesirable way.

1

The Human Genome Project brings to a focus many of these public concerns. As the most ambitious research agenda ever to be initiated in biological science, it has brought together developments in biotechnology, biophysics, information theory, computerization, and genetic research. It has also created a national and international organization of research directed toward specific targeted goals to be accomplished within a stated time frame and within a defined budget. Its completion will shape the face of medical science, and many other areas of biological science, for centuries to come.

Outside of the scientific community, however, there is a striking lack of awareness of the character or purpose of the HGP. Any reader of the newspapers and weekly magazines is aware that it is a major research enterprise, and that there is now a "race" of some kind taking place to complete it (Wade 1999a), but the reasons for such a competition, and for the substantial funding behind it, are not transparent. If it is to the credit of the organizers of this research enterprise that they considered the advice of ethicists, philosophers, legal scholars, and other humanists important from the beginning of the HGP through the creation of the ELSI program, the need they sensed for such a component also indicates an awareness by the scientific elites themselves that the work they were to embark upon had significant ethical and humanistic significance, and carried potential social dangers.

In the face of such an organized effort, it seems appropriate to look beyond the immediate questions and issues that are presented to society by the initiation of such research. We must also to ask about the deeper questions being posed by such massive control of the processes of life, what Philip Pauly has called the "engineering ideal" in modern biology (Pauly 1987, chap. One), the turn of life science from description and analysis to a conscious effort to control living things by the combination of scientific and technological knowledge. To pose such questions, it seems useful to look beyond the proximate origins and issues in genetics, molecular biology, informatics, and the politics of modern medical research, to some issues that underlie the "biotechnological program" in the life sciences more generally.

To examine this question, I return directly to the roots of this enterprise in the speculations of Réne Descartes in the seventeenth century. More than any other "prophet" of the new science, Descartes laid out a philosophical agenda that defined the initial framework for a reductionist and "mechanistic" approach to life sciences.[1] Reworked, revised, and reinterpreted, there are still strong elements of this program surviving within the conceptual structure of modern life science as attitudes and styles of research that need to be excavated and re-examined as we embark on this new adventure in the realm of living beings.

RETURNING TO THE TREE OF DESCARTES

I shall define the "Cartesian" research program to embody at least three axiomatic claims of relevance to the subsequent development of the life sciences: first, the life sciences, like all other sciences, are to be approached by the method of analysis and synthesis, meaning that the complexities of organisms are to be understood by breaking them down into their most elementary parts and processes. The operations of the whole are then to be understood as a result of the synthesis of these elementary ingredients. Second, the explanatory entities are those to be found at the lowest level of this analysis, rather than being located at higher levels of organization, teleological purposiveness, or governing vital agencies. Third, all autonomous forces, special agencies, faculties, vital properties, novel emergent powers, and animating souls are abolished from the explanatory framework of biology.[2] Although it would be incorrect to emphasize the exclusive role of Cartesianism in formulating any one of these ingredients, nonetheless the clarity with which Descartes conceptualized the questions of the life sciences and the way in which he unified all three of these ingredients in a programmatic philosophical enterprise, rendered his project more directly relevant historically to the development of the life sciences than the competing methodological projects of Bacon, Harvey, Gassendi, Galileo or Newton. Through his very early endorsement of Harvey's theory of the circulation of the blood, Descartes made circulatory theory a central component of his mechanistic theory of the organism. Descartes was also the only one of the prophets of the new science to make a new mechanistic medicine,

---

1 This is not to deny the importance of a Baconian thrust toward a technology of life, as argued by Lily Kay (Kay 1996, 88). But I conclude that Descartes' program is both more coherently articulated and that it much more deeply and explicitly affected the life sciences in their concrete development.

2 I am of course simplifying a very complex set of issues here. For a penetrating brief discussion of some of the complexities see Canguilhem 1994, chap. Ten.

based directly upon a reformed physics, the primary *goal* of his philosophical enterprise (Descartes to William Cavendish, October 1645, Adam and Tannery 1897, IV, 325).[3]

Furthermore, Descartes explicitly conceptualized a novel unification of epistemology, mechanics, physics, anatomy, medicine, ethics, and theology. Although these relationships were never made the subject of a specific treatise, they were sketched out in the author's letter to the Abbé Claude Picot that prefaces the second edition of his *Principles of Philosophy*, published in 1647. Through the metaphor of a tree, Descartes envisioned a total connection of the sciences that was both a classification of the sciences and also a programmatic agenda for the future, a promissory note for the development of the sciences that would eventually flow from his reform of philosophy.[4]

The roots of the Cartesian tree were to be supplied by his comprehensive reconstruction of metaphysics and traditional natural philosophy. These new Cartesian foundations rested upon three primary philosophical principles, proven to his satisfaction with complete certainty by rigorous philosophical argument of the *Meditations* of 1641: (1) the existence of God; (2) the immateriality of the soul and the separation of soul from body; and (3) the clear and distinct conception of matter as extension.

The trunk of this tree was formed by the new mechanistic physics that he proposed to build directly upon these foundations, particularly upon the theory of matter as extended substance. This new physics included his definition of three main laws of nature, later fundamental for Newton's developments, and the subordinate mechanical principles that followed from these physical laws. These physical principles formed the foundation for the development of a speculative proto-evolutionary cosmology, geology and terrestrial mechanics sketched out in Part III of the *Principles of Philosophy*.

The third portion of this tree, yielding its eventual fruits, was only sketched out in the *Principles* and was never fully developed in his writings, due to Descartes' premature death in 1650. Descartes

---

3 For discussions of the significance of this "medical" vision see Canguilhem 1994, chap. Ten; Carter 1983; and Martial Guéroult ([1952] 1985, 2: chaps. 19–20. I have discussed some of the historical background in my Sloan 1977.

4 As Roger Ariew has shown, this metaphor was not novel with Descartes. Charles de Raconis and Francis Bacon had both used this analogy previously (Ariew 1992, 1998). Nonetheless, there are important differences in Descartes' use of this image that I will develop in this essay.

envisioned   in the letter to Picot   the production of   three primary
pro-ducts of his reform of philosophy:

> Thus Philosophy as a whole is like a tree; of which the roots are
> Metaphysics, the trunk is Physics, and the branches emerging from this
> trunk are all the other branches of knowledge. These branches can be
> reduced to three principal ones, namely, Medicine, Mechanics and Ethics
> (by which I mean the highest and most perfect Ethics, which
> presupposes a complete knowledge of the other branches of knowledge
> and is the final stage of Wisdom.) (Descartes [1647] 1983, xxiv)

Remarkable in this vision of the unification of knowledge was the
place of medicine and applied technology—his apparent meaning of
méchanique in this context. Both were conceived to flow directly
from his new physics, itself based on matter, motion, and
foundational scientific laws.

The medical dimensions of the tree followed along the lines of the
mechanistic physiology that Descartes first sketched out in Part Five
of the *Discourse on Method*. These views were then elaborated in
detail in the posthumous *Traité de l'homme*, published first in Latin
in 1662, and in French in 1664. In this treatise Descartes developed a
speculative physiology based upon a model of the organism as a
hydraulic machine, driven by a "fire without light" residing in the
pores of the septum of the heart. Using this model and the theory of
the circulation of the blood, Descartes elaborated a mechanistic
account of the senses, muscular action, nervous system, and reflex
action. All of this was integrated into a model that was striking in its
ambitious claims:

> all the functions that I have attributed to this machine, such as the
> digestion of food, the beating of the heart and arteries, the nutrition and
> growth of the members, respiration, waking and sleep, reception of light,
> sound, smell, taste, heat and such qualities by the external sense organs. .
> . follow completely naturally in this machine solely from the disposition
> of the organs, no more nor less than those of a clock or other automaton
> from its counterwights and wheels. . . ; it is not necessary to conceive on
> this account any other vegetative soul, nor sensitive one, nor any other
> principle of motion and life, than its blood and animal spirits, agitated
> by the heat of the continually burning fire in the heart, which is of the
> same nature as those fires found in inanimate bodies. (Descartes [1664]
> 1972, 106–07, trans. Sloan)

These speculations supplied the first coherent model of a reductionist biological program in the history of modern science. As Cartesian theoretical medicine and physiology developed over the next century, it literally sought to build medical science upon a reduction of organic functions to mechanics, in which the body was conceived as composed of wheels, tubes, sieves, pumps and levers (Mendelsohn 1964, chap. Four; Duchesneau 1982, chap. Two; Gariepy 1990).[5] The iatromechanical tradition continued for a half-century to pursue a modified version of the Cartesian reductive ideal along the lines of a blunt mechanistic physics.[6]

As historical studies have shown, the iatromechanical program failed historically by the 1770s. It was replaced by vitalistic physiological and medical programs identified with the names of John Hunter, François Xavier Bichat, the Montpellier vitalists, Albrecht von Haller, Robert Whytt, Caspar Friederich Wolff, John Brown and William Cullen (Canguilhem 1994, chap. Thirteen; Duchesneau 1982, 1985; Brown 1974; Risse 1971). The history of functional life science since the 1770s has in many respects been a story of the way in which this "vital" science of the late Enlightenment was itself replaced by a new form of reductionism in the nineteenth. Although this new reductive science had no immediate historical connection with the original Cartesian project, it pursued the ideal of a reductionist and analytical biology that retained many of the conceptual ingredients of the Cartesian program in a surprisingly intact degree.

---

5 Descartes himself never seemed to have envisioned such analogies as anything more than metaphors, and the opening language of the *Traité de l'homme* seems to recognize the man-machine only as a heuristic device. In this respect, many of his iatromechanical successors were more literal mechanists than was Descartes himself. More recent efforts to utilize AI and computer analog models of biological systems are closer to this literalist extension of the original Cartesian program than are the physico-chemical models that I would claim constitute the novel version of reductionism to emerge in the nineteenth century.

6 I am intentionally not making a sharp distinction between properly Cartesian developments of this program, which maintained a strict reliance on contact forces and physical causation in physiology, as might be found in Boerhaave, Archibald Pitcairne, and Friederich Hoffmann, and the specifically "Newtonian" versions, such as were articulated by Stephen Hales, which introduced forces in addition to contact actions. In a general framework, most of the early Newtonian variants were only refurbishings of earlier Cartesian models. Later "etherial" variants of the Newtonian program led away from mechanism in the direction of vitalist physiology (Schofield 1970, chap. Nine).

REGROUNDING THE MECHANISTIC PROGRAM

The more recent biophysical and reductive approaches to living phenomena that emerged in the nineteenth-century, represented the result of important social developments of science, and also conceptual reformulations of crucial questions. Institutionally, these more recent "mechanistic" approaches to the life sciences were a product of the incursion of physical science into the work of medical faculties in the early decades of the nineteenth century, especially in the German universities and research institutes of the period (Lenoir 1988, 1992). Conceptually, these developments specifically represented a connection of life science and medical physiology with the new analytical chemistry that emerged at the end of the eighteenth century. These research agendas also applied experimental physics and physical instrumentation to physiological study in new ways. In ways not quite as envisioned by Descartes, the life sciences from this period have increasingly been drawn into a more and more intimate relation with the researches in the physical sciences.

The foundations of this historical development can be most readily, if not exclusively, dated from the landmark work of the French chemist Antoine Lavoisier in his revolutionary experimental work of the 1770s and 80s on respiration in animals and on the cause of animal heat (Lavoisier [1777] 1780; Lavoisier and Laplace [1780] 1784; Mendelsohn 1964; Holmes 1985). Lavoisier was at least partially successful in illustrating in these papers the close similarity in the reactions and in the resultant products yielded by the combustion of charcoal and animal respiration. The long-mysterious phenomenon that had seemed to define the autonomy of life—the ability of animals to generate and maintain their own body heat—apparently yielded to this combination of chemistry, quantitative analysis, and experimental physics. Extending Lavoisier's approach, and obtaining by these methods a satisfactory detailed solution to the issues of general physiology and intermediate metabolism with a similar quantitative rigor defined much of the project of general physiology in the nineteenth century (Culotta 1968; 1972), with the foundational work of Julius Robert Mayer and Hermann von Helmholtz on the principle of the conservation of force a crucial component in this development (Caneva 1993, chap. Three, Elkana 1974, chap. Four).

Attention of historians of science has been drawn to the young group of students surrounding Johannes Müller and the Berlin Physical Society—Emil DuBois-Reymond, Hermann von Helmholtz, Carl Ludwig, and Ernst Brücke—who formed the vanguard of a generation of researchers in the Germanies around the middle of the nineteenth century, and who sought to exploit the methodology and the conceptual tools in this new form of biophysics. The spirit of this movement was captured by an almost solemn pledge to pursue this approach in biology. As one of these workers, Emil DuBois-Reymond, put this in his retrospective memoirs:

> Brücke and I pledged a solemn oath to put into power this truth: no other forces than the common physical-chemical ones are active within the organism. In those cases which cannot at the time be explained by these forces one has either to find the specific way or form of their action by means of the physical-mathematical method, or to assume new forces equal in dignity to the chemical-physical forces inherent in matter, reducible to the force of attraction and repulsion. (DuBois-Reymond, 1918:108 trans. Bernfield as in Sulloway 1992, 14)

Researches into the new chemistry, physical quantification, and technological instrumentation of living functions pursued by Helmholtz, Ludwig, DuBois Reymond, and many other biophysically-minded medical physiologists, achieved the rigorous biophysical explanation of a limited body of fundamental properties of life by the end of the nineteenth century. They also supplied a model for further efforts to reduce the biological to the physical in a form that has been ascendant since the late nineteenth century. The modern reductionist program in the life sciences has been expanding in scope ever since this period, exported by dedicated disciples of this biophysical approach into other national contexts.

MODERN BIOPHYSICS, REDUCTIONISM, AND THE HGP

The historical connections of modern "molecular" biology to these nineteenth-century developments in German biophysics is a topic still in need of detailed exploration, and it requires analysis of the impact of foreign study and the interaction of national traditions on individuals who later were important in setting up programs that carried out biophysical analyses of living forms in a form that came to be called in the 1930s "molecular" biology.[7] In the original definition of this new approach to biology, "molecular" biology was to constitute "a new field . . . in which delicate modern techniques are being used to investigate ever more minute details of certain life processes" (Weaver 1970, 582; Kay 1996). "Molecular" could, and did, mean many things to different workers. In some contexts it relates to the form of collaborative biomedical research created by physical scientists, mathematicians and information theorists who migrated into the life sciences in the post World-War II era. It also related to the level and scale of research and the degree to which it emphasized topics at a minute scale. Beyond this, it is not easy to point to any larger unifying theory behind molecular biology, and as Richard Burian has argued, it is better understood as an "immensely powerful battery of techniques for getting at the interrelated families of complex mechanisms found in all sorts of organisms" (Burian 1996, 80), than as a body of theory. This form of scientific practice accomplished a broad institutional and conceptual conquest of traditional biology departments in much of the western world in the 1960s, and is now being extended to human and medical genetics through the Human Genome Project.

Philosophically, the reductive and analytic *methods* in molecular biology have commonly been accompanied by strongly reductionist philosophical programs that have had several implications for modern life science and its humanistic relations. An interest in completing such philosophical agendas was a strong personal motivation for many of the personnel who first created "molecular" genetics. As Francis Crick summarizes his own motivations in an interview with Horace F. Judson:

My impression is that other people have gone into molecular biology with the same *general motives* that I had, but sometimes with a difference in point of view. . . .I went into it, for example, to try to show that you can explain all these phenomena—the term molecular biology

7 Quantitative data on the contacts of Americans with German physiological schools and methods is to be found in Frank 1987. The translation of German styles of biological inquiry to the United States, especially by Jacques Loeb, has been explored by Pauly (1987, esp. chaps. One-Two). Some useful remarks are also to be found in Turney (1998, chap. Four).

wasn't common then, certainly I didn't know it, but the phrase I had in my mind was 'the borderline between the living and the dead'—that you could explain these phenomena just by the laws of ordinary physics and chemistry. (Judson 1996, 586–87)

The linkage of this modern philosophical reductionism to the historical "Cartesian" program is, to be sure, analogical rather than causal-historical. Nonetheless, to the degree that the modern biophysical program has successfully completed an explanatory unification of biology with the underlying physical and chemical sciences through a strongly analytic biology, it has fulfilled at least the general programmatic dimensions of Descartes' efforts to derive a reductionist biology from more elementary physical foundations and principles. Presumably this reduction also does away with the appeal to special explanatory functions or higher properties of living beings that cannot be dealt with by these methods.

THE GENOME PROJECT

This brings us to the genome project itself. Contemporary molecular genetics, as a sub-species of this larger enterprise of molecular biology, represents a complex historical synthesis of biochemistry and medical physiology with specialized developments in plant and animal genetics. It also added the use of informatics and other components drawn from mathematical and early computer sciences, although the degree to which these were intimately involved is a subject of some debate (Kay this volume, Sarkar 1996a). It has also extended these methods into the domain of human and medical genetics. As we learn from the essays of both John Opitz and Jean-Paul Gaudillière in this volume, clinical human genetics had primarily developed as the creation of pediatricians who were concerned with the lesions and abnormalities encountered in young children rather than with questions of theoretical biology or biophysics. With tensions alluded to by John Opitz in his essay, the field has now been occupied by research scientists who may have obtained their primary scientific socialization and higher education in such fields as chemistry, physics, microbiology, biophysics and mathematics rather than in traditional biology departments or in clinical medicine. In practical terms, we see in modern genomics an extension of basic science up into the clinical world in ways that are proving transformative.

It originated through many complex stages of development, developing on the development of microbial genetics that followed upon the work of André Lwoff, Jacques Monod and Joshua Lederberg, the development of DNA genetics that followed the work of Watson and Crick in 1953, the unraveling of the coding problem in the 1960s, and the development of Polymerase Chain Reaction techniques in the 1970s. It was also pursued by individuals whose training was typically in biochemistry, medical microbiology or one of the basic sciences instutionally removed from the departments of biology and agricultural science that typically taught genetics. With it came a new collaborative style of biological research that was familiar to the physics community and some medical research groups, but had not been one common to traditional biology departments. A similar intellectual and institutional conquest of traditional human and medical genetics is now in the process of being achieved through the developments in molecular genetics.

This entry of research styles drawn from the physical sciences, and the pursuit of new research goals by personnel trained either in the physical sciences, biochemistry, or in the biology of very elementary forms of life, has had immense impact on the style of doing biology in the last four decades, and has dramatically altered the conception of the field of genetics. Biological research in genetics that at one time might have been carried out by a single professor and a few graduate students working in a small laboratory with modest or non-existent funding—elaborations of Thomas Hunt Morgan's "fly-room" in many respects—, has been replaced by group-research employing advanced technology, and displaying a remarkable degree of interdisciplinary collaboration between genetic research, government support, commercial biotechnology, theoretical molecular biology, and traditional clinical genetics. It has created a paradigm example of "technoscience" at work, in which the boundaries between theoretical science and applied technology have become impossible to define.

The importance of this style of research for the character of the genome project is not only related to its particularly organizational form of research. It is also connected to strategic decisions that were made in the origins of the genome project, particularly the mass-sequencing goal, that is at the root of many of the subsequent ethical dilemmas the HGP has posed.

John Beatty has given in this volume a basic overview of the remote and proximate history of the HGP. I wish here to elaborate on one aspect of this recent history that involves the Alta, Utah meeting, held on 9–12 December of 1984, and jointly organized by Raymond White of the Howard Hughes Medical Institute of the University of Utah and Mortimer Mendelsohn of the Lawrence Livermore National Laboratory.[8]

As Beatty tells us, the Alta meeting was sponsored by the Health Effects Research Division of the Department of Energy, and was principally called to review the results of the long-term study by the Atomic Bomb Casualty Commission on the effects of radiation on the populations of Hiroshima and Nagasaki, carried out under the direction of one of the founders of human genetics, James V. Neel. It was to be a small gathering of invited molecular geneticists with a few human geneticists at a local ski lodge with the specific focus to be the discussion of the ABCC results and a review of the prospects for a more effective direct detection of human genetic mutations through the application of new recombinant DNA technologies (White 1984; Mendelsohn 1985, 1986; Delehanty et al., 1986; Cook-Deegan 1989).

The ABCC study, summarized for the conference the opening morning by James V. Neel and John Mulvihill, set the stage for the discussion of newer methods of detecting DNA alternations. Molecular geneticist David Botstein summarized the RFLP technologies and reported on the first applications of these to study human mutations in Nagasaki and Hiroshima. The possibility of more massive analysis of human mutation occupied the second day of the meeting, beginning with a discussion by Sherman Weissman on genes for which there was good DNA structural data. This was followed by Raymond White's presentation on polymorphisms and mutations and then a discussion by Maynard Olson on the use of cloned DNA and PCR methods on yeast clones for analysis of the genome, using automated imaging of the gels. Edwin Southern, inventor of the Southern Blot method, followed Olson with the suggestion of using a combination of RFLP and probe methods (Anon. 1984).

---

[8] The second was the Santa Cruz meeting of 24–25 May 1985, organized by molecular biologist turned University Chancellor, Robert Sinsheimer, eager to find a biological field that would initiate "big biology" on a scale similar to that of major projects in space and high-energy physics (Sinsheimer 1989; Cook-Deegan 1994, chap. Five).

DRAFT AGENDA

"DNA METHODS FOR MUTATION DETECTION"[9]

RUSTLER LODGE, ALTA UTAH
10-13 DECEMBER 1984

Monday *December 10*	Tuesday *December 11*	Wednesday *December 12*	Thursday *December 13*
Raymond White *Prefatory Comments*	Sherman Weissman *Mutation in HLA*	George Church *Direct Sequencing*	*Report* *Development*
Mortimer Mendelsohn *Background*	*Leroy Hood *Mutation in Gene* *Families*	Charles Cantor *Big DNA*	*Report* *Development*
J.V. Neel *Current Status*	John Roth *Discussion*	Sherman Weissman *Discussion*	
John Mulvihill *Current Status*	Ray White *Mutations and Pm's*	Leonard Lerman *Gels*	
Elliot Branscomb *Somatic Methods*	Maynard Olson *Big Measurements*	Richard Myers *Gels*	
Thomas Caskey *Mutation at HPRT* *Locus*	*Leroy Hood *Discussion*	Raymond Gesteland *Discussion*	
David Botstein *Calculation for* *Rxn Enzymes*	*Clyde Hutchison *New Genetic Methods*	Mendelsohn- White *General Discussion*	
	Edwin Southern *Fragments*		
	General Discussion		

*unable to attend meeting

---

[9] I thank Mortimer Mendelsohn for a copy of the draft agenda and a list of the invited participants (personal communication). The meeting opened on 9 December with preregistration and social events.

ALTA SUMMIT PARTICIPANTS[10]

Name	Affiliation 1984	Ph.D	PhD. Topic
David Botstein	Biology, MIT	U. Michigan 1967 (Human Genetics)	Synthesis and Maturation of DNA by the Salmonella Phage 22
Elbert Branscomb	Biomed-Science, Lawrence-Livermore	Syracuse 1964 (Nuclear Physics)	Internal Symmetries in S-Matrix Theory
Charles R. Cantor	Human Genetics, U Columbia	UC Berkeley 1966 (P-Chem)	Sequence Dependent Sequence of Oligonucleotides
C. Thomas Caskey	Medical Genetics, Baylor Med.	MD Duke 1963	None (Intern NIH 65-67 with Marshall Nirenberg)
George Church	Biogen Research, Cambridge MA	Harvard 1984 (Mol. Bio/Biochem)	Genetic Elements in Yeast Mitochondria and Mouse Introns
John D. Delahanty	Boroughs-Wellcome	Florida St./Oakridge Nat'l Lab 1980	Sequence Isoacception of TRNA from Euglena
Charles Edington	Health & Environment, DOE	U. Tenn. 1956 (Zoology)	Recessive Lethal Genes Induced by Radiation
Raymond Gesteland	Hughes Med Inst, U Utah	Harvard 1966 (Biochemistry)	E. coli Ribosomes and Ribonuclease
Michael Gough	OTA, Wash.	Brown 1966 (Bacteriology)	DNA Relationships in E. Coli
Leonard Lerman	Biology, SUNY Albany	Caltech 1950 (Chemistry)	Slow Contraction of Frog Muscles and Antibody Reaction
Mortimer Mendelsohn	Biomed-Science, Lawrence-Livermore	MD Harvard, 1948, Cambridge 1958 (Biochem)	Microspectrophotometry and Cytochemistry of Nucleic Acids
John Mulvihill	Nat'l Cancer Inst. NIH, MD	MD U Wash., 1969	None
Richard H. Myers	Biochem-Mol. Bio. Harvard	Georgia State 1979 (Psychology)	Psychological Distress in Early Detection Study for Huntington's Disease
James V. Neel	Genetics, U Michigan	U Rochester 1939 (Genetics)	Bristle Patterns on Drosophila
Maynard Olson	Genetics, Wash. U	Stanford 1970 (Chemistry) 1970	Oxygen-17 Magnetic Resonance of Vanadium and Chromium Solutions
David A. Smith	Heath & Envr. Res., DOE	USC 1964 (Biochem)	Metabolism of 5-Hydroxyuracil Compounds
Edwin Southern	MRC Mammalian Genome Unit, Edinburgh	U. Glasgow 1964	Synthetic and Naturally-Occurring Enzyme Metabolites
Sherman Weissman	Human Genetics, Yale	MS U Chicago (Math)	MD Harvard 1955 Internship NIH
Raymond L. White	Hughes Med. Inst. U Utah	MIT 1972 (Microbiology)	Recombination in Bacteriophage Lambda

---

10 Table prepared from standard reference sources. I also wish to thank Mortimer Mendelsohn, Frank Church, Sherman Weissman, David Botstein, C. T. Caskey, Edwin Southern and John Delehanty for valuable assistance with some important details of this table.

This was developed by further group discussion into an informal proposal for a means to obtain the sequence of the *entire* genome by a high-technology application of the new techniques that had been developed in microbial and yeast genetics.[11]

This extension of purview from simple biological systems to human beings suggested to the participants, and to the DOE, some dramatic new vistas that prefigured much of what was to become the Human Genome Project. In the words of Mortimer Mendelsohn's internal report:

> This extraordinarily stimulating and successful meeting has precipitously broadened the research frontier for heritable mutation testing, offering hope for soon to be available, powerful, new methods and raising innumerable research and strategic questions. An inevitable consequence will be the demand for new, large-budget activity for federal research, and a new focus of scientific activity bridging the DNA, genetic and epidemiologic fields. It carries an exciting and humbling set of responsibilities. (Mendelsohn 1985, 10)

Viewed conceptually as well as institutionally, this move away from the study of atomic bomb effects toward "epidemiology"—the more general study of diseases—displays how the basic-science approach of molecular biology was turned toward questions of medical genetics.

GENOMICS AND REDUCTIONISM

In the extension of basic science, reductionist methodologies and biotechnology into issues of relevance to human genetics, illustrated in microcosm by the Alta meeting, we can see illustrated the way in

---

[11] The procedure emerged with the ungainly label of "Synthetic-Oligonucleotide Heteroduplex Tetramethylammonium-Chloride Melting Method." It would identify mutations by separating out abnormal sections of the DNA of either parent using ammonium chloride to discriminate single-pair mismatches. As the report summarizes: "The beauty of this method is the efficient, one-pass parallel processing of a genome's worth of DNA. Conceivably the measurement of mutation rate could be made in a single triad with only a few days of technician time." (Mendelsohn 1985, 5). I thank Dr. Mortimer Mendelsohn of the Lawrence Livermore National Laboratory for supplying a copy of this internal report and many other assistances with the details of this meeting that considerably extend the understanding of its importance beyond the information available in the early account of Cook-Deegan 1989.

which contemporary biology has turned the powerful techniques of molecular biology onto the questions of human inheritance.

The consequence of the decision to sequence the entire three-billion base pairs residing on the DNA in the forty-six chromosome pairs of the human genome, as it has been fleshed out by the formal development of the HGP, both domestically and internationally, has the potential for medical benefits that will develop curative medicine in many ways in the next decades. It also has more unsettling possibilities that center on the conjunction of the injection of this information into a pervasive reductionism that has accompanied the growth of modern life science. If we are to avoid some of the unfortunate consequences of this new biological inquiry, I would suggest that it requires careful attention to the meaning of the reductionist program in the life-sciences.

Since Descartes, reductionist and "analytic" strategies have formed a powerful and successful methodology for exploring questions in the life sciences. Francis Crick has spoken in a passage quoted above of how he was drawn to biology by his interest in "explaining" the living by the non-living. Philosophically, however, much depends on the meaning of "explain" in such statements, and this in turn requires a sensitivity to subtle issues that have been explored in detail by philosophers of science (Grene 1967, 1971; Rosenberg 1985, chap. Four; Sarkar 1992; 1996a; 1998; Schaffner 1993; Burian 1996; Peacocke 1985). If 'explanation' is taken in a weaker methodological sense, meaning the specification of the necessary conditions of life in terms of the principles of physico-chemical science, or as a heuristic research strategy that employs simplifying idealizations, there is little to question in the pursuit of this kind of analytical biology. It can be argued that for the most part, reductionist and analytical methodologies have proven to be the most highly creative approaches in the history of life science (Roll-Hansen 1969; 1976; Bechtel and Richardson 1993). Methodological fertility is not, however, equivalent to a compelling argument for the metaphysical and epistemic adequacy of reductionist philosophical agendas, nor does heuristic reductionism imply metaphysical materialism and ontological "eliminativism." These are extensions of methodological reductionism that do not rest upon empirical science, but upon additional philosophical premises that must be argued for on philosophical grounds. Generally, these extensions are not being made by philosophers of biology, who have been careful about technical distinctions, and who discriminate in detail between epistemic, intertheoretic, and ontological meanings of reductionism (Sarkar 1992, 1998). Richard Burian, for example, has argued that molecular biology is "far less reductionistic than biochemistry," because it must consider the integration of multiple levels of function (Burian 1996, 80).

The difficulties created by the *social* extension of the insights and methodologies of molecular biology hinge on the loss of these subtle distinctions as the empirical findings of the life sciences enter popular discussion. Without attention to these issues, the consequence is very often either the explicit or implicit advocacy of strong ontological reductionism and metaphysical materialism by individuals who claim this to be a consequence of science. Genetics, as the science that deals with the linkages of generations, with concerns for posterity, and with the explanations of physical properties of organisms, necessarily acquires socially-significant implications when interpreted in terms of metaphysical reductionism.

This more popular reductionist program, in its genetic expressions, has depended heavily on a realistic interpretation of the "gene" as a material entity which can be mapped, cloned, and perpetuated essentially unchanged from one generation to the next as the "determinant" of life. Conceived of as the underlying "true cause" of expressed phenotypic traits, the biophysical "structural" understanding of the gene in terms of a specific base sequence within regions of the DNA molecule has fleshed out Crick's sense of "explain" in highly realistic terms.

Historians of biology who have looked closely at the history of the gene concept have detailed the "hardening" of the thesis of genetic causation by underlying structural genes over the past seventy years. They have displayed how an original "heuristic" understanding of the gene that recognized the complex differences between simple Mendelian trait assortment and the factors that are involved in the emergence of phenotypic traits in the embryological process, was been replaced by a more literal preformationist concept (Moss 1998, chap. 1; Portin 1993). Due to the success of the Morgan chromosome theory, the gene came to be conceptualized primarily as a nuclear and chromosomal structure. (Sapp 1987; see also Harwood 1993). As aspects of this history are elaborated by Evelyn Fox Keller, Jean Gayon and John Opitz in this volume, these developments have served either to obliterate or sideline a host of residual issues surrounding developmental and organismic biology.

The result has been the establishment of a gene concept that plays a significant role in metaphysically reductionist analyses of organic life. By extension, it also features centrally in much of contemporary discourse about the way human beings are regarded in relation to this organic world. Whether these interpretations are even warranted on empirical grounds has been recently questioned (Moss 1998). But even if the structural DNA gene concept employed in much of scientific discourse and the textbooks of biology best suits a genetic approach to diseases, this does not itself justify, without further argument, the philosophical agendas that claim to derive their warrant from this empirical fertility.

Criticisms of strong genetic reductionism from philosophical, developmental, and linguistic directions have been substantial, as we can find sampled in the essay by Evelyn Fox Keller in this volume. It is not, however, evident that these critiques will alter genetic research at a practical level, nor that they have succeeded at a philosophical level (Sarkar 1996a). The development of effective DNA-based medical therapies that can give solutions to AIDS, tuberculosis, and cancer promises to be one of the great benefits of the HGP, and the likely success of reductionist strategies in solving these problems in the next century will only reinforce the power of the structural gene concept and the implicit reductionism that accompanies it. For this reason I suggest that the reductionist challenge needs to be addressed within a framework that accepts reductionism as a highly successful strategy in the life sciences, but reexamines the location of the human being in this enterprise.

Here again I return for perspective to the tree of Descartes. The historic Cartesian tree acknowledged, however problem-laden the Cartesian solution turned out to be, the reality of the "human phenomenon" at the base of the construction of a reductionist program in modern science. In Descartes' systematic philosophy, this was cast in the form of a substance dualism between thought and reflection, with thought presumed to reside in an immaterial substance, *res cogitans,* separated from the material body, a material and mechanical entity within the domain of *res extensa.* It is unnecessary to pursue the complex issues and debates surrounding the adequacy of classic Cartesian dualism except to extract from it one important insight.

Descartes' location of reflection and thought at the *base* of his reductive program, essentially as its primary root, has too often been ignored or forgotten in the historical development of the life sciences. Instead we have seen a series of historical proposals since La Mettrie's *L'homme machine* in the eighteenth-century that have proffered to *explain*, in a strong metaphysical sense of this term, the world of life and eventually of reflection, consciousness, and even ethics (Loeb 1912, 31), by reducing it to the materialism and mechanism governing the domain of matter. Contemporary expressions of this project are manifested in assertions of a material reduction of consciousness and life to biophysics, computer informatics, and neurophysiology in some schools of contemporary philosophy of mind.

Engaging this issue in its full dimensions would require a careful examination of the presuppositions of several forms of naturalistic epistemology that cannot be attempted here. We can, however, examine some of these questions within the historical framework I have been developing in this essay. As we have seen, strong reductionism formed a central thesis in the historic Cartesian program, implied in the linkages envisioned by Descartes between physics, biology and medicine. At the same time, this did not imply a reduction of the human being to the physical domain. To explain the thinking self by the same reductive analysis that Descartes employed to analyze the actions of the senses or the physiology of the body would have undermined the entire attack on epistemological scepticism that I have suggested elsewhere generated the agenda of the Cartesian program in physiology (Sloan 1977).

Nonetheless, the turning of reductionist analysis upon the human observer, with the elimination of a priviliged status to an independent consciousness, was the express development of the Enlightenment materialists and the philosophical naturalists influenced by their work—La Mettrie, Diderot, D'Holbach, Hartley, Priestley, Hume. Problems in Cartesian dualism suggested to these *philosophes* that the transcendental subject was superfluous and that a fully naturalistic epistemology, one that denied the transcendence of thought over the material conditions of life, was both possible and compelling.

The difficulties in this project were diagnosed at the time with penetration by Immanuel Kant, and I accept many of his conclusions as providing at least a route to a more satisfactory analysis of the issues. From his position as a philosopher concerned with the full range of questions of ethics, epistemology, natural science, and even medicine and biology, Kant was well-positioned to engage the claims of his contemporaries at a fundamental level and articulate in a more satisfactory way than had Descartes the epistemic relations between the transcendental observer and the nature explored by scientific understanding.

Fundamental to Kant's complex analysis of biological and epistemological topics in the first and third *Critiques* (1781; 1787; 1790) was his recognition of the contribution of the human knower to the input of sensory experience, elaborated in the *Critique of Pure Reason* in terms of the contributions of the faculties of Sensibility (space and time), Understanding (categories), and Reason (ideas). As this framework was extended to the science of living nature, Kant recognized the unusual character of living phenomena that prevented it from being simply assimilated under the categories of physics.

Kant's contemporary scientific and medical context presented him with the two alternative options in the life sciences of his day we have spoken of above—the reductive and mechanistic life science that had developed institutionally from Cartesianism, and the new vitalistic medical theories that emerged across many national traditions in the latter decades of the century. This new vitalism also was accompanied by the revival of long-defunct Greek ideas of teleology, organization, form, and vital agency in a new form (McLaughlin 1990, chap. One; Lohff and Englehardt in Cimino and Duchesneau 1997).

Kant's great contribution to the theory of the life sciences follows from his acceptance of the insights contained in this new "vital" science expounded by Blumenbach, Haller, and Caspar Friederich Wolff, while at the same time preserving the intentions and methodological strategies of the reductionist program. This complex move was accomplished by making the intentional, organizational and teleological properties of organisms primary *epistemic preconditions* of biological inquiry, intimately connected to human intentionality and the noumenal freedom of the self, rather than employing them as explanations of biological activity. A quote from Kant illustrates my point:

> We may and should explain all products and events of nature, even the most purposive, so far as in our power lies, on mechanical lines—and it is impossible for us to assign the limits of our powers when confined to the pursuit of inquiries of this kind. But in so doing we must never lose

sight of the fact that among such products there are those which we cannot even subject to investigation except under the conception of an end of reason. (Kant [1790] 1963, Pt. Two, 74)

In other words, *explanations* in the life sciences were to be pursued along causal-deterministic lines, not vitalistic or teleological ones. Traditional explanatory appeals of the Aristotelian tradition and its heirs to formal and final causation were not to be employed within life science, even though they could not be eliminated as guiding ends of reason in its regulative function. But this concession to the mechanistic-reductionist program also imposed significant restrictions on the applicability of this same causal-deterministic analysis to the human subject. If the empirical subject can be made the focus of a causal science, the transcendental self conducting this scientific inquiry cannot. It was precisely the antinomy involved in this puzzle—how one could reconcile the universal determinism of nature with the obvious experience of inner freedom and moral choice—that Kant's philosophical program claimed to resolve (Kant [1787] 1963, B xxviii, 28). This enabled him to accept the teleological, intentional, and organizational aspects of living beings as properties that are impositions of our empirical experience of the living world upon consciousness. Such principles of the reflexive judgment in turn provide the necessary conditions for the very conduct of reductive life science. Put briefly, we can pursue our reductive analysis of life to the most extreme levels, but this inquiry is always within the structure supplied by the pre-reflective experience of organisms as more than physical mechanisms.

Kant's transcendental idealism has had many critics and this is not the place to try to deal with these. At least some who have argued for the possibility of a fully adequate "naturalistic" view have at least seen the force of Kant's arguments and have tried to engage them in the name of a more realistic interpretation of science and non-transcendental naturalism (Shimony 1992, chap. Two). I would agree with those who defend a stronger scientific realism than Kant provides on this issue. But the difficulty with many realist responses to Kant is that they advocate in turn a strong reductionism that fails to recognize the importance of the insights Kant has drawn to the surface in the domain of biology.

Some combination of these Kantian insights with a more robust scientific realism can be at least sketched out here. My own arguments are drawn from some of the reflections of philosophers of science within the phenomenological tradition (Kocklemans 1966) and the writings of Marjorie Grene and Michael Polanyi (Grene 1967; 1971; 1987; 1988; Polanyi 1964; 1968).[12]

The issue on which I will focus in this discussion is the importance in the life sciences of the epistemic activity of selecting out or "thematizing" certain features and processes of organisms in order to subject living beings to systematic scientific study. Analytic biology engages in this activity by breaking down the complex functioning of whole living beings into various components that can be studied by intensely reductive methods. By making aspects of living beings the objects of primary focal interest or attention, this implies that other properties of living beings can, and even must, be "backgrounded" or rendered subsidiary objects of attention. To employ an example, a cellular biologist works within a domain of study that might be defined by pure cultures of specific cells derived at one time from multicellular organisms. As these studies are pursued more deeply, one might render not only the whole organism, but even the complex organizational and hierarchical properties of the cell, or even those of the nucleus of the cell, as a pre-given background set of conditions and structures within which one pursues the reductive analysis of the cellular and biochemical mechanisms to increasingly refined levels. A cell can be isolated, microprobed, and macerated. Its components can be analyzed out biochemically or by ultracentrifuging. Its membrane can be dissected out, fixed, and subjected to electron microscopy. There are no in-principle limits to such analysis of the cell within the specified zone of attention it defines as a holistic structure.

We also can, by a similar process of selection, pursue this reductionism even further by making the complex structure of the DNA molecule a subsidiary background within which biochemical and physical analysis of its components can be pursued to the atomistic

---

12 In bringing these authors together, I am not claiming that Polanyi or Grene are endorsing the main themes of the Phenomenological tradition. But I do see Polanyi in particular as developing in different terminology some of the same themes as can be derived from some dimensions of Husserl and Merleau-Ponty as these have been applied to scientific questions by Kocklemans. Marjorie Grene's positions have a much stronger tie to Aristotelian realism, and she argues in several of her writings for the role of teleological purposiveness in organic life. I have explored these issues in my essay "Teleology and Form Revisited" (Sloan 2000) to appear in the Marjorie Grene festschrift (Burian and Gayon 2000).

and quantum-mechanical level. But nothing in this analysis eliminates or "reduces" these complex organizational properties to their material conditions. As Michael Polanyi argued in a perceptive short essay (Polanyi 1968), the complex processes of life occur within "control hierarchies" that impose limiting conditions on the specific ordering and directionality of the biophysical processes occuring within the cell or its subordinate structures. The claim that one can "explain" these organizational properties by the specification of the terminal results of this analysis is always a claim that presupposes the prior recognition of organized and hierarchical properties.

The importance of these points has primary bearing when we situate human beings within this pursuit of the reductive project of modern biology. For much of modern biological and biomedical inquiry, the intentional and self-conscious human being who is conducting scientific inquiry has simply been made subsidiary and backgrounded in the analytic pursuit of a scientific understanding of life and consciousness itself. Within the focal domain then delimited, there seems to be no limit to the reductive explanation of any human activity or property, either physical or mental. But such reductive analysis is only one *interest* among many possible interests of the scientific observer. Human beings as intentional, conscious agents can also be made the object of our focal attention, in which case the whole biological apparatus and functioning of the human body is made subsidiary to this inquiry, with the concern now with the domain that constitutes language, art, culture, ethics, religion, friendship, and love—in other words, the domain of *human* existence.

These points may seem obvious. But it is remarkable how they have either been forgotten or ignored in the literature of modern life science, not only in the writings of the popularizers and proselytizers of strong sociobiology, genetic reductionism, and explanatory cognitive neuroscience, but also in the textbooks upon which are educated future physicians, nurses, genetic counselors, and biological scientists. The problem presented by the success of the contemporary reductionist program in biology, as it has subtly shifted from a methodology to a set of claims about biological reality, is that it easily loses sight of the fact that reductive analysis of life takes place within a framework that has merely backgrounded the self-reflective, intentional and moral dimensions of human existence. It in no way has eliminated or "explained away" these interests or the intentionality of consciousness any more than my structural "genes" determine the writing of this sentence or the intentionality behind its composition.

I do not offer this as a defense of some kind of hidden vitalism or a dogmatic assertion of "top-down" causation. My point is more epistemic than metaphysical. Recognition of the ability of human attention to shift focus from level to level, rendering at each of these some aspects focal, and others subsidiary, requires us to take seriously the role of the human knower in science and the reality of the human interests behind our biological science. Hence one need not question the value of the methodological reductionist program, but only its slide into a set of metaphysical claims made in the name of science. It demands in particular that we look more closely at the claims of this program when it is turned upon human beings.

However one wishes to account for the distinction of knower and known to which I have drawn attention is not as important as the recognition that reflective consciousness and human freedom stand historically as *preconditions of* the Cartesian tree rather than its terminal fruits. Recognition of this simple, but crucial, point offers some hope for reorienting the relationship of the cognitive, vital, physical, and ethical issues that at present seem deeply dislocated by the success of the reductionist revolution in biology. If we are to unlock the giant of a new biotechnological genetics and apply this to human beings with more tolerable consequences than those that resulted from the utopian hopes of the eugenics movement of the 1920s and 30s, it seems imperative that we achieve some better balance between reflective human interests and the applications of technological knowledge. Consideration of this difference will also free us to oppose, when necessary, our scientific interests on grounds that reflect other human interests and concerns.

One further dimension of the Cartesian tree seems relevant here. This centers on those interests we might term theological or religious in character. In formulating his prophetic vision of the interrelations of knowledge, Descartes also located theological issues in a novel position with relation to the new science, placing them at the level of an a-priori *precondition* of his mechanistic and reductive science, rather than making these issues to be dealt with at a later stage in the development of the Cartesian tree. This theological foundation alone ensured that conclusions of human reason about the natural world, at least when it was understood in its reductive and mechanical dimensions, corresponded with the true structure of nature.[13] Because he placed theological questions at this foundational epistemic level, Descartes made no venture along the road of "design-contrivance" natural theologies that were to prove the nemesis of the English-language solution to the science-theology relationship. In the Cartesian ordering of the sciences, theological questions were not to be usefully entangled with scientific inquiry, nor did theological conclusions flow immediately from scientific investigation. This did not, however, deny the validity of theological and religious interests of human beings in relation to the new science.

The great conflicts of the past between science and religion, first over Copernicanism, and later over Darwinism, have involved what have seemed to be insoluble conflicts between two competing explanations of the same body of phenomena—the motions of the heavens and earth; the origin, distribution and development of living beings in relation to the history of the earth. I would suggest, however, that the issues presented by contemporary life science are not of the same character as represented by these classic cases. In our present context, a new level of conflict between theology and science is being generated not by any single issue or theory—it can be argued that molecular biology is not even governed by a unifying idealizing theory (Schaffner 1996)—but by the convergence of a wide range of inquiries—evolutionary biology, molecular genetics, reductive physiology, naturalistic scientific cosmology and cognitive neuroscience—in a totalizing naturalistic world view that claims to give a comprehensive explanation of all aspects of existence.

Viewed against the backdrop of the historic Cartesian project, the attainment of such an eventual "consilience" of the sciences on naturalistic and reductive grounds was surely one of the desired goals of Descartes' vision of the new science. But from this perspective, such a consilience depended on a warrant for the realistic knowledge claims being advanced. Descartes perceived with penetrating insight that such claims could not justify themselves. The strong reductionist

---

13 This point is most clearly made in *Meditation Five*: "Thus I see plainly that the certainty and truth of all knowledge depends uniquely on my knowledge of the true God, to such an extent that I was incapable of perfect knowledge about anything else until I knew him. And now it is possible for me to achieve full and certain knowledge of countless matters, both concerning God himself and other things whose nature is intellectual, and also concerning the whole of that corporeal nature which is the subject-matter of pure mathematics" (Descartes [1641] 1984, 1, 49).

and naturalistic program has taken the scientific realism of Descartes for granted, but has ignored the foundational questions that justified it in the first place. Recognizing once again the Cartesian location of theological issues at the *base* of the tree, as the grounding for the realistic knowledge claims of the new science, at least opens the door for a more productive discussion between adherents of modern scientific naturalism and traditional theism than has generally been true in the past. At the very least it can refocus these questions away from a "warfare" that pits competing explanations of the world against one another, and redirects them to a deeper recognition of the various interests of human beings, some of which are theological in character. Ideally from this can emerge a productive "interpenetrating dialogue" of science and religion rather than a fruitless confrontation.

The return to the historical presuppositions of contemporary life science I have attempted in this essay surely does not supply immediate solutions to any of the practical ethical questions which face parents, clinical geneticists, pediatricians, genetic counselors, lawyers, legislators, and insurance actuaries in the face of the genetic revolution upon us. But solutions to these practical issues at many points involve, either explicitly or implicitly, answers to more fundamental questions about the ethical relations of generations to one another, about human nature, about the value of human life, about the knowledge claims upon which our ethical decisions rest. It requires that some position be taken on basic moral issues that involve more ultimate philosophical positions.

As we continue to open the "book of life," to borrow from the title of Lily Kay's essay, my appeal is for a concerted effort to recover the "human" phenomenon in the scientific enterprise itself, a deep and substantial recovery that can enable us, in the name of ethical principles derived from our reflective human concerns, to resist and even refuse to pursue some of the technological options that the genome project and its related enterprises in reproductive science now present. Ideally we can then use the genome project for worthy human ends, while avoiding the grim scenarios that many have projected as its inevitable social product. This suggests that the full completion of the Cartesian tree implies, contrary to Descartes' own sketch, that wisdom emerges not out of our physics and biophysics, but from our recognition of our science as a product of human consciousness reflecting upon itself.

Part One

Origins of the Genome Project
Introductory Comments

The opening historical section of the volume explores the origins of the genome project from several different points of view. The opening essay by Timothy Lenoir and Marguerite Hays displays the way in which the "big science" of the Manhattan Project was extended in the post World War II period into the field of biomedicine, forming a collaborative research effort of clinical medicine and biophysics that reached considerably beyond research into human genetics. Jean-Paul Gaudillière then explores the developments in pediatric medicine and its connection with genetics in a French context in a similar historical period, displaying some of the ways in which medicine and genetics were associated in the French clinical setting to form a strong association of human genetics and pediatrics. A commentary by Robert Bud then helps bring into focus some of the important issues in the Gaudillière paper.

Following this is a more theoretically-oriented essay by Lily Kay, exploring the role of metaphors of "information" and "code" and the "language" of genetics that grew out of the important convergence of information theory and genetics. It also extends beyond these historical questions into an exploration of role of metaphors of code and information in constructing the way we conceptualize the living realm. Such interactions of theory, metaphor, informatics and genetics has formed one of the key components of the genome project. Hans Jörg Rheinberger's commentary further illuminates the role of language in the work of contemporary genetics.

John Beatty's essay then provides historical detail on the stages by which the "Manhattan Project" came to be connected to genetics, and the stages by which this eventually involved human genetics. His paper gives us both a long-term perspective, and a short-term history, and develops some of the connections of the HGP and the older Atomic Bomb Casualty Commission studies.

The essay by Alice Dreger on the use of metaphors of mapping in the presentation of the HGP to Congress underlines some of the political issues that surrounded the origins of the mass-sequencing project. In this provocative essay, she illustrates how nationalistic rivalries entered into the definition of a project that originally was to display the cosmopolitan dimensions of science.

The various historical perspectives provided by these papers displays with striking force the complex historical roots of modern "genomics," and the nature of the interpenetration of disciplines and different forms of training and socialization of the scientific community that has produced the dynamic enterprise we now see before us. They also illuminate the importance of the World War II period in bringing about the collaboration of government, politics and clinical medicine with the theoretical developments in biophysics and molecular biology, all of which have become involved in the HGP.

2

The Manhattan Project for Biomedicine

Timothy Lenoir and Marguerite Hays

A topic of central concern to policy makers since the close of the Cold War has been assessing the importance of federal investment in scientific research. With economic competitiveness replacing concerns about military security as a rationale for national funding priorities, there have been calls for a new contract between science and society establishing a closer working relationship between academe, industry and the national laboratories, and creating a supportive environment for the commercialization of innovative technologies generated through public research funds. A model for the new contract between science and society has been the Human Genome Initiative. Indeed the HGI has been heralded as a potential Manhattan Project for biology through the possibilities it affords of stimulating collaborative work among physicists, engineers, computer scientists and geneticists in the national labs which may prime the pump for commercially viable biomedical and information technologies analogously to the way in which the World War II Manhattan Project was the technological incubator for important Cold War electronics, computer science, military, and aerospace technologies (DeLisi 1988, 489; Kotz 1995).

While the call for launching a Manhattan Project for biomedicine may have the rhetorical ring of newness about it—and its formulation in terms of mapping the human genome certainly is new—the fact is that since the days immediately following the Hiroshima and Nagasaki bombings there always has been a Manhattan Project for biology and medicine. Our aim in this paper is to show how, well before the end of the War, the chief medical officers for the Manhattan Project began to contemplate plans for adapting the work they had done under extremely close security conditions to the postwar world. They not only contemplated how to continue the promising paths of research they had opened during the war but also how to carry their research outside the confines of military classification into the civilian world. We discuss their plans to establish academic and medical disciplines in health physics, biophysics, and nuclear medicine; we also explore programs they initiated to train medical personnel in the use of radioactive materials, as well as a strategic program to subsidize the development of radioisotopes, radiopharmaceuticals, and biomedical instrumentation as part of an effort to create the infrastructure of a self-sustaining biomedical nuclear industry. We argue that within roughly a decade, by 1960, this program of discipline building was essentially complete, and with it perhaps the most successful example of government technology transfer in the history of American industry.

ATOMIC MEDICINE

Initial efforts to use radionucleides for physiological studies and the first glimmerings of a medical research program into the effectiveness of using selectively localizing radioactive isotopes to destroy cancer cells preceded the Manhattan Project. In 1923 George Hevesy, working with Hans Geiger and Ernest Rutherford in Manchester, experimented with thorium^B to study the absorption and localization of lead in plants. He continued this line of exploration with naturally radioactive elements, but these early tracer studies were limited to heavy elements and very slow sampling techniques. In order to examine physiological function, radioisotopes of the lighter biologically active elements were needed. The first of these, heavy water, was made by Harold Urey in 1932, and a number of other radioactive isotopes followed in rapid succession. Ernest O. Lawrence's construction of the Berkeley cyclotron in 1931 and subsequent production of radiosodium obtained by bombarding sodium with deuterons in 1934 opened the path to physiologic tracers (Lawrence 1934). Cyclotron-produced radioactive phosphorus, P^{32}, was used in a number of pathbreaking investigations of phosphorous metabolism by Hevesy, Otto Chiewitz, Hardin Jones, Waldo Cohn, and John Lawrence. John Lawrence seized upon the preferential absorption of inorganic phosphorous in hematopoietic tissues and rapidly multiplying cells, such as malignancies, and used P^{32} to treat leukemia. This first therapeutic use of artificially produced radioisotopes occurred on Christmas Eve 1936 (Lawrence et al. 1939; 1940). Also about the same time, beginning in 1936, first investigations with radioactive iodine, I^{128}, were undertaken by Robley Evans, Arthur Roberts, and Saul Hertz in the thyroid clinic at Massachusetts General Hospital, but its 25-minute half-life rendered it difficult to use (Hertz et al. 1938; Evans 1969, 105). Stimulated by their Berkeley colleague Joseph Hamilton to construct an isotope of iodine with a longer half-life, Glenn Seaborg and J. J. Livingood devised a method of manipulating the cyclotron to produce a mixture of twelve hour and eight-day isotopes of iodine, I^{130} and I^{131} (Livingood and Seaborg 1938). This development not only suggested that radioisotopes might be tailor-made for specific needs, but also, given the specificity of iodine for thyroid tissue, a number of researchers, including Saul Hertz, Arthur Roberts, Robley Evans, Earle Chapman, John Lawrence, and Joseph Hamilton, pursued the localized therapeutic use of radioiodine in hyperthyroidism. In November of 1940 MIT physicists produced their first sample of the I^{130}—I^{131} mixture in their new cyclotron, using the Berkeley methodology. And in January 1941 Saul Hertz at the Massachusetts General Hospital and Roberts at MIT gave a patient the first therapeutic dose of I^{130}—I^{131} mixture, and started a program of study with thirty patients. In late 1941, the Berkeley group also began treating patients with hyperthyroidism with I^{130}—I^{131}, and, before the end of the war, several other groups were using this treatment. Marinelli and Oshry proved the therapeutic value of I^{130}—I^{131} between 1943–46 in their treatment of a patient with thyroid cancer (Seidlin et al 1946). Thus, by 1943 when the Manhattan Project got fully underway—the first nuclear pile went critical in Chicago on 2 December 1942—there were already a number of major successes and several well demarcated lines of research and therapy had been opened up by a small core group of biophysicist-physician teams in nuclear medicine.[1]

From its inception the Manhattan Project incorporated a substantial medical research program (Anon.1945). Beyond the need for medical care for project workers at Hanford, Washington, and Oak Ridge,

---

1 See Hertz et al. 1938; Hamilton and Soley 1939; Lawrence 1940; Hamilton 1942.

Tennessee, appropriate programs of industrial hygiene had to be instituted in Manhattan District laboratories and plants working with the unusual chemicals required for processing uranium. It was already well understood that the materials involved in these facilities were hazardous and that determining how to protect workers would require a considerable research effort.[2] Investigation was required to determine what effect absorbed uranium and plutonium compounds would have on the human body; what effect absorbed products of uranium (fission products from the pile process) and radioactive products produced in the actual explosion of the uranium and plutonium bomb would have; and what hazards might be encountered from absorption of other chemicals developed for use by the District. In addition, although the dangers of work with radium were fairly well known, it was discovered that the knowledge concerning the maximum safe exposures to various radiations was not well-founded, and that considerable research would be required to establish safe levels with certainty (Lawrence 1938, 1.1–1.2).[3] A natural corollary

---

2 John Lawrence called attention to the potential hazards for laboratory workers following his early work on isotope production following neutron bombardment with brother Ernest's accelerator: ". . . At the present time physicists rather than biologists are exposed to this potentially dangerous form of radiation; and it is imperative that those workers become familiar with the changes which take place in tissues of the body after sufficient doses.

We have as yet no knowledge of the possible delayed effects of exposure to small doses over a long period of time. However, the changes that occur after relatively large doses make it imperative that workers in laboratories where neutrons are generated protect themselves, by screening, from exposure. . . .The daily dose to those working with neutrons should not exceed one-fourth that accepted as the tolerance dose for x-rays. Whether daily doses of this magnitude over a long period of time will result in damage is not known, nor is information as yet available concerning the effects of neutrons on the skin" (Lawrence 1938, n.p.).

3 In his 1948 address to the Pharmaceutical Association, Stafford L. Warren, the Chief Medical Officer for the Manhattan Project, later provided a revealing sense of the uncertainties regarding exposure levels. Warren stated that for years dosage exposure to x-rays had been set at one-tenth of a roentgen by the Bureau of Standards. When Warren checked into the accuracy of this he found that the standard had been set out of 'thin air.' It was set a hundred times lower than the Bureau thought hospitals could achieve at the time: "In order to qualify this tenth of a roentgen as a reasonable figure and supply some experimental data against the time in the postwar period when the Project would be disbanded and we would have these thousands of employees who might claim compensation, we set up a dog, rat, monkey, mouse experimental program in which these animals had blood counts. . . . It has shown up the fact that a tenth of a roentgen is not safe" (Warren 1948, 3).

to these investigations was the need for an intensive research program designed to establish early signs of toxic effects from these chemicals or from radiation, and to develop specific measures useful for treating over-exposures.

From 1942–1945 the work of the Medical Division of the Manhattan Project was conducted at laboratories at the universities of Chicago, Rochester, U.C. Berkeley, Columbia, and Washington, at the Los Alamos laboratory and at Clinton Laboratories in Oak Ridge (Brundage 1946a; 1946b). The research programs conducted at these sites were extensive and wide-ranging, aimed at "the diagnosis and control of effects produced by exposure to radiations emitted by radioactive materials during experimental or processing operations as well as the chemical toxicity or localized radiation from such materials deposited within the body"(Warren 1946, n.p.). Among the programs organized and begun during the three-year Manhattan Project were studies of methods for the physical measurement of radiations of various types with an aim toward standardizing dosages of radiation to be used in biological experimentation and in measurement of radiation which might be found in a plant area (ibid., 4). Among the types of biological effects investigated were survival times or percentages that a given radiation dose will reduce the life span of different animal species, studies of genetic effects of radiation as manifested in the development of abnormal individual types from changes in the hereditary mechanism, histopathological changes in the makeup of various body tissues, physiological changes produced by the alteration of normal function of animal tissues following radiation, biochemical, and enzymatic disturbances presumed to be the potential source of these physiological abnormalities. Studies were also undertaken of the hazards due to handling the specific toxic materials used in processing uranium, plutonium, and a variety of fission products. Given the pre-war record of research in leukemia and polycythemia, efforts to understand the mechanisms by which radiation suppresses the hematopoetic system received priority, and investigations were made of therapeutic measures for treating acute radiation damage, including possibilities of replacement of hematopoetic elements destroyed by severe radiation damage (ibid., 5–6). Research and development of precautionary methods to be taken in the handling of special materials, the removal of hazardous dusts, reduction of skin contact, prevention of ingestion, and studies of methods for shielding against radioactive materials, ventilating and exhaust systems, and the development of remote control methods for processing radioactive materials received primary attention by the Medical Division.

The members of the Medical Division of the Manhattan Project were convinced by their wartime experiences that a continuation and expansion of their work would be essential to public health, and to worker health and safety in the nuclear age. Whatever the future military prospects of nuclear energy—and they seemed considerable—the potential of nuclear power and the application of other aspects of nuclear science for civilian industry seemed unlimited. Manhattan Project medical personnel were also convinced that their work in the Manhattan Project was the harbinger of a revolution in biology and medicine. Since the great discoveries in bacteriology and the introduction of instrumentation into medicine in the late nineteenth century, medicine had taken on the ideology and aura of "science," but medical practice remained very much art- and craft-driven. Of course, the war effort had resulted in the production and widespread use of the "miracle drug," penicillin, but the Manhattan Project opened up more expansive vistas for creating science-based medicine and for basing medicine on its own foundations of research and experiment than anyone had previously contemplated. In his address to the American Association for the Advancement of Science in 1948 Stafford Warren gave a sense of the excitement and sense of the opportunities he envisioned awaiting the implementation of the lessons learned from the Manhattan Project in a peacetime civilian setting:

> It was biological research at an entirely different tempo than we had ever experienced before: generous financial support, an organized program with men working in teams, a specific narrow goal for each group, pressure for speed, immediate application of results. . . . Even though discoveries in fundamental science were not pursued in any program, those discoveries were either made or were tantalizingly visualized for peacetime research. (Warren 1950, 143)

* * *

> Before the war, and even during the early part of the war, one occasionally heard the statement in research committees that "biological research ideas could not be bought with money," or that, "even with money, medical research was a sterile field in certain areas." It is evident from the war experience that the previous opinion was the result of frustration and isolation, for such statements are not heard any longer....

The best minds were not directed into research channels because there was no possibility of making a living in a research career. Medical and biological research was tolerated only as an offshoot of teaching.

In summary then modern developments in nuclear energy have influenced public health problems by showing the benefits and accomplishments that can be achieved by team work in correlated, large-scale biological and medical research; by devising new techniques and new instruments, and furnishing a copious supply of radioactive isotopes for general research purposes; by training many young men in new research fields; by aiding in bringing about a rejuvenation of biological and medical research; by creating a new field in industrial hygiene called radiological safety and a new profession, the health physicist (or biophysicist); by creating the greatest public health hazard of all time; i.e., the possible fantastic contamination that would result from an all-out atomic war, and by posing the problem of how to combat mass fear and how to prevent another war. In every phase of this development there is an obvious lack of trained men, not excluding the fields of social and political sciences. If war can be prevented there is no doubt that we can look forward to the solving of world-wide problems in health and welfare in both the strictest and the broadest terms. We all know the goals and we have the tools. (Ibid., 148–49)

These statements were an early public expression of the transformation of medicine Warren and his colleagues envisioned following in the wake of the Manhattan Project, a vision already acquiring institutional momentum required for its growth. A new era of biological and medical research populated by an entirely new breed of medical professionals, the biophysicists, health physicists, and practitioners of nuclear medicine, could be inaugurated by implementing the Project's organizational structure in the civilian biomedical sector. The new research medicine would be multidisciplinary, involving close working relationships among physicians, physicists, engineers, physiologists, and other biomedical specialties working in teams at newly minted medical research centers with well-equipped research laboratories. It was necessary to create the conditions for retaining the medical personnel gathered during the war and to establish programs of training and recruitment suited to reproduce the medical profile of this new breed of medical scientist. To nurture the new field of medical research, it was necessary to maintain the atmosphere of emergency that had enabled its wartime success, to retain and recruit academically trained medical and scientific personnel, and to remove secrecy surrounding the medical projects and publish useful wartime research results. Warren and his colleagues envisioned a hybrid setting, fulfilling the necessary security requirements of the military with respect to nuclear technology but facilitating a cooperative structure of research and training facilities in government, industry and academe.

Steps toward realizing this vision were taken through the founding of the Atomic Energy Commission, charged specifically with the responsibility to conduct research and development activities relating to, among others, utilization of fissionable and radioactive materials for medical, biological, health or military purposes, and the protection of health during research and production activities (U.S. Congress 1946, Section 3a). An interim Advisory Committee to the Medical Division of the Manhattan District which first met on 5–6 September 1946 in the Manhattan District offices in New York City, was established with Warren as chairman. It was continued as a permanent committee under the new AEC. At meetings of this committee in 1946 and 1947, Warren urged a general strategic plan for transforming the work of the Medical Division of the Manhattan Project into a launching pad for a wide-ranging program of medical discipline-building. A key element of Warren's vision for attracting and retaining appropriately trained scientific personnel was the creation of faculty and research appointments at universities with provisions for tenure (Warren 1946, 10).[4] Believing that men with scientific aspirations would only be attracted by an open community, a second element of Warren's plan was to declassify all Manhattan Project reports in the medical, biological, and health-physics fields, and organize a national meeting at an easily accessible central location open to scientific researchers from all parts of the country, at which Manhattan Project researchers would present information on medical aspects of nuclear energy.[5] In

---

4 In Warren's view the desirability of this proposal from the AEC's perspective was that: "teaching institutions may be utilized and work stimulated in all parts of the US; these men and their laboratories also offer facilities of use to the Manhattan District in emergencies and for medico-legal consultation" (Warren 1946, 9).

5 On the issue of declassification of medical documents the AEC was more guarded in its cooperation. A memorandum dated 17 April 1947, from Colonel O.G. Haywood, Jr., of the Atomic Engergy Commission to Dr. Fidler, U.S. Atomic Energy Commission, Oak Ridge addressed the subject of medical experiments on humans as follows:

"1. It is desired that no document be released which refers to experiments with humans and might have adverse effect on public opinion or result in legal suits.

connection with such a meeting, the third element of Warren's vision included creation of a new society for radiobiology as related to medical problems. A fourth, crucial element of Warren's proposal was that in connection with declassifying Manhattan Project research reports encouragement be provided to manufacturers of instrumentation for radiation measurement (Warren 1947a).

In the AEC's reply Carroll Wilson explicitly acknowledged the importance of fostering the disciplinary aims Warren had outlined by situating the proposed medical research units in universities and creating conditions for them to pursue independent research. Warren's proposed budget for continuing the interim work of the medical division was approved, and it reflected the commitment to locating the main research effort in universities.[6]

(contd.)
Documents covering such work should be classified "secret". Further work in this field in the future has been prohibited by the General Manager. It is understood that three documents in this field have been submitted for declassification and are now classified "restricted". It is desired that these documents be reclassified "secret" and that a check be made to insure that no distribution has inadvertently been made to the Department of Commerce, or other off-Project personnel or agencies.

2. These instructions do not pertain to documents regarding clinical or therapeutic uses of radioisotopes and similar materials beneficial to human disorders and diseases" (Haywood to Fidler 1947).

Other documents suggest that managing the image problem related to human experimentation continued, as suggested in the correspondence of Shields Warren with Albert Holland, Director of the Oak Ridge facility:

"Reference is made to your memorandum dated January 28, 1948, which requested reconsideration of the classification on the following documents:

CH-3592, entitled, "Uranium Excretion Studies."

CH-3607, entitled, "Distribution and Excretion of Plutonium in Two Human Subjects."

CH-3696, entitled "An Introduction to the Toxicology of Uranium."

MUC-ERR-209, entitled, "Distribution and Excretion of Plutonium."

It is the feeling of the Division of Biology and Medicine, the Advisory Committee for Biology and Medicine and the Technical Information Office of the Atomic Energy Commission that these documents should not be declassified" (Shields Warren to Holland 1948).

6 Wilson to Warren, 30 April 1947: ". . . .The Commission also hopes to strengthen the position of the Universities to attract personnel of high calibre to the medical research work, by authorizing them, with the approval of the commission to enter into contracts of employment with key personnel for periods up to three years.

The Commission is entirely sympathetic with the view of your Committee that research personnel engaged in medical projects should be encouraged to exercise their own initiative and should be given an opportunity to devote part of their time to

	1945–46	1946–47
University of Chicago	$2,500,000	$1,000,000
University of Rochester	1,700,000	1,200,000
University of California	250,000	250,000
Biochemical Research Foundation	120,000	25,000
Columbia University	80,000	75,000
University of Washington	60,000	30,000
Los Alamos	100,000	100,000
Clinton Laboratories		200,000
Other Installations for Miscellaneous Problems as Appropriate	100,000	1,000,000
Totals	**$4,910,000**	**$3,880,000**

TABLE ONE

(contd.)
pursuing lines of research which appear fruitful to them, even though not immediately related to specific items in the approved program for the particular project. Accordingly, the Commission is authorizing its Area Managers to approve such research, up to twenty percent of the time of the research personnel engaged on such medical projects. When such approval is given, the Director of the medical project will be required to certify to the Commission (1) that the research is useful and is not outside the general scope of the Commission's research interests, and (2) that the research will not unduly interfere with the progress of the work on the approved program at the project. The director also will be required to indicate separately in his reports to the Commission what research has been done under this authorization."

The memo went on to discuss guidelines for obtaining medical data:

"It is understood that your committee has recommended a program for obtaining medical data of interest to the commission in the course of treatment of patients, which may involve clinical testing. The Commission wishes to make clear to your Committee its understanding of the program which is being approved. The Commission understands that in the course of the approved program:

a. treatment (which may involve clinical testing) will be administered to a patient only when there is expectation that it may have therapeutic effect;

b. the decision as to the advisability of the treatment will be made by the doctor concerned."

Table One gives a breakdown of the approximate 1 July 1945–30 June 1946 budget. These funds were for fundamental, applied medical and biological research over a one-year period.

Although the budget reflected a twenty-percent reduction to remain consistent with the postwar downsizing of military expenditures, Warren concluded that "It is believed that approximately $5,000,000 per annum is an appropriate budget for a program as difficult and broad as this must be" (Warren 1946, 9)—a figure several times less than what Warren thought was actually needed[7]—and in his summary statement Warren went on to recommend a sustained commitment to continuing the research effort of the medical division well beyond the transition period for at least 10 years (ibid., 10).

Warren and his colleagues made use of the resources of the Manhattan District, and after January 1947 the AEC, to achieve this full-scale disciplinary program. A first step in this direction was the issuing of AEC contracts to universities for projects in nuclear medicine and areas related to civil defense. One of the contracts authorized was AEC contract GEN-12 to UCLA. This project, funded initially at $250,000 annually, was the nucleus of major developments in the fields of biophysics, radiology, and nuclear medicine at UCLA, and also the core of a cancer research institute at UCLA. Warren's assignment after the war had been to head up the collection of data from the Bikini tests as part of Operation Crossroads. While preparing to launch this effort he was approached by Robert Sproul, President of the University of California, to consider the deanship of the new medical school being considered by the Governor and the California legislature for siting in Southern California. Upon his return from the Bikini tests Warren agreed to move to UCLA, "because this was an opportunity to really organize a medical school from scratch. . . , to put the clinical and the basic science departments together. . .on a bigger scale"(Warren 1983, 994–95).

---

7 In his oral history interviews with Adelaide Tusler Warren later recalled:
"Well, when the committee and I resigned during this interim period late in 1947, we recommended that a budget of $28 million be given to the division each year because there was so much needed for the instruments and for the full support of the programs at Oak Ridge, Brookhaven, Argonne, Rochester, Berkeley, Los Alamos, and the new one at Los Angeles. We felt it was warranted. Shields Warren didn't believe this was so. Anyway, the recommendations on his side, I think, were only $6 million or so. Fortunately this did not cut either Rochester or Los Angeles, but it didn't do much for Oak Ridge" (Warren 1983, 1074).

The Manhattan Project was a crucial resource for these developments. Three of the original five founder faculty members of the UCLA Medical School—Warren, John S. Lawrence and Andrew Dowdy—were all colleagues and had worked closely together in building the Manhattan District facilities at Rochester. The new AEC projects were also a key resource for building up his new faculty:

> But now the chief problem was to get some faculty, hopefully who would be able to start their research in the Atomic Energy Project. I couldn't think of a field, except maybe psychiatry, where this couldn't be done, because the effect of fallout and radiation was so catholic. It entered every field and every discipline. For instance, Dr. John Lawrence worked in Rochester on the effects of radiation on the blood and had Dr. William Valentine as his assistant. Dr. Valentine had just finished his residency at that time. It was quite simple, when I picked Dr. Lawrence to be the chairman of medicine, to have him bring Dr. Valentine and later Dr. William Adams, both of whom were working in this same field in Rochester during the war on Manhattan money. All three could be cleared top secret, no problms, and I got one of the best clinicians in the country. I could set them down in the lab right away. So this made it easy for them to come, too; no hiatus—their work could go right on. (Ibid., 1075–76)

Consistent with Warren's strategy of maximizing available government and university resources, recruitment of Valentine and Adams also illustrate the importance to the new school of the nearby Wadsworth VA Hospital. Adams received his salary from the VA for several years, and both Adams and Valentine worked in research laboratories at the VA. The AEC project was crucial to building the research facilities Warren envisioned and for recruiting physicists, engineers, and other scientists without medical degrees into a medical environment. Contract GEN-12, provided a valuable source of contracts needed to train personnel and develop instrumentation for biomedical research (Warren 1983, 777–79, 1066–75).

Rapidly growing from a small group in 1947 funded at $250,000 to a roster of 145 persons in 1949,[8] the AEC project was from the very beginning a large support organization for Warren's design of a new research-based medicine fusing basic and clinical sciences modelled after the Manhattan Project. In square footage of its fa-

---

8 Due to security arrangements there was no official listing of persons employed in the AEC project. An internal phone directory dated 15 October 1949 shows a roster of 145 persons (Warren 1949).

cilities alone, the AEC Project occupied 33,600 square feet, whereas the space allocated initially to medicine and surgery totalled 31,000 square feet.[9] While the AEC Project was housed in facilities separated a short distance from the university medical center, projects such as the summer course "The Application of Nuclear Physics to the Biological and Medical Sciences,"[10] taught by Staff Warren, Andrew Dowdy, and Robert Buettner were typical of the integration sought between the AEC Project and the new medical school. The budget for the AEC Project incorporated separate line items for support of work in various basic science departments of the medical school. Numerous forms of collaboration developed within different sections of the AEC project and University clinical investigators. Typical was an important early project to develop scintillation counters and radiation flourimeters. Ben Cassen and Fred Bryan reported to Warren in 1948 that:

> Sufficient progress has been made recently by the Medical Physics section of this [the AEC] project in collaboration with the Industrial Hygiene division on development of prototype instrumentation for the measurement of roentgen rays, alpha particles, and beta activity in tissue in vivo, that it is now becoming desirable to make arrangements for the future clinical application and testing of these devices.
>
> These instruments are called tissue radiation fluorimeters and essentially consist of a small diameter light pipe in or insertable in a hypodermic needle. Any fluorescence or scintillation of a small amount of screen material placed on the end of the light pipe is transmitted to a photomultiplier tube outside the body. They can be used for measuring roentgen ray intensity in tissue or counting scintillation produced by the presence of radioisotopes or alpha emitters in the tissues. (Cassen and Bryan to S. L. Warren, 2 December 1948)

To carry their project forward Cassen and Bryan sought permission from Warren and the AEC to engage in clinical trials with

---

9 Arthur (1992, 92) reports as space allocations: 15,500 for medicine, 15,500 for surgery, physiological chemistry 18,000, physiology 17,000, pharmacology 5,400, obstetrics 9,000, pediatrics 9,000, and infectious disease 22,795 square feet. Building plans and space allocation for the AEC Project are discussed in (Buettner and Warren, 8 December 1948).

10 Taught 2–20 August 1948, see AEC Project annual report for 1948, Administrative Papers of Stafford L. Warren, UCLA University Archives, Collection 300, Subseries 600, Box 30, Fldr. "AEC Project UCLA."

investigators at the Birmingham Veterans Hospital affiliated with the UCLA medical school teaching program (more on this below). Stating his view that such research did not appear to fall under the AEC ban against human experimentation which was directed primarily toward the use of toxic materials in the human body, Warren urged pushing forward on this project at the earliest possible moment (Warren to Cassen and Bryan, 31 December 1948b). Such success in integrating the research projects of the AEC group with the clinical and basic science faculties led Warren to appoint a committee in 1950 to work out the formal integration of the AEC Project with the UCLA Medical School (Warren to Bryan et al., 11 August 1950). Both parties, the AEC and the UCLA Medical School, were pleased with these arrangements, and in 1953 when reporting to Warren on the approval of plans by the AEC to expand the facilities to twice the size (63,200 square feet) of the original building, Robert Buettner noted that the AEC was extremely pleased with the graduate research training program Warren had constructed at UCLA, particularly the academic influence on the program. The new AEC resources were to be used to expand the graduate training program in the Division of Biology and Medicine from 12 to 24 (Buettner to Warren, 8 December 1953, esp. 2, point [7]).

Staff Warren's visionary efforts to create a multidisciplinary medical center integrating clinical and basic science research by utilizing the resources made available to him through his connections with the Manhattan Project and his leading role on the Medical Advisory Board to the AEC was perhaps the most spectacular but by no means the only effort to recreate the institutional infrastructure of the Manhattan Project after the war. A similar vision was pursued by John Lawrence at Berkeley. Prior to the war in 1941, William H. Donner, president of the International Cancer Research Foundation, later named the Donner Foundation, donated $165,000 for the construction of a laboratory for medical physics. The Donner Lab became one of the sites after the war for the continuation of AEC support for nuclear medicine and biophysics. But well before the end of the Manhattan Project, Berkeley pioneers in nuclear medicine and medical biophysics were working to establish the institutional arrangements necessary for continuing their work after the war. In August of 1944 the Regents of the University of California approved the appointment of four faculty members, John Lawrence, Joseph

Hamilton, Cornelius Tobias, and Hardin Jones to the Physics Department, where they formed the Division of Medical Physics, the rationale being that "no distinction was made, at first, between the two types of work—physical and biological" (Westwick 1996, 7). The official approval granted by the Regents of the University of California on 1 July 1945 explicitly acknowledged the desirability of linking the appointments of these new physicists with the Medical School:

> There should be a nucleus of men including Drs. John Lawrence, Joseph Hamilton, [Paul] Aebersold and [Cornelius] Tobias who would hold joint appointments in the Medical School and in the Department of Physics, and then a second group of people who would play important parts in the development of medical physics such as Doctors Miller, Chaikoff, Hamilton, Anderson, Robert Aird, Strait, Low-Beer, Althausen, David Greenberg, C. L. A. Schitt and Soley. This second group would be carrying out experimental studies and therapeutic studies through the Medical School. (Budinger 1987, 9–10)

In addition to the links between Donner Lab medical physicists and clinical personnel in the University Medical School, the guidelines for the new appointments also stipulated that the Donner Lab have direct control of its own clinical research facility, effectively rendering it an outpatient department of the university hospital:

> there should be no limitation or "ham-stringing" of the freedom of the members of the subdivision of Medical Physics or others in the medical school to carry on treatment or investigations at Berkeley if research were the prime interest. (Ibid., 10)

BUILDING THE INDUSTRIAL INFRASTRUCTURE: THE ISOTOPE DISTRIBUTION PROGRAM

The most important of the components necessary for the nascent medical field was the support it received from the Atomic Energy Commission's program for distributing radioisotopes. Paul Aebersold who had completed his Ph.D. with Ernest Lawrence at the Berkeley cyclotron with a dissertation on the collimation of fast neutrons, became the director of the isotope program, transferring from Los Alamos Lab to Oak Ridge in January of 1946. Aebersold successfully headed the program for twenty years, until his retirement in 1965.

Isotope Program Milestones [11]

• 1 January 1947	Jurisdiction of the program turned over to civilian Atomic Energy Commission.
• 1 March 1947	Inititation of service whereby the radioisotope user may submit own sample for irradiation or neutron bombardment in the nuclear reactor.
• July 1947	First compound labeled with radiocarbon (methyl alcohol) available from the Commission.
• 25 January 1948	Enactment of regulation by the Interstate Commerce Commission for shipment of radioisotopes by common carrier (rail and truck).
• 1 April 1948	Initiation of a program making available, free of production costs, three isotopes—radiosodium, radiophosphorus, and radioiodine—for research, diagnosis, and therapy of cancer and allied diseases.
• June 1948	Opening of first Commission-sponsored training course in radioisotope techniques by Oak Ridge Institute of Nuclear Studies
• July 1948	Production and distribution of radioisotope labeled compounds by commercial firms.
• 25 February 1949	Extension of Commission's support of cancer research program to include the availability of all normally distributed radioisotopes free of production costs for use in cancer research.
• June 1949	Initiation of a program for commission support of commercial development and synthesis of selected radioisotope labeled compounds.
• June 1949	Initiation of a program for cyclotron production and distibution of certain long-lived radioisotopes not producible in the nuclear reactor.
• 20 July 1949	Enactment of regulations by Civil Aeronautics Board for shipment of radioisotopes by commercial aircraft.

TABLE TWO

The program for distributing reactor-produced radioisotopes was formulated between January and August of 1946 by a committee ap-

---

11 Source: Oak Ridge National Laboratory 1949, Appendix 1-A, 30–31.

pointed by the National Academy of Sciences (Atomic Energy Commission 1946). The enactment of the Atomic Energy Act on 1 August 1946 provided for the distribution of radioactive byproduct materials from nuclear reactors. The first shipment was on 2 August 1946 (Atomic Energy Commission 1946a, 1). Among the milestones of the early isotope distribution program, the items listed in Table Two are most significant for our purposes.

As authorized by the Atomic Energy Act, the directors of the Isotopes Division interpreted their mission in its broadest terms as "improving the public welfare, increasing the standard of living, strengthening free competition in private enterprise, and promoting world peace" (Aebersold 1949, 1). This mission could best be accomplished by "assisting and fostering private research and development to encourage maximum scientific progress" (ibid., 2). How best to transition the facilities at Oak Ridge, which had operated under maximum security conditions throughout its brief existence, raised challenging scientific, technical and managerial issues,[12] In the postwar effort to downsize and consolidate the nation's military research enterprise, consideration was given to transferring the basic research programs at Oak Ridge to Chicago. Such a move would have resulted in Oak Ridge losing its status as a national laboratory under the AEC's newly forming guidelines on national labs, and this loss of lab status was bitterly protested through a letter writing campaign to President Truman and AEC officials by the directors of the Oak Ridge facility, who feared their facility was being demoted to a chemical processing factory.[13] David Lillienthal, the Chairman of the Atomic Energy Commission, intended to achieve this objective of uniting government sponsored academic research and industrial prototype development through the establishment of the Oak Ridge Institute of Nuclear Studies.

From the earliest phases of its establishment the Isotopes Division at Oak Ridge was chartered with the explicit aim of performing re-

---

12 See Johnson and Schaffer (1994, 50–55) for a thorough discussion of the relationship of Monsanto to Clinton Laboratories and the transfer of plant operations from Monsanto to Union Carbide in 1947.

13 The AEC did not release a precise definition of a national laboratory. It granted the title, however, only to laboratories engaged in broad programs of fundamental scientific research, that had facilities open to scientists outside the laboratories and cooperated with regional universities in extensive science education efforts.

search and development that would stimulate industrial and medical uses of atomic energy.[14] In an era when American science and the free enterprise system had won the war, the projects generated under the aegis of the Atomic Energy Commission could not interfere or compete with American industrial interests. The goal of the Isotopes Division was to promote the use of radioisotopes as widely as possible through dissemination of information, through training programs, and through the distribution of radioisotopes to researchers, and at the same time the Division was to encourage private commercial interests to develop the industry. The projects generated by the AEC were to be handed off to private commercial interests as quickly as possible. It was decided that a reasonable policy for the Commission to follow in the synthesis and distribution of isotope-labelled compounds would be that the Commission only distribute materials which commercial firms were not prepared to produce and distribute in the foreseeable future and that the Commission withdraw from production as soon as a commercial firm demonstrated a satisfactory synthesis. Syntheses developed in Commission laboratories should be made freely available, and isotope-labelled compounds which were not of interest to commercial firms but which were of significance to scientific research should be made by Commission labs (Anon. AEC 1947, "Minutes," 9).

While Aebersold and the Isotopes Division were eager to put the entire industrial production of radioisotopes in the hands of commercial vendors, including reactor production as soon as feasible, they repeatedly argued that the cost of building the reactors and other facilities would make radioisotopes prohibitively expensive. Thus from the beginning they followed the policy established within the Manhattan Project of government ownership of the reactor and operation by management contractors from private industry. Citing separate studies by Bendix Aviation and Tracerlab on the feasibility of private radioisotope production in 1951, Aebersold urged, "It is obvious that the Commission must supply radioisotopes until private industry is ready and willing to produce them"(Aebersold 1951, 13–14) and he doubted that reactor production of radioisotopes would become a truly comptetitive enterprise in the near future. One of the approaches taken to commercialize production of radioisotopes was to

14 These concerns are clearly evident in the early memos of Paul Aebersold. See Aebersold 1946.

encourage the entry of intermediate producers of isotope-labeled compounds. Early entrants into this field were Tracerlab of Boston and Abbott Laboratories of North Chicago. Tracerlab submitted a proposal to synthesize C^{14} labeled compounds for medical research use (Tracerlab 1947). According to the conditions of the arrangement worked out in December 1947, Tracerlab received 100 millicurie allocations of C^{14} from Clinton Laboratories in Oak Ridge which it could stockpile and process into a variety of synthesized intermediates. Abbott worked under a similar arrangement but quickly pursued the lease of a site in Oak Ridge for a radiopharmaceuticals processing plant. By 1952 Abbott had built a plant in Oak Ridge adjacent to one of the main reactors, and by 1955 announced plans to double the floor-space and triple the capacity of its radiopharmaceutical processing plant (Aebersold 1951, 9). Although firms like Abbott and Tracerlab were beginning to get into the isotope processing business, in the early stages of the isotope program they could not meet all the needs of the scientific and medical researchers Aebersold was attempting to stimulate, nor could they generate a synthesis program on a scale adequate to the needs of the experienced research groups already within the AEC labs. Considerable delay in the research programs at these facilities (by as much as 6 months according to one Commission estimate—memo 3 September 1947—of delays caused for Memorial Hospital, New York, Western Reserve Medical School and others in medical and cancer research) would have been caused by enforced dependence on commercial producers. Moreover commercial producers were not in a position to handle very large quantities of radioactive material (in the hundreds of millicuries) safely.

In light of such concerns another strategy developed by the Isotopes Division was to encourage AEC labs to develop the production and distribution stages of processes for synthesizing radioisotope-labeled intermediates they might use for their own research purposes. It was argued to be more cost-effective to add the additional capcity required to increase the batch size of isotopes beyond the needs for immediate experimental work and distribute them to off-commission researchers than it would be for commercial producers to enter the market. An effective example of this strategy was implemented in the Berkeley Radiation Laboratory, where Dr. Melvin Calvin had developed numerous effective, high-yield syntheses of C^{14} labelled compounds. Concerns were raised about the advisability of entering the business of providing intermediate compounds by Carroll L.Wilson, the General Manager of the Oak Ridge Laboratory, urging that "the making of compounds may well represent the basis for a useful business in the manufacture of chemical intermediates or final products. Unless the Commission considers very carefully the various steps in this direction, it may seriously prejudice the possibility of building a small but useful industry in the production and sale of compounds containing radioisotopes." Such concerns were countered by the view that:

> The program will actually improve the opportunity for such entrance by immediately helping to build up the market and by accumulating data on synthesis procedures and costs as well as on degree of demand. This should enable the Commission within 12 to 18 months to offer private concerns a definite industry and market. It will save private firms a considerable inititial investment in C-14, in facilities and development costs without their having even an approximate knowledge of the market. It will allow the Commission time to perfect the techniques employed in the syntheses and to aid in the training or dissemination of information that will be required for larger scale off-Commission operations with C-14. (Williams 1947, 6)

As a guarantee to industry that the Isotope Distribution Program was intended to promote scientific research and facilitate rather than compete with private industry in promoting the commercial use of nuclear isotopes the agreements reached in the 30 October 1947 meeting between AEC officials and industry representatives specified that syntheis processes which had been developed in Commission laboratories should be made freely available, and that the Commission would make isotope-labeled compounds which were not of interest or profit to commercial firms but were important for scientific research (Anon. AEC 1947, "Minutes," 6).

In addition to contracting within AEC laboratories for the development of economical syntheses and their initial scale-up to production level, the Isotope Division also operated a program for purchasing on open bids from private laboratories special compounds of special interest for biological or medical research in cases where the synthesis was difficult and the demand uncertain. Examples of compounds developed in this manner were labeled folic acid, thiouracil and hormones needed for cancer research, labeled vitamins, and labeled DDT (Aebersold 1951, 12). These programs for research and development of syntheses as encouragement to the commercial production of medically relevant isotopes succeeded in generating more than 225 compounds by June of 1952 (ibid., 13).

In the 1949 formal review of the distribution program, Paul Aebersold reported that Oak Ridge distributed radioisotopes for cancer research free of production costs. The total expenditure for production and free distribution of radioisotopes for medical research in 1949 was $300,000. Aebersold projected that the cost of this free distribution program for 1950 would be $450,000 (Aebersold 1949, 8–9). In addition the AEC budgeted $100,000 in 1949 for the production of isotope-labeled compounds, including $25,000 for research on new synthesis methods, and provided an additional $100,000 for cyclotron-produced radioisotopes. In his general address to the AEC for the 4 March 1949 meeting, Aebersold defended the level of promotional effort as essential, and he urged even larger future government support for these programs:

> The Government can go further than above and even distribute the irradiated materials and basic products at less than the already inherently low production costs. This is because of the expectation of great returns to the general national health and welfare from the wide scale use of the materials in medicine, science, agriculture and industry. Although the cost to the govenment of the isotope distribution program is not over one million dollars a year each for radioactive and stable isotopes, the benefits resulting from this investment, through all the investigations and applications made possbible by the progam, could in not too many years hence easily be worth 10 to 100 times the original investment. (Ibid., 3)

Aebersold assured his AEC colleagues that such Government promotion of isotope usage would not interfere with private enterprise, "but actually encourages new areas of enterprise for industry as well as profitable improvements in existing industry" (ibid., 6). The early success of this program in launching the medical and industrial uses of radioisotopes was considerable, and every indication suggested it would continue to grow. Thus in a "Memorandum of Information" on the program filed in 1951, Aebersold projected the total sales volume would double within the next five years, and he reported that radioisotopes had become so widely adopted in medical research that their use would continue despite removal of government subsidy, although at a retarded rate (Aebersold 1951, 30). Industry interest in radioisotopes, while initially small, was increasing rapidly after the first three years of the program (commercial distribution was authorized by AEC-108, 2 June 1948, see Table Two above). Aebersold noted that 183 of the 800 institutions receiving radioisotopes in 1951, or approximately 23%, were industrial organizations. This figure represented a 50% increase over the previous year. Industrial use of cobalt 60 was especially strong, growing by 552%, due principally to its importance in radiographic testing in several industries. But the importance of government subsidy for the new field was undeniable. As Aebersold reported, while Union Carbide projected "sales" of radioisotopes exceeded $1-million for 1952, the $450,000 AEC subsidy for the free distribution of medical isotopes, combined with the approximately $200,000 per year of the Oak Ridge National Laboratory's "sales" of isotopes to other AEC installations meant that approximately 65% of the costs for producing and distributing isotopes were subsidized by the AEC. Furthermore, when it was considered that many of the institutions purchasing isotopes did so with funding supplied by contracts through a variety of AEC divisions, the subsidy was considerably higher (ibid., 31).

THE OAK RIDGE RADIOISOTOPE
TRAINING PROGRAM

From the beginning of Paul Aebersold's career as director of the Isotope Distribution Program it was clear that the major obstacle to overcome in creating conditions for greater commercial participation in production and distribution as well as the utilization of radioisotopes in industry was the lack of personnel trained in radioactivity techniques. The lack of training was considered far more of a problem for industry than either the fields of medical or scientific research. Aebersold and the Isotopes Division were the primary forces in setting up training courses in radioisotope techniques by the Oak Ridge Institute of Nuclear Studies beginning in June of 1948. Through publication in scientific and technical journals, and by distribution of fliers sent to universities and medical schools and companies encouragement to attend these courses was given to a wide range of potential candidates in medicine and industry (Oak Ridge

Institute 1947, 1).

In its first year the training school accepted forty-eight students at a time into one-month training courses. The courses were intended primarily for technicians in university and industrial laboratories, agricultural experiment stations, medical schools and other organizations who might desire to employ radioactive isotopes in their research programs but did not have the appropriate personnel or equipment for such work. The goal of the course was to develop familiarity with the uses of radioactive isotopes as a technique and research tool in tracer studies. Trainees were selected on the basis of previous training and experience of the individual applicant, but also on the basis of evidence presented that the institution they represented intended to establish a laboratory for isotope tracer work. Selection was also made to ensure wide geographical participation and diversity of the type of institution and research activity (Brucer 1951). Trainees paid a $25 fee for the course and had to defray their own travel and living expenses.

The instructional program was divided equally between laboratory and lecture courses from 9–12 noon Monday through Friday. Monday and Friday afternoons were reserved for seminars. Occasional afternoon demonstrations of equipment by commercial manufacturers and conferences were also part of the program. The laboratory was partly done by demonstration and partly direct experiments performed by trainees. A demonstration lab was scheduled on Wednesdays from 1:30 to 4 PM. Individual lab work was done in smaller groups of twelve trainees. The lecture topics included basic nuclear physics, instruments (such as ionization chambers, Geiger-Mueller counter tubes, preamplifiers, electroscopes and photographic detection, and analysis by mass spectrometry), design of tracer experiments, radiochemistry, design of small radioisotope tracer laboratories, health physics and precautions, production of isotopes, and sources and procurement of isotopes and equipment. There were "special lectures on radiation effects on gene mutations, chromosome breaks, and mitosis; nature of radiation burns, radiation induced cancers, electron microscope techniques and general biophysics" (Brucer 1951).

The isotope tracer technique school was extremely important for the development of nuclear medicine and for the nuclear industry generally. In the first formal review of the Isotope Distribution

Program in 1949, Paul Aebersold reported that practical considerations related to the recruitment of qualified teaching staff in the program had limited the number of students that could be accepted and trained in the program to 200 per year, but that more than three times that number of applications was received from qualified applicants (AEC 1949, 7). In response to this demand the Navy opened a training course similar to the Oak Ridge school at Bethesda, Maryland. The contributions of both the Bethesda and Oak Ridge schools to the growth of the field was impressive; in its second annual report to the AEC on the progress of the school, the Oak Ridge Institute of Nuclear Studies reported that the total number of participants who had attended the basic course to date was 963. Another 120 persons had attended the advanced biochemical research sessions (Brucer 1952). But in spite of the apparent success in attracting trainees to the isotope tracer technique program, Aebersold reported that while medical interest in isotopes was strong, "Industry has been slow in utilizing radioisotopes in research, in process control and in new products. . . . Industrial uses will not increase rapidly until it can be demonstrated that the overall costs of procurement, shielding, instrumentation, and special handling will permit economic advantage to the user" (AEC 1949, 8–9). There seemed to be no shortcut to alleviating the problem other than to continue and possibly intensify the promotional programs.

BUILDING A CLINICAL INFRASTRUCTURE FOR NUCLEAR MEDICINE: THE VETERANS ADMINISTRATION (VA) HOSPITALS

A crucial component evident in the discussion of the programs at Berkeley and at UCLA discussed above was establishment of links from the AEC projects both to academic science disciplines and to clinical facilities for medical training and research. The Donner Lab group worked persistently at increasing its immediate access to patients for clinical research, unrestricted by the necessity of having their research initiated and approved by the Medical School.[15] The

15 See Lawrence to Deutsch, 23 April 1947. Lawrence urged Deutsch to expand clinical access of the Donner group: "The suggested revision of paragraph one would result in interference with academic freedom to initiate research, which would be a very dangerous situation in the University. Furthermore, this revised paragraph

problem was resolved eventually by construction of the Donner Pavilion, a special two-floor wing costing $191,000 added to the Cowell Hospital student health center in 1954 for research in radio-biology under supervision of the Donner Laboratory. At UCLA Staff Warren faced the difficulty that no hospital facilities yet existed and building completely new ones required financial resources exceeding funds available to him from the allocation of the state legislature. The solution for this problem confronting efforts to transition the work of the Medical Division of the Manhattan Project to the postwar civilian medical enviroment was provided by the VA hospitals.

A set of developments with direct implications for the clinical in-frastructure of nuclear medicine within the VA began with the atomic bomb tests at Bikini in 1946. After Japanese forces had been driven from the Marshall Islands in 1944, the islands and atolls, Bikini among them, came under the administration of the U.S. Navy. In 1946 Bikini was selected as the site of Operation Crossroads, a vast military-scientific experiment to determine the impact of atomic bombs on naval vessels (Weisgall 1994). The world's first peacetime atomic-weapons test was conducted at Bikini on 1 July 1946. A twenty-kiloton atomic bomb was dropped from an airplane and exploded in the air over a fleet of about eighty obsolete World War II naval vessels, among them battleships and aircraft carriers, all of them unmanned. The second test, on 25 July, was the world's first underwater atomic explosion; it raised an enormous column of radioactive water that sank nine ships. Stafford L. Warren at that time still in his position as Medical Director for the Manhattan Engineering District had been assigned in April 1946 to serve the

---

(contd.)

would prevent members of the Division of Medical Physics who are not licensed physicians and indeed others in the University, (such as Doctor Hardin Jones, Doctor Cornelius Tobias, Professor C.F. Cook and numerous others on this campus) from carrying out any study on a human being unless it were initiated and approved by the chairman of a division of the Medical School, even though the particular investigation were a safe one." Lawrence went on to suggest: "With reference to the revised paragraph five. . .we would suggest the following revision: It is recognized that in order to have a continuing supply of patients who need the types of investigation that will be proposed, contact with the hospital and out-patient department with large clinical facilities is advantageous. It is to be recommended therefore that the University of California Out-Patient Department and Hospital can be used for obtaining certain specific patients for study. The Medical Physics work can be considered an Out-Patient Unit of the University of California Hospital."

Joint Task Force One for the Operation, charged by Leslie Groves with setting up a radiological civil and military defense to protect the troops and then to create the information which would enable a force to occupy an area whether it was military or civilian (Warren 1983, 821–22). Dr. George M. Lyon, a pediatrician from Huntington, West Virginia, was Warren's chief medical officer on Operation Crossroads. Lyon, who had also served during the Manhattan Project as an observer at the Alamogordo test, assisted in designing the gathering of data on fallout and related exposure effects of the Bikini detonations. Lyon's next assignment was as Special Assistant for Atomic Medicine in the Office of the Chief Medical Director of the Veterans Administration.

The Atomic Medicine Division of the Veterans Administration was initiated out of concerns raised about potential future legal claims arising from Operation Crossroads, similar future atomic tests, and from human exposure to ionizing radiation in laboratories and processing plants. In a summary history of the program dated 8 December 1952 Lyon states that at the meeting convened with General Leslie Groves and others by General Hawley in August of 1947, it was agreed that a Central Advisory Committee be formed consisting of Stafford Warren, Hymer Friedell, Shields Warren, Hugh Morgan, and Perrin Long, all of whom had participated in either the Manhattan Project or Operation Crossroads, to advise on steps to be taken for dealing with potential alleged service-connected disability claims associated with atomic energy. The Central Advisory Committee held its first meeting on 5 September 1947. According to Lyon, no one on the Committee was interested in drawing public scrutiny to problems with nuclear energy, and they were concerned that creating alarm by drawing attention to suspected grounds for disability claims might jeopardize the positive development of nuclear medicine. It was decided to downplay the negative concerns behind the formation of the program by characterizing it primarily in terms of research aimed at bringing veterans the benefits of medical breakthroughs connected with the use of radioisotopes carried on during the War. Lyon wrote:

At this time (September 5, 1947) the objectives of the Atomic Medicine Program were formulated and the broad aspects of a scheme for establishing a radioisotope program to support the more inclusive Atomic Medicine Program were drafted. The advisory committee was given the name, "Central Advisory Committee on Radioisotopes", as it was not desired at this time to publicize the fact that the Veterans Administration might have any problems in connection with atomic medicine especially the fact that there might be problems in connection with alleged service-connected disability claims. The committee recommended, (a) the establishment of an Atomic Medicine Division within the Department of Medicine and Surgery and the appointment of a Special Assistant for Atomic Medicine to head up the Division and to represent the Chief Medical Director in the handling of atomic medicine matters, and (b) the establishment of a Radioisotope Section to implement a Radioisotope Program. It further recommended that, for the time being, the existence of an Atomic Medicine Division be classified as "confidential" and that publicity be given instead to the existence of a Radioisotope Program administered through the Radioisotope Section. General Hawley took affirmative actions on these recommendations and it was in the manner described that the Radioisotope Program was initiated in the Fall of 1947. (Lyon 1952, 554)

A related mission of the Division was to prepare for the possible role of the Veteran's Administration in connection with civil defense. Specifically the VA was charged with developing a program of training and education in civil defense:

A *major objective* of the Radioisotope Program was, from the very beginning, to provide the Veterans Administration with qualified professional, scientific, and technical personnel, as well as the specialized facilities required, (a) to meet the varied and unique problems of atomic energy that might be of concern to the Veterans Administration, particularly in respect to problems associated with the study and analysis of alleged service-connected disability claims, and (b) in connection with certain responsibilities that the Veterans Administration may have in civil defense. (Ibid., 554–55)

Thus, it was decided to emphasize the VA Radioisotope Program in the public eye, describing it primarily in terms of its contribution as research in support of civil defense programs and in terms of the potential of research with radioisotopes for improving clinical medicine affecting veterans, particularly through the development of diagnostic aids. The Division's original mission to collect data and conduct research for evaluating potential disability claims became a minor effort (United States Advisory Committee on Human Radiation Experiments 1995, 477–79).

The emphasis on the basic biological and clinical research character of the radioisotope program was the principal message in the letters inviting Stafford Warren, Hymer Friedell, Shields Warren, Hugh Morgan, and Perrin Long to serve on the Central Advisory Committee sent by E. H. Cushing, the Assistant Chief Medical Director for Research and Education of the Veterans Administration (Cushing to S. L. Warren, 15 August 1947). Cushing also emphasized that benefits the contemplated program would offer the larger medical research community:

> Within the hospitals of the Veterans Administration there is an important reservoir of valuable clinical material. The administrative facilities and the professional organization of the Veterans Administration in large measure tend to favor the harvesting of valuable information through the conduct of investigations of an approved nature within these hospitals. The close association of the staff of many of these hospitals with outstanding teaching and research centers through the Dean's Committees favors the conduct of investigations and the appropriate utilization of radioisotopes within certain hospitals of the Veterans Administration on a high plane of medical acceptance. (Ibid., 2)

Pointing to the use of P^{32} in the treatment of patients with chronic leukemia and with Hodgkin's disease as an example, Cushing observed that the VA offered controlled conditions permitting statistical evaluation on a larger scale than anywhere else. "As time goes on," he noted, "there will be continuous opportunity for similar contributions" (ibid.).

Summarizing these developments we can observe that the "research ethos" in the immediate postwar period nurtured a fruitful convergence of interests among the Veterans Administration and discipline-building physicians, such as Warren. For their part physicians like Warren seeking to transform clinical medicine into a science-based experimental research field required new hospital training and research facilities to accommodate their plans of expansion. In the immediate postwar era, traditional academic positions were scarce and did not always fit the profile of the emerging new biomedical researcher. The construction of the radioisotope research units established within VA hospitals provided crucial enabling components for this program in terms of positions, facilities, and "clinical material." For its own part, through the creation of research facilities in its hospitals for radioisotope work as well as other research, the Veterans Administration benefitted by establishing close working relations with academic medical programs that could aid in solving its desparate shortage of qualified medical staff, intensified by the increasing numbers of veterans requiring medical care.

At the end of the first year of the program eight Veterans Administration hospitals had been equipped with Radioisotope Units.[16] Each of these units was established in close association with local medical schools having active radioisotope programs. The five year history of the program demonstrates a steadfast effort to expand the number of radioisotope facilities: in 1949 there were twelve VA Radioisotope Units, fourteen in 1951; and by the end of 1953 there were thirty-three units employing 202 full-time staff (Committee on Veteran's Medical Problems 1954, 622). These labs occupied spaces between 1,200–3,000 square feet in their hospital units (Central Advisory Committee 1949, 817). During the first four years of its existence, the VA Radioisotope Program was funded at roughly $500,000 annually (Table Three).

The progress reports of the Radioisotope Units filed annually for the meetings of the Central Advisory Committee provide evidence of the sophisticated assembly of equipment and experimental research being undertaken in these labs.[18] Typical VA-Radioisotope Units

---

16 These were VA hospitals in Framingham, Massachusetts; Bronx, New York; Cleveland, Ohio; Hines, Illinois; Minneapolis, Minnesota; Van Nuys, California; Los Angeles, California; and Dallas, Texas. Two additional Radioisotope Units were authorized for San Francisco, California, and Fort Howard, Maryland at the end of the first year (VA Information Service 1948).

17 In its preliminary recommendation of 15 September 1948, the Central Advisory Committee recommended that at least 2,000 square feet be allocated in new hospital constructions for the radioisotope labs (Central Advisory Committee 1948, 2).

18 We are relying here on the reports filed on equipment and facilities at the Meeting of Representatives of VA Radioisotope Units, 21–22 November 1948, Washington, D.C. The early reports are quite extensive, since they sought to elaborate the full range of equipment being installed in the labs and the programs contemplated. Typical equipment in the VA Radioisotope units included Geiger-Mueller Scaling Units (usually two) such as those produced by Technical Associates

	Equipment and Supplies[19]	Personnel
FY 1948	$220,500	$ 31,604
FY 1949	157,970	290,260
FY 1950	216,052	299,838
FY 1951	225,480	183,978
Totals	**$820,002**	**$805,680**

TABLE THREE

consisted of an administrative office, separate biochemical and biophysical laboratories with specially constructed hoods, chemical benches, etc., and a clinical room or rooms where patients were brought for diagnostic studies involving administration of radioisotope materials. The staff of a radioisotope laboratory consisted initially of five-eight full-time individual professional, scientific, and technical personnel. In addition usually two or more faculty members or laboratory staff persons from a local university medical school served on a part-time basis in the unit. University faculty also served

---

(contd.)
and accessory equipment such as lead shields, all glass G.M. tubes for beta and gamma ray counting, an oscillograph, preamplifiers and a counting rate meter. Each lab had a range of monitoring equipment, including Victoreen miometors, Potter decade scalers, an Autoscaler, and a Berkeley decade scaler. Also on hand were Tracerlab sample changers (both manual and automatic), Lauritsen Quartz Fiber Electroscopes and a variety of radiation reference sources. Physical instruments included specal equiment such as Beckman photoelectric spectrophotometers, mass spectrometers (particularly at the Van Nuys unit). The Wadsworth VA in Los Angeles had a number of state-of-the-art instruments, including a mass spectrometer, an infrared spectrometer, an electron microscope, and a visible spectrometer with ultra violet range attachment (see Raymond Libby in Central Advisory Committee 1949, 22). Isotope units also had chemical instrumentation useful in immunological work such as Beckman pH meters, analytical balances, drying ovens, a centrifuge, a bacteriological bench, autoclave, 370 degree incubator, and a refrigerator. Apparatus for ultramicrochemical studies included an electrophoresis apparatus, a Chambers micromanipulator, an air-driven microcentrifuge, and a mercury microburet. A Spencer research microscope and a circular Barcroft-Warburg apparatus for work with the micromaniplator was also a standard lab component. VA labs were also equipped with a high vacuum line glass apparatus for C^{14} studies.
    19 Exclusive of construction and alterations.

as consultants (Lyon 1950, 1).[20] Oversight of the Radioisotope Program in an individual hospital was administered by a Director of the unit, usually an MD in internal medicine, or on occasion a radiologist. The Director of the radioisotope unit was directly responsible to the Chief of Medical Service of the hospital.

The salient feature of the VA radioisotope units was their close relationship to a local medical school or university. In fact, VA radioisotope laboratories were established only in VA hospitals having affiliation with medical schools (Lyon 1952, 555). Thus, the radioisotope units of the VA hospitals were crucial enabling structures for nuclear medicine and for efforts to place clinical medicine on experimental foundations in the post-War era. The VA radioisotope units played at least two essential roles: first they were crucial sites for developing therapies, instrumentation, training programs, and the production of basic biomedical research conducted in close connection with both the AEC labs and with university departments of basic science. Secondly, by designing and implementing practical laboratory spaces and clinical wards integrated seamlessly within the structure of the normal hospital, the architects of these radioisotope units—George Lyon and his staff—not only created structures that were implemented throughout the large number of VA hospitals (by 1965 there were eighty-six VA hospitals licenced by the AEC to use isotopes). These clinical laboratory facilities were the prototypes of all such clinical-laboratory facilities in the U.S. and as such were major sources of transformation in the American hospital system.

An excellent example of the manner in which the radioisotope units served as major research centers in the nascent field of nuclear medicine is suggested by the work of the VA Center (VA Wadsworth) in Los Angeles. As we have indicated above, this center developed in close association with Staff Warren's program for building the UCLA Medical Center. He saw the unit as a means for creating positions for physicists, radiologists, and physicians working with radioisotopes that he could not get appointed on the regular faculty at UCLA. In addition, the VA provided the crucial clinical wards needed for his fledgling medical school, and it served as an important training center for potential residents in nuclear medicine. By 1955, when the formative stages of the VA radioisotope program can be considered to

---

20 By 1955, however, the staff of these facilities had greatly expanded. See below.

have been completed, the radioisotope unit of the VA Wadsworth had a staff of nineteen persons including three consulting physicians and radiologists, and a budget of $66,000, of which $59,000 went for salaries of support staff and consultant fees (Bauer 1956). Members of the radioisotope unit made twenty presentations of work at scientific meetings and produced fourteen publications in 1955 on a range of topics including studies of skin collagen with the electron microscope (Linden et al. 1955), use of radioactive phosphorus in the diagnosis of skin tumors (Bauer and Steffen 1955), and uses of radioisotopes in studying neuromuscular diseases (Blahd et al. 1955). Members of the unit were working on eight different long-term research projects, particularly in areas of muscular distrophy, the use of rubidium[86], potassium[42], and sodium[22] in electrolyte disorders, peripheral clearance of radioisotopes in patients with heart disease, and several projects involving electron microscopy. In addition to projects initiated and directed by members of the radioisotope unit, six different research projects directed by UCLA medical faculty were being carried out with collaboration of the Radioisotope Service. A similar record of high-level research achievement would be documentable at most of the other thirty-three radioisotope units. No less significant was the clinical work being done in the unit. Seventy-five patients received a variety of radioisotope treatments and 1100 patients received radioisotopes in diagnostic procedures in 1955. Two residents on the Medical Service and one resident in Radiology were assigned to the Radioisotope Service on a three-month rotation.[21]

CONCLUSION

The trademark of today's clinical nuclear medicine services, the "scan," or image of the distribution of radioactivity in a patient's body, is the product of work begun at the Wadsworth VA Hospital in the late 1940s. Herbert Allen, who assumed direction of the VA Radioisotope Service at Wadsworth in 1950 after having set up the first six radioisotope units under George Lyon, was seeking a way to localize radioactivity precisely in the patient's body. He used a gamma

---

21 This emphasis on training was a key part of the VA radioisotope effort. In 1967 the VA established a formal two year training program for physicians in nuclear medicine which became a residency program after the Board of Nuclear Medicine was founded in 1972.

radiation detector which Benedict Cassen and his colleagues at the UCLA AEC laboratories had developed (Cassen et al. 1950). This detector was collimated so that when it was placed directly over an anatomic area, it could measure the activity present in that area without much interference from surrounding areas. Allen used this device to map out the pattern of radioactivity in patients' thyroid glands, laboriously counting the activity at each point as the detector was manually moved through a grid pattern (Allen et al. 1951). Cassen and Raymond Libby together with engineers Clifton Reed and Lawrence Curtis, extended this concept to develop an electrically driven device which would move the collimated detector back and forth to "scan" the anatomic area of interest and so to map out the pattern of its radioactivity (Cassen et al., 1951). This device was first demonstrated at a national meeting of VA radioisotope unit personnel in 1950 (Lyon 1951). Its earliest use was in imaging the normal and diseased thyroid gland, taking advantage of the fact that the thyroid gland concentrates radioiodine in its simple iodide form.

This radioisotope scanner greatly expanded nuclear medicine. Soon radioactive compounds were developed to image many organs of medical importance. The clinical usefulness and commercial profitability of scanning and the agents which made it possible led to commercial interest, and to rapid expansion of clinical nuclear medicine applications, not only in university and veteran's hospitals but in community hospitals as well. Within a few years, by the mid-1960s the Anger scintillation camera, developed in the AEC Donner Lab at Berkeley, began to come into use. The Anger Camera, which provides a simultaneous image of all parts of the field of view, gradually supplanted the scanner, though the term "scan" persists to this day. Modern tomographic methods use principles inherent to both the camera and the scanner.

We have described the first Manhattan Project for biomedicine as the efforts to build the infrastructure of research facilities in the basic and clinical sciences as well as the industrial base needed to transition the work of the Medical Division of the Manhattan Project to the commercial sector. These efforts were successful beyond anyone's wildest imagining. In 1976, thirty years after the legislation establishing the Atomic Energy Act an ad hoc committee headed by James Potchen of Michigan could report to the Energy Research and Development Agency (ERDA):

The effect that nuclear medicine has had on the practice of medicine can be demonstrated in at least two ways. The first relates to the use of nuclear medicine procedures in the clinical practice of medicine. For example, in 1973, some 7.5 million Americans received *in vivo* nuclear medicine procedures. This represents approximately one procedure for every 4.4 hospital admissions. These procedures were performed in some 3,000 hospitals and 2,000 independent clinical laboratories. The related nuclear medicine industry is now growing between 20 percent and 25 percent per year according to recent market research analyses.

The second area of major impact relates to the effect of nuclear medicine as a scientific discipline with regard to careers in health care. The Society of Nuclear Medicine now has some 8,000 members and the American Board of Nuclear Medicine has certified 2,070 physicians as specialists in nuclear medicine since its inception on July 28, 1971. These numbers represent individuals who spend a major portion of their careers in nuclear medicine, and do not include the large number of other users of nuclear medicine techniques in fields such as endocrinology. Careers range from the physician-practitioner to the research radio pharmaceutical chemist. That this field is well recognized as a medical discipline is shown by the establishment of the American Board of Nuclear Medicine and the recent formation of a Section on Nuclear Medicine in the American Medical Association. Additional testimony can be gleaned from the 1975 Report of the Joint Congressional Committee on Atomic Energy which concludes that "nuclear medicine represents one of the most successful applications of the peaceful use of atomic energy."(ERDA 1976, 10)

Modern nuclear medicine, carried out by physicians, scientists and technicians certified after formal training in the field, is now a requirement for hospital certification. It depends upon commercial availability of radioactive tracers tailored to the needs of the patient, on highly sophisticated equipment and on modern knowledge of the principles of radiation safety. We have told in this paper how all of these elements began in the programs of the AEC and the VA, programs which were the direct descendents of the Manhattan Project.

3

Whose Work Shall We Trust? Genetics, Pediatrics, and Hereditary Diseases in Postwar France*

Jean-Paul Gaudillière

In 1942, in the midst of World War II, the French pediatrician Maurice Lamy started to lecture students at the Paris Medical Faculty on human inheritance. This was the first course of its kind given in the country. A few years later, in 1951, Lamy was appointed professor of medical genetics. His inauguration conference *programmatically* defined human genetics as a science based on "Mendelo-morganism" and biometrics. Thus, Lamy explained:

> The most urgent task (for geneticists specializing in the study of human pathologies) is to establish a general catalogue of truly inherited diseases. The second task is to elucidate their mode of transmission. The third task is to collect statistics and to compute exact frequencies. By so doing, it will be easy to determine the number of individuals from the general population who carry a transmissible defect. In the end, we shall expose these secret carriers. (Lamy 1951, 1676)

Observers and practitioners of contemporary human genome research may think that this crude form of medical genetics was rapidly superseded by the growth of molecular biology and the rise of genetic

---

* I am indebted to Phillip Sloan for help in revising this paper. Helpful comments on a previous version of this study have been provided by Judith Melki, Marie-Anne Bach, Robert Bud, Soraya de Chadarevian, and Ilana Löwy.

counseling. Histories of medical genetics have often distinguished three stages in the postwar development of the field. In the 1940s and 1950s, medical genetics departed from authoritarian prewar eugenics by focusing on Mendelian disorders and on counseling. In the 1960s and 1970s, the discovery of important chromosomal disorders, including trisomy-21, originated in cytogenetics and resulted in new diagnostic opportunities. Finally, during the past ten years, genomics and human genome research have developed highly specific means of investigation and have given birth to perspectives of genetic therapy.

It seems to me that a survey of medical genetics in the postwar era should, however, depart from this vision by questioning both notions that genetic knowledge of humans was first established in the laboratory to be later applied in the clinics, and that collaboration between geneticists and clinicians was a straight-forward process. The point of departure taken in this paper is that geneticists and clinicians operate in different social worlds. Consequently, both groups do not structure (and do not use) either genetic concepts or clinical observations in the same way, although they may investigate and speak of seemingly common diseases. One way to analyze the coordination between these different systems of practice is to focus on the changing ways in which genetic and medical features have been associated, within specific sites, in order to diagnose hereditary diseases.

A decade ago, the sociologist Harry Collins introduced the notion of experimenter's regress as a typical problem of bench workers who must decide whether an unknown entity is a natural phenomena or an artifact (Collins 1985). When radical novelty is at stake, scientists face a methodological loophole, since both the procedures they use, and the data they assess are new. If they believe the phenomenon they observe is natural, they will accept the technique employed for its determination as reliable. If they don't believe the results are true, they may emphasize the method's uncertainties. These circumstances would result in endless controversies between experimentalists, except for the use of shared resources, either material, social, or cognitive. Similarly, medical geneticists have to handle laboratory techniques, biological data, clinical examination and medical knowledge. Consequently, they face more complex but analogous situations. Claims for genetic causality commonly depend on the proper and simultaneous usage of laboratory tools and diagnostic procedures. Both poor technical work or erroneous diagnosis may, for instance, account for controversial results or weak correlation between genotypes and medical phenotypes. New knowledge in medical genetics therefore relies on the co-evolution of biological techniques and medical categories. This may result in two regimes of collective assessment. One is akin to the experimenter's regress. By analogy, the second may be called a clinician's regress. In the context of medical genetics, the former is characteristic of debates focusing on the work done in the genetic laboratories, while the latter is characteristic of controversies centered on the bedside activities of physicians, especially on the definition of disease (nosology).

This paper traces instances of both the experimenter's and the clinician's regress. It is based on three episodes in the development of medical genetics in France. Historians have strongly emphasized the peculiar fate of both genetics and eugenics in France. On the one hand, Richard Burian, Jean Gayon, and Doris Zallen have convincingly argued that the fact that many French biologists working during the interwar period did not show much interest in the sort of genetics practiced by Morgan and the students of Mendelian factors illustrated a physiological approach to heredity rather than a conservative commitment to "Lamarckianism" (Burian, Gayon, and Zallen 1988; Burian and Gayon 1990). On the other hand, William Schneider has shown that the role of physicians differentiates French eugenics from its counterparts in Britain or in the United States (Schneider 1990). His analysis of the eugenicist's struggle against social scourges has been reinforced by Anne Carol's recent study of the French medical discourses on inheritance (Carol 1995). Accordingly, the typical French eugenicist was a pediatrician who did not care much about Mendelian diseases. He was more concerned with the declining birth rate and plagues of medical importance, such as syphilis, tuberculosis, and alcoholism.

The first part of the paper concentrates on the transformation of "mongolism" into a chromosomal disorder in the late 1950s and early 1960s. The trajectory of the French human geneticist, Jerome Lejeune, sheds some light on the processes which channeled debates about the diagnosis and the etiology of Down's syndrome into the standardization of karyotyping practices. Lejeune's work at *Hôpital Trousseau* in Paris highlights the legacy of French pediatric eugenics. This local culture is contrasted by looking at the alternative form of research and counseling which emerged in the nearby genetic service headed by Maurice Lamy at the *Hôpital des enfants malades*. Finally, the third part of the paper describes some of the conditions which made possible the articulation of a true clinical regress by presenting discussions on the molecular causes and clinical heterogeneity of a series of muscular atrophies which took place at the *Hôpital des enfants malades* in the 1980s.

DEFINING MONGOLISM:THE PROBLEM OF DOWN'S PATIENTS WITH NORMAL CHROMOSOME COUNTS

In 1956, cytogeneticists J. H. Tjio and A. Levan published what became a landmark article on the number of chromosomes in normal human cells (Kevles 1986, chap. Sixteen). Tjio and Levan opposed the then prevailing view that the correct chromosome count was forty-eight, and claimed the proper number to be forty-six. Three years later, a team working at the *Hôpital Trousseau* in Paris announced that cells from nine Down's patients contained forty-seven chromosomes. Although a priority dispute later surfaced, the discovery that Down's syndrome is caused by trisomy-21 has usually been attributed to the authors of this paper—Jérome Lejeune, Marthe Gauthier and Raymond Turpin (1959). Logically, the latter achievement relied on the former: without the knowledge of the correct chromosome number, Down's abnormality could not be traced. Lejeune and his colleagues commented that their data included only "good" microscopic in cytological studies as the origin of the discovery of trisomy-21. This simplified account undermines the technical and medical uncertainties which underlay the discussion of the etiology of Down's syndrome.

As several authors have noticed, when Turpin and Lejeune issued their report, there was no consensus on what was a good technique for chromosomal examination, and the exact chromosome count in humans was still equivocal.[1] Moreover, although the results of Tjio and Levan were confirmed by several groups of cytogeneticists, their technique remained questionable. The French workers, for instance, argued against the techniques used by British and American cytogeneticists. The former repeatedly explained that the use of colchicine to increase the number of dividing cells, the use of hypotonic salt solu-

---

1 For a discussion of the chromosome count problem see Kottler 1974. See also Kevles 1986 on Lejeune's doubts about the chromosomal count.

tions to disrupt the cell, and the use of squashing procedures that were employed by the Anglo-American groups resulted in artifacts. Lejeune thought these techniques were changing the structure of chromosomes, inducing breaks, and increasing the counts (Lejeune, Turpin, and Gauthier 1959b). The French team favored culture synchronization in order to increase the number of dividing cells. They employed hypotonic serum solutions, and simply dried the preparations in the air (Lejeune, Turpin, and Gauthier 1960).

This disagreement over procedures had long-range consequences beyond the obvious fact that chromosome analysis could only be statistical work. One effect was that the forty-six number remained fragile for a long time. One approach to the problem consisted in emphasizing history and the fact that observers had for a long time seen forty-eight chromosomes in human cells. Thus, the new karyotyping techniques might well be producing nice pictures that were only artefacts. This position was never seriously considered within the small group of biologists who investigated human chromosomes since the end of W.W. II. For the "cytogeneticists" like Levan, Hsu, Ford, or Lejeune who participated in the search for major changes in human genetic material induced by radiation, the forty-six count was the most serious candidate. But this was not enough to settle the matter. It could well be that humans did not have a definite number of chromosomes. One year after his first report on Down's syndrome, Lejeune was thus commenting: "One should recall that thorough studies in animal species have shown quantitative and morphological variation of chromosomes. Rational expectation is therefore that the same kind of analysis will reveal similar variability in humans"(Lejeune 1960a, 342). In other words, the chromosome count could well vary between forty-six and forty-eight in different individuals or different tissues. This was actually supported by a few experimental claims (Kodani 1957).

In addition to biological uncertainties, clinical uncertainties originated in the description of Down's syndrome.[2] By the late 1950s, a majority of clinicians thought mongolism to be a single disease with homogenous etiology. Nonetheless, reports of mongoloid symptoms associated with hypothyroidism, and other diseases were occasionally used to question this clinical homogeneity (Warkany 1960). The most discussed issue was, however, the problem of diagnosis, especially the

---

2 On this distinction see Gaudillière 1995.

diagnosis of Down's syndrome in babies and young children. Clinical judgment was usually based on physical examination, i.e., the search for hypotonia, epicanthus, abnormal palm lines including the famous "four-fingers line," and furrowed tongue. Morphological analysis was complemented by intelligence testing (Down's patients normally showed IQ values ranging from 50 to 75). Variability of association of these signs was, however, the rule. The American specialist Theodore Kushnick thus warned that "the diagnosis of mongolism should not be made on the basis of one or two physical characteristics but on a multiplicity of these features. One or two of the so-called findings can be found in many normal individuals"(Kushnick 1961, 151). As a French psychiatrists phrased the rule of thumb: an individual would qualify as "Mongolian" as soon as he or she would show "both mental deficiency (IQ above 65) and at least half the signs included in the Mongolian series" (Lecuyer 1958, 46). Pediatricians and psychiatrists, however, thought that mongolism could not be mistaken for long by experienced clinicians.

As the karyotyping of Down's patients became more frequent, troublesome findings emerged and reinforced these uncertainties. A few months after the publication of Lejeune's analysis of the karyotypes of nine "mongols" showing forty-seven chromosomes, oral claims were made that if many Down's patients have forty-seven chromosomes, others have only forty-six. As the British geneticist Lionel Penrose put it: "Just now there is an epidemic of mongols with chromosomes stuck together in the wrong way. I suppose this phenomenon will become elucidated eventually. Ford has one, Fraccaro has one and we have one" (L. Penrose to J. Frézal, 21 January 1960, Penrose Papers file 133/7). Other cases soon followed.

This could theoretically result either in changes in the definition of the disease, or in changes in the karyotyping procedures. Several alternate possibilities were at hand. One was to play down the notion that the forty-seven-count was the cause of Down's syndrome. Accordingly, alternative origins of the disease should be investigated and the karyotyping technique ought not be employed in clinical practice as the reference tool for making diagnosis. The alternative route was to go on with the genetic explanation, either by considering that the troublesome patients had been misdiagnosed, or by looking for unnoticed alteration of the chromosomes. Commenting on the case of "a mongol girl with 46 chromosomes," the British cytogeneticists

P. E. Polani and C. E. Ford, who had both confirmed Lejeune's conclusion, thus wrote: "On present knowledge it is difficult to believe that there might be a genetic mechanism causing mongolism that did not involve the presence in triplicate of at least part of chromosome 22. The clinical diagnosis of mongolism in the present case is undoubted. . . .The most likely interpretation of the finding is an unequal reciprocal translocation" (Polani et al. 1960, 722). Difficulties were also raised by clinically-diagnosed Down's patients showing variable numbers of chromosomes (Clarke et al. 1961). Swedish pediatricians discussing the "provocative" case of a "Down's syndrome with normal chromosomes" captured the problem in the following terms:

> The finding of a normal karyotype in a typical case may be explained in several ways. . . .The possibility exists that the whole or part of the Down's syndrome chromosome is hidden somewhere in the karyotype. Nor is another possibility precluded—namely that a genic mutation may have the same final effect as the extra chromosome 21. The vast majority of cases with Down's syndrome, in which the chromosomes have been analysed, have been trisomic for chromosome 21. For practical purposes this establishes its usefulness as a diagnostic sign for the syndrome. That the correlation, and possibly the etiology, may not be an absolute one is demonstrated by the present case. (Hall 1962, 1027)

The most important feature in the discussion was not that such reports existed but the fact that they neither led to changes of the clinical work nor jeopardized the use of karyotypes. Clinically-defined Down's patients remained Down's patients. In a few years most cases were linked to chromosomal abnormalities, either as trisomy-21, or as translocations and mosaics which were frequent in cases of "familial" mongolism (Carter et al. 1960). A reasonably small number of typical Down's syndrome remained problematic.

The association between genotypes and medical phenotypes which was attained in the early 1960s rested on the following claims: 1) Down's syndrome is usually a disease of chromosomal origin; 2) karyotypes generally provide a useful sign for pediatricians; 3) if clinical and laboratory criteria conflate, the former prevail for all medical purposes.[3] This may be exemplified by means of the book on

---

3 Only then did chromosomal examination became a means to settle debates about clinical cases *previously* recognized as clinically problematic. The first report

Down's syndrome by the American specialist Clemens Benda published in 1969. The chapter on diagnosis was entirely devoted to the description of clinical signs beginning with a classical trilogy: hypotonia, epicanthus, and the "four finger line" in the palm print. One sentence only dealt with karyotypes: "In all cases where there is a question with Down's syndrome, a chromosomal test can dispel all doubts" (Benda 1969, 22). Benda further added: "If the stigmatization is very conspicuous and yet the chromosomal analysis negative, one has spoken of paramongolism" (ibid., 128).

How was the linkage between a theoretical commitment to genotypes and a practical reliance on phenotypes stabilized? The above-mentioned notion that the attribution of a chromosomal origins to the Down's syndrome requested a stable chromosome count in humans, points to the relationship between the standardization of karyotypes and the closure of the etiologic debate. By 1959, most cytogeneticists thought that homogenous techniques and standards would dissolve many problems associated with chromosomal analysis, including the medical anomalies. Lejeune thus wrote: "It would be highly desirable to establish a complete description of the normal human karyotype agreed upon by all geneticists who have been working on this topic. Unfortunately, it is, for the time being, difficult to compare samples prepared through heterogeneous procedures"(Lejeune 1960a, 342). The first standardization meeting was organized in 1960 in Denver with the support of the American Cancer Society. The final document focused on the classification and measurement of normal chromosomes.

One unstated but obvious by-product of the normalization of karyotypes was to establish officially the chromosome count in somatic cells at forty-six (Anon. 1960). Down's syndrome was not mentioned in the final document. Yet, the debates hinged upon the problem of chromosomal abnormalities. Norms of chromosomal size were actually decisive in making decisions about the transfer of chromosome parts associated with the putative translocations shown by Down's patients with normal chromosome count. Accordingly, the Denver agreement on the measurement of human chromosomes provided "firm ground" for the detection of translocations. This in turn favored the "genetic" normalization of most Down's patients with

(contd.)
of this sort was Lee et al. 1961.

forty-six chromosomes. In other words, debates about the problematic relationship between genotype and phenotype rapidly evolved into a typical "experimenter's regress." This is further evidenced by the fact that there were only isolated attempts to distinguish authentic Down's patients and "mongoloids" on the basis of chromosome studies and laboratory surveys: clinically they seemed all alike and this was enough.[4]

Attention to the composition of the standardization workshops provide some hints at the conditions for this technical response to diagnostic uncertainties. Beside basic geneticists and radiation geneticists, the Denver workshop gathered a small group of medical geneticists actually working on Down syndrome: Lejeune, Ford, Böök, Fraccaro. Thus, a few researchers who had rapidly been convinced that Down was caused by trisomy-21, and that Down's patients "truly" showing forty-six chromosomes were cases of undetected chromosomal alterations, played a decisive part in the debates. This "core-set" actually mediated between the cytological laboratory on the one hand, and the pediatric ward or the mental deficiency institution on the other hand. Developments in Paris illustrate their motives as well as reasons why the technical response to diagnostic uncertainties did not affect the clinical vision of the disease.

MONGOLISM BETWEEN KARYOTYPES AND PEDIATRICS: LEJEUNE AND THE LEGACY OF FRENCH EUGENICS

In 1959, Jerome Lejeune was a young researcher working in a pediatric unit headed by Raymond Turpin, a prominent character in French medicine. Professor of therapeutics at the *Faculté de médecine* in Paris, Turpin was head of a pediatric service at *Hôpital Trousseau*. As recounted by William Schneider, Turpin had been involved in the French eugenic movement in the 1930s (Schneider 1990).[5] His vision of familial diseases then did not depart from the views of other "baby-docs" involved in the eugenics movement. Turpin was more concerned by the declining French birth rate than interested in the inheritance of insanity or criminality. He viewed eugenics as a branch

---

4 The credibility of this approach is attested by a few attempts reaching the publication stage. See for instance Gibson et al. 1965.

5 See Turpin's articles on the control of (social) diseases (Turpin 1932; 1939; 1941a).

of social medicine focusing on the family as living unit and on reproduction, in other words on the making healthy babies by healthy
mothers. Although Turpin thought that diseases displaying a
Mendelian pattern of transmission to be theoretically important, he
considered the physician's duty to study heredity as a system of vertical transmission which included gene transmission, infection, intoxication, and other biological relationships between parents and children. This was legitimated through references to public health concerns. From the clinical vantage point, rare Mendelian disorders did
not matter much, in contrast to causes such as tuberculosis, syphilis or
alcoholism (Turpin 1941b).

In the early 1930s, Turpin engaged in the study of mongolism. His
understanding of the etiology of the disease included the hypothesis of
a change in the hereditary makeup. The research perspective he
adopted, however, focused on predispositions. The aim of Turpin's
collection of clinical descriptions of mongols was the correlation between the disease and: 1) signs of a weak constitution in the family
such as small malformations, tuberculosis or mental deficiencies; 2)
environmental "influences" such as poor living conditions; 3) factors
characteristics of the mother-child relationship such as the age of the
mother or the birth rank (Turpin and Caratzali 1934). Before World
War II, the main outcome of Turpin's research on Down's syndrome
was to emphasize increasing incidence with the advancing age of the
mother, and with the possession of a furrowed tongue by the parents
of the mongoloid child, a physical abnormality transmitted as a dominant trait.

Given this background, one may understand why, in contrast to the
classical chronology of medical genetics, World War II and the decreasing legitimacy of racial and authoritative eugenics had little impact on Turpin's work as pediatrician interested in heredity. After
1945, his continued commitments to the betterment of the constitution
of French babies were illustrated by his establishment of the *Centre
de progénèse*. One aim of this small, but publicly supported, foundation was to promote research and conferences on the medical control
of human reproduction. Generation, not heredity, was the problem of
interest. Turpin then distinguished between hereditary and environmental factors. Yet, his ambition was to evaluate and enhance reproduction as a single process. *Consequently,* Turpin worked out the dif-

ference between "progénèse" and "metagénèse," rather than isolating environment and heredity:

> Hereditary and environmental forces influence human generation in two ways: before and after fertilization. The first period includes the part played by the hereditary makeup of the reproductive couple and by the influence of environmental factors on the formation of gametes and reproductive organs. . . .The second period consists in the interplay between the hereditary factors inherited by an individual, of the intrauterine environment where the embryo will develop, and of the extrauterine environment. (Turpin 1955, 4)

Similarity of inheritance prevailed in his analysis of hereditary disorders. His textbook, *L'Hérédité des prédispositions morbides*, published in 1951, was typical in this respect. Turpin contrasted biochemical diseases like phenylketonuria that could be described with simple Mendelian patterns, with the complex relations between genetic factors (*la constitution*) and environmental factors (*l'ambiance*) that characterized the really important diseases. Thus, the book mainly discussed hereditary predispositions to allergy, infectious disorders, cancer, congenital malformations, mental retardation.[6]

After the war, research on Down's syndrome resumed when Lejeune joined Turpin's medical unit. There, he occupied an odd position. Lejeune was a general practitioner who could not pass the internship examination. Lejeune had no hope for a position at the Paris Medical School because of this failure. He was nonetheless nominated resident by Turpin. This promotion however meant that the possibility to work in the hospital as a pediatrician continuously depended on Turpin's protection. Administratively, Lejeune was left no choice other than to apply for a position at the *Centre national de la recherche scientifique*, a newly created state research agency.

Benefiting from decreasing incidence of congenital syphilis in the country, Lejeune gradually transformed the consulting room for venereal diseases into a place for the examination of Down's patients. Taking over Turpin's prewar survey of "mongoliens," he started to gather data about features such as sex-ratio or palm prints that might

---

6 "Les graves problèmes médico-sociaux sont posés par le cancer, les infections, la tuberculose, les troubles mentaux, les accidents allergiques, les maladies de la nutrition. Le public attend que le médecin le renseigne sur l'importance d'une prédisposition héréditaire à ces maladies" (Turpin 1951, 43).

be characteristics of families including a mongoloid child (Lejeune to Penrose, 19 January 1953, Penrose Papers, file 174/4). In 1954, Turpin and Lejeune reported that palm lines of normal human beings did not resemble those of inferior monkeys like macaques. But that the palm lines of Down's patients did. To the authors, this fact of regressive change was a new argument in favor of the genetic causation of the disease.

While following the morphological track, Lejeune established a second line of inquiry which linked Turpin's interest in the environmental factors affecting human reproductive ability with opportunities originating in the development of collaborative research on the effects of radiation. In the late 1950s, Lejeune was receiving funds from the American NIH and the French *Institut national d'hygiène* for investigating the medical consequences of X-ray exposure. The introduction of cytogenetics into the service was part of this program. In 1958, Lejeune's collaborator Marthe Gauthier actually traveled to the United States in order to learn karyotyping procedures presumably with Tjio (Lejeune, Gauthier, and Turpin 1959c).[7]

Working with Turpin, Lejeune may have rapidly been convinced that "mongolism" was a defect of the genetic constitution at an early stage of his research. In any case, in the summer 1958, as soon the first forty-seven count in Down's patient was obtained, Lejeune began to argue for a causal relationship.[8] Doubts about the significance of these early results, however, seem to have vanished only in the fall of the same year when di-zygotic twins, one affected and one healthy, were correlated with forty-seven and forty-six chromosome counts respectively (Lejeune, Gauthier, and Turpin 1959a).[9] Then results like Kodani's cases, which suggested a variable chromosome count in normal humans, became artefacts which, according to Lejeune, originated in the poor quality of the pictures, or in the analysis of testis from unnoticed Down's patients, or in peculiarities of cells from the germ line (Lejeune 1960). The correspondence between Penrose and Lejeune similarly documents the fact that instances of "Down's patient

7 Also based on an interview of Mme. Lejeune by the author, January 1996.

8 A preliminary report of this case was actually published in the Danish press following an interview given by Lejeune during the summer 1958 as he was travelling with his wife. (Mme. Lejeune, interview with the author, January 1996).

9 This, in addition to rumors that British scientists were on the same track, triggered decision to publish the first observations (Lejeune, Gauthier, and Turpin 1959a).

with normal chromosomes" also became a motive for suspecting misdiagnosis, in private correspondence if not publicly. For instance, in February 1962, Penrose wrote about a case published in the Australian medical press: "as I did not feel that the evidence for the diagnosis was necessarily conclusive, I wrote to Sidney to ask for palm prints of the child. I have now learned. . . that, in fact, the prints were sent to you"(Penrose to Lejeune, 24 February 1962, Penrose Papers File 62/7). Following the reception of Lejeune's drawings of the palm prints, Penrose once again articulated his doubts about the diagnosis, and the observation was ignored.

In the early 1960s, Lejeune's scientific credentials were based on the forty-seven chromosome hypothesis and on his participation in the network of cytogeneticists. Yet the peculiar niche Lejeune had found for himself and his work made medical practices more influential than they were in other research settings. Lejeune was eager to show the medical utility of his work. Like Turpin, he viewed medical genetics as a form of "preventive medicine" concerned with pregnant women and newborns. In other words, it was a form of medicine that would pay sufficient attention to the environmental and hereditary factors which influence human generation before fertilization.

Within this context, Down's syndrome remained embedded in the science of influences. Rather than being the first cause leading to illness, chromosomal abnormalities were viewed as "passage points" connecting the constitution of the mother, the conditions of pregnancy, and the abnormal child. Lejeune focused on the most abundant type of Down's syndrome: the patients showing a trisomy-21 and no familial transmission. These mutations were not hazardous accidents, but outcomes triggered by specific circumstances and facilitated by unknown predispositions.[10] The mechanism could barely be specific, but Lejeune kept mentioning the signs echoing the disease which had been noticed by Turpin in the unaffected relatives of Down's patients. In this respect, Lejeune was not isolated. The psychiatrist Clemens Benda, the most prominent specialist of Down's syndrome in the

---

10 "All statements about the effect of heredo-syphilis and fetal trauma have been proved wrong. The incidence of the disease, however, dramatically increases with the age of the mother. Similar features have been observed with dominant mutations. In the latter cases, however, the magnitude of the phenomenon was much smaller. Peculiar mechanisms must account for the mutation causing mongolism" (Lejeune 1960, 11).

United States, thus wrote in 1969: "While the mechanics of trisomy can be traced to non-disjunction, the etiology of nondisjunction has not been established and can only be suspected in very general terms. We see, however, at this point that mongolism is not due to so-called hereditary factors but the error of cell metabolism must be traced back to the environmental conditions which produce a peristatic constellation which interferes with normal cell division" (Benda 1969, 218–19). Benda then mentioned three factors: a virus disease of the mother, radiation, and nutrition.

By contrast, the British scientist Lionel Penrose—then a geneticist without medical duties—focused on genetically more interesting cases: patients showing mosaics, translocations and familial transmission (Penrose 1964; Penrose and Smith 1966, § 11, Aetiology). He reclassified Down's patients according to epidemiological data and genetic features. One group gathered sporadic and age-dependent cases, mostly trisomy-21. A second group gathered familial and age-independent cases analyzed in terms of chromosomal structure and pattern of transmission—translocated Down's, mosaic Down's, etc. The latter were particularly valued by Penrose since they made possible a form of genetic prognosis. Thus, Penrose defined a pseudo-mendelian Down's syndrome while Lejeune maintained a pseudo-constitutional disease.

In terms of medical practice, the pediatric context in which Lejeunes researches were evolving meant particular attention was given to the fate of families affected with the disease. Lejeune had the responsibility for the "consultation des mongoliens." Consequently, he had to advise the families about care and institutionalization. By the late 1950s, the main issues in French medical genetics were not risks of recurrence or reproductive choices, but medical care. This was rooted in the use of antibiotics which radically altered prognosis and extended life expectancy for Down's patients from childhood to adolescence and adulthood. Parents of Down's patients then fought for the creation of new institutions and opposed institutionalization in general asylums. During the decade which followed the end of the war, the number of Down's patients living in specialized centers increased from a few thousand to roughly one hundred fifty thousand (Pinell and Zafiropoulos 1983). The main role of pediatricians working in hospitals was therefore to help a family decide where, when and how would care be provided.[11] Lejeune responded to this issue by establishing early links with the *Union nationale des associations de parents d'enfants inadaptés* which was established by the families of French mentally retarded patients.

Lejeune's work as a geneticist was directly affected by the supremacy of clinical aims. First, karyotyping analysis lead to few changes in clinical practices. As a French specialist in mongolism put it: "it is exceedingly rare that one hesitates to give a clinical diagnosis of mongolism. If it happens to be that the first examination raises some doubts, then further examination, a few weeks later will make the case" (Mallet 1964, 10).[12] Cell culture and microscopic examination could be helpful when assessing rare clinical puzzles, for instance that of "mongoloid blacks," but they would not matter much for normal clinical work. In Turpin's service, Lejeune practiced routine diagnosis, but he rarely addressed the clinical status of his patients. For instance, he did not try to correlate the description of different chromosomal abnormalities with variable clinical symptoms. In contrast to Penrose, who described different forms of the disease and even looked for correlating different genotypes and palm prints, Lejeune did not challenge the homogeneity of Down's syndrome. In a similar way, he did not lean toward the description of "fascinating" cases which would shed light on chromosomal dynamics, but he rapidly focused on therapeutic issues. From 1960 onward, Lejeune advocated a treatment based on the discovery of an anomaly of the amino-acid metabolism which he viewed as a metabolic characteristic of the syndrome (Lejeune and Turpin 1960).

The pediatric background finally sheds light on the fact that Lejeune contributed little to the rise of genetics as medical specialty. In the early 1970s, following the development of amniocentesis, the prenatal diagnosis of Down's syndrome became feasible. Lejeune *now* supported chromosome examination after birth, but fought prenatal diagnosis and medical abortion. He advocated against the generalization of (prenatal) chromosomal analysis and against the new form taken by genetic counseling (Lejeune 1970). Although this choice has

---

11 Lecuyer 1958, § 12, "Conduite à tenir du médecin en présence d'un arriéré mongolien."

12 This may be compared with the rule articulated in recent textbooks in pediatrics. Though clinical criteria did not significantly change, the karyotype analysis has become an obligatory rite-of-passage.

generally been viewed as the consequence of Lejeune's deep commitments to Catholicism, and religious beliefs certainly played a part in his positions, his stance should be interpreted in view of the fact that Lejeune's vision of medical genetics was centered on mother-child relationships and "progénèse."

Our analysis of the transformation of Down's syndrome from an environmental consequence into a chromosomal disorder reveals two issues in the definition of genetic disease. First, the new vision of the links between genotype and phenotype was highly constrained by pediatric aims. The fact that most uncertainties regarding the diagnosis of the disease were articulated in terms of adjusting the karyotyping techniques to the clinical definition of Down's syndrome reflected this situation. The strength of clinical medicine thus channeled the discussions of the period 1959–1962 into an experimenter's regress. This in turn explains why the attempts to distribute or classify Down's patients on the basis of family patterns and cytogenetic data emerged in settings devoted to human genetics such as Penrose's laboratory and had little influence on pediatric practices. In other words, the genetic heterogeneity of Down's patients did not result in a clinical heterogeneity.

Secondly, this pattern was especially visible in France. The most prominent team of researchers interested in Down's syndrome gathered pediatricians with roots in the local tradition of medical eugenics. Moreover, in the 1950s, the matter of importance was to decide where and how an increasing number of surviving Down's patients would be cared for. Both circumstances favored the construction of a constitutional Down's syndrome opposed to what may be regarded as the pseudo-mendelian version of the disease which could be found in England. This hypothesis may be checked by looking at a second French location where practices echoed developments in the British Isles.

LOCAL ALTERNATIVE: LAMY, MENDELISM, AND MEDICAL GENETICS

One peculiar aspect of the fate medical genetics in postwar France is the small number of service centers and their centralized location. In 1965, Turpin and Lejeune moved from *Hôpital Trousseau* to the *Hôpital des enfants malades*. From that year onward, the top laboratories gathering together pediatricians interested in genetics were located in one single setting. This arrangement was paradoxical, since two groups in the children's hospital were simultaneously examining patients affected with hereditary disorders, offering karyotyping service, and advising families. Styles of work, scientific culture, and professional interests, however, opposed Lejeune's cytogenetic unit and the large clinic and research laboratory headed by Maurice Lamy.

Genetics at the hospital for sick children in Paris started during World War II within the department of pediatrics when Lamy was asked to give a course on hereditary disorders by the professor of general pediatrics, his mentor Robert Debré. At the same time, small-scale research on heredity was conducted within the Lamy's specialized out-patient service (Lamy 1951). In contrast to most French physicians, Lamy sojourned in the United States as a medical researcher. From the 1930s onward, he became the translator of Mendelian genetics into a French medical language. This may be perceived in Lamy's (and Debré's) fierce opposition to the widespread use of "hérédo-contagion" for explaining the transmission of syphilis or tuberculosis. The translator's role also surfaced in Lamy's choice of a research strategy, one emphasizing the collection of pedigrees and the Mendelian analysis of patterns of transmission, as represented in the investigation of hemolytic diseases (Debré, Lamay, See, and St. Schramek 1936).

Lamy's early commitment to Mendelo-morganism was clearly reflected in his war-time teaching. Opposing French pediatric eugenics as well as German racial hygiene, Lamy's description of heredity was built on the Mendelo-Morganian understanding of human genetics. His course at the Hospital for Sick Children began with classical patterns of transmission illustrated with iconic disorders such as hemophilia and color blindness (Lamy 1943). Predisposition to medical disorders was hardly mentioned. Discussing research prospects, Lamy considered twin studies to be the best *dispositif* for sorting out the respective influence of genes and environmental factors. During the war years, Lamy actually worked out an original bond between research and clinical medicine. While routinely practicing pediatrics in Debré's department, Lamy established a specialized out-patient service for twins. In the latter, care was not an issue. Lamy provided general advice about education while conducting measurements, physical examination, inquiries, and psychological tests. The main product was research material.

In the aftermath of World War II, Lamy rapidly benefited from the fact that Debré started to build a biomedical empire at the *Hôpital des enfants malades*. (Debré 1974; Picard 1992; Gaudillière 1992). As mentioned above, Lamy was nominated as the first professor of medical genetics in France. Two fields of expertise emerged in an expanded laboratory: statistics and biochemistry (Lamy 1972; Debré 1976; Frézal 1959a; 1959b). On the one hand, statistical studies of muscular, skeletal or neural malformations were launched in order to measure the incidence of these disorders and to quantify the risks of recurrence. On the other hand, following Penrose's emphasis on the exemplar of phenylketonuria, the study of Mendelian metabolic disorders was thought to be a natural target. By the early 1960s, Lamy's student Jean Frézal, who had spent some time in Penrose's laboratory, was organizing a small research team which would focus on "inborn errors of metabolism," i.e., PKU, diabetes, obesity, galactosemia, defects in amino-acid metabolism, and lipid metabolism deficiencies (Lamy, Royer and Frézal 1959; Frézal to Penrose, 17 January 1960, Penrose Papers file 132/5). Routine practice consisted in collecting "fascinoma": clinical observations of interest to the geneticists. This could mean cases of rare disorders, families showing complete and informative pedigrees, or groups of patients characterized by similar biochemical signs of putative diagnostic value. The increasing part played by teaching and laboratory bench work resulted in loosened links with the clinics.

The use of karyotyping techniques in the laboratory may illustrate this tendency and highlights similarities with the study of "fascinating" exemplars of chromosomal disorders pursued by the British human geneticists. Karyotyping was introduced in 1959 by a student of Lamy, Jean de Grouchy, who did not learn the procedures with Lejeune but went to England where he visited a series of laboratories, including Penrose's Galton Laboratory (de Grouchy and Lamy 1961). Although de Grouchy did practice chromosome analysis for local pediatricians, in the early 1960s, Down's syndrome was not a research topic. Competition with Lejeune certainly contributed to this choice. Nonetheless, it is worth noting that studies focused on other chromosomal abnormalities, namely sex disorders of chromosomal origins, which offered more direct insights into genetic determinism.

Turner's and Klinefelter's syndromes thus provided opportunities for combining karyotype, hormone measurement, and morphological analysis. In contrast to the service role Lejeune adopted, de Grouchy centered his contribution on case studies which aimed at nosologic and etiologic innovations. Thus, in contrast to Lejeune, de Grouchy repeatedly defended the classification of clinical entities on the basis of genetic evidence.

The work on Turner's syndrome is typical of this pattern. By 1962, following the exemplar of Down's syndrome, Turner's syndrome had been associated with four features: 1) dwarfism and malformations; 2) low development of the ovaries; 3) high amounts of pituitary hormones in urine; 4) a forty-five chromosome count with one single X and no Y. In 1963, Lamy and collaborators published observations of familial Turner's syndrome, which included two sisters *referred* by nearby internists. Clinical tests and diagnosis had been completed ten years earlier, but the new perception of Turner's syndrome made a new study of a putative familial case interesting. The chromosomal analysis led de Grouchy to conclude that karyotypes of both sisters were normal. The clinical features were then re-evaluated. Commenting that these patients did not show significant malformations and very little dwarfism, the geneticists suggested that cases which had occasionally been described as "familial Turner's syndrome" were not really Turner's syndrome. They should instead be viewed as instances of a different hereditary disease possibly caused by a recessive mutation and characterized by underdeveloped ovaries (Josso, de Grouchy, Frézal, and Lamy 1963).

In this case, as in many others, patients were dealt with in a clinical department. They were sent or transferred to Lamy's genetic service for a short period of biological and clinical examination. The reinforcement of this division of labor gradually led the geneticists to practice a new form of medical work which mainly consisted in counseling. In the 1950s, it was current custom in Lamy's unit to distinguish "genetic" activities centered on the study of incidence and transmission, and "eugenic" work, meaning the whole range of medical activities (Lamy 1956).[13] From the late 1950s onward, however, examination sessions were viewed as meetings to inform families about transmission, recurrence, severity, and penetrance of truly

---

13 For a practical exemplar, see the study of families with children suffering from muscular distrophy in Lamy and de Grouchy 1954.

hereditary diseases (Frézal 1959). The local emphasis on Mendelian inheritance encouraged two aspects which opposed this form of counseling and pediatric eugenics. First, the establishment of family trees and medical pedigrees became an obligatory stage in establishing transmission patterns and in computing risks. This reinforced the tendency to focus medical work on couples as reproductive units rather than on individuals or on families as living units. Second, as mentioned in the introduction, concerns about population genetics and public health were reframed in terms of "carrier" detection (Frézal 1967). Accordingly, when the Guthrie test made the diagnosis of PKU in newborns feasible, Lamy's team provided the driving force behind systematic post-natal screening in France.[14]

By the late 1960s, this form of genetic counseling was greatly affected by the development of amniocentesis and the rise of prenatal diagnosis. Increased demands for the diagnosis of chromosomal abnormalities in turn facilitated the generalization of what was still a local development and triggered debates about the professionalization of medical genetics. Thus, in 1970, the French *Académie de médecine* held its first hearing on genetic counseling. Raymond Turpin advocated for the creation of new centers for premarital or postnatal counseling, with a special eye on the role of environmental mutagens and on the prevention of chromosomal changes. The author of the main report was Maurice Lamy. Once again, he stressed the importance of inherited disorders which followed Mendelian modes of transmission. Lamy called these diseases "maladies authentiquement héréditaires" (Lamy 1970). His report put a special emphasis on inherited metabolic disorders which could be diagnosed through biochemical or physiological assays and treated by means of specific diets. According to Lamy, laboratory procedures could, in a few instances, be employed to screen families for heterozygous individuals, or to achieve prenatal diagnosis following amnio-centesis. For diseases like PKU, sickle-cell anemia, thalassemia, and galactosemia, anticipations of reduced incidence and/or limited effects were reasonable. Lamy's report was endorsed by the *Académie* and became the first official statement of what genetic counseling should be: a source of non-directive information about the genetic status of families with newborns, fetuses, or relatives affected with "truly" genetic disorders.

---

14 J. Frézal, interview with the author, June 1995. See also Frézal and Rey 1964.

BETWEEN COUNSELING AND GENOMICS: THE MOLECULAR CLASSIFICATION OF SPINAL MUSCULAR ATROPHIES (SMA)

Given this background, it is no wonder that the genetic department at the *Hôpital des enfants malades.* was, fifteen years later, among the main centers benefiting from the growth of human genome research. In order to contrast the early days of chromosomal examination, I shall in conclusion discuss the search for the genes causing spinal muscular atrophies (SMAs) that was launched in the 1980s in this very same setting. Following the development of this SMA project will highlight the changes that made possible the articulation of a typical clinician's regress: the rise of genetics as a medical specialty centered on counseling and the creation of a lay demand for a molecular biology-based medical research.

In 1987, the French muscular dystrophy association, the *Association française contre les myopathies* (AFM), imported the Telethon format, modeled on the TV-based fund raising campaign organized by the Muscular Dystrophy Association in the United States. The first attempt at presenting muscular dystrophy to a wide popular audience resulted in a huge collection of 200 million French francs. This success transformed a small-scale support association into a major patron of French biomedical research. AFM 1988 research budget (roughly 160 million FF) compared favorably with the funds allocated by large charities such as the *Association pour la recherche contre le cancer.*[15] AFM had been established with the patronage of a few specialists of neuromuscular disorders (Paterson and Barral 1994). The Telethon provided the means to modify this form of dependency. A three-stage program emerged aimed first to create a mapping infrastructure and develop sequencing tools; second, to identify the mutated genes causing muscular disorders and other hereditary defects; and third, to invent the procedures for gene therapy (Barataud 1992). Following debates between AFM officials and Daniel Cohen from the *Centre d'études du polymorphisme humain* (CEPH)—a privately funded research center which had been established to investigate the correlations between special sets of histocompatibility markers and chronic diseases such as diabetes,

---

15 On cancer charities in France see Pinell 1992. Between 1980 and 1987, ARC collected roughly 150 million French francs per year for French biomedical laboratories.

myocardial infraction or atherosclerosis—special emphasis was placed on the systematic mapping of the entire human genome. The decision was made to implement this project within the framework of a new AFM center for genetic research , Genethon, which would be built in a Paris suburb (Cohen 1993; Gaudillière 1990).[16]

Beside this commitment to a large scale and centralized approach, AFM supported many local programs. Muscular disorders were supplemented with a wide range of genetic disorders, and by 1990 the fund raising campaign was run under the claim that AFM goal was to win the war against genetic disorders. The first research project concerned with the identification of pathological genes dealt with spinal muscular atrophies. The DMD gene involved in the Duchene-type muscular dystrophy had been cloned in 1986 and some members of the scientific advisory board rapidly advocated an analogous strategy-for other genetic disorders. At that time, the genetic department at the Hospital for Sick Children was already involved in mapping work, and Jean Frézal, the former student of Maurice Lamy who took over the service when Lamy retired, secured the SMA contract.

By the spring of 1988, AFM allocated funds for the cost of equipping a laboratory of molecular genetics which would be closely linked to the pediatric genetic service. Following internal reshuffling, SMA work was to be completed under the leadership of a young neuropediatrician with training in molecular biology. The strategy consisted in the search for significant correlation between the transmission of the disease within affected families and the transmission of genetic markers of known chromosomal location. The project depended upon the coordination of three major resources: DNA samples from patients diagnosed with SMA and from their relatives; Restriction Fragment Linked Polymorphism (RFLP) probes recognizing specific polymorphic segments of the human genome; and statistical tools for linkage analysis.

Problems related to the recruitment of SMA patients are worth mentioning because they illustrate the complex relationship between the physicians, the geneticists, and the patient association which underlay the development of a "French Spinal Muscular Atrophy Investigators." Collaboration with the pediatricians, neurologists and

---

16 It should be mentioned that the mapping project was not invented by AFM: CEPH scientists were already discussing the issue with Collaborative Research and other players in the American HGP.

geneticists involved in the care of SMA patients in France was facilitated by putting a neuropediatrician in charge. Letters requesting clinical histories, pedigree information and blood samples from families with SMA were sent to relevant specialists while the head of the localisation project started to visit clinical services all over the country in order to urge the physicians to participate in the study, and to collect samples and records. By the fall 1988, however, a dozen samples only had been sent to the research team. A large-scale data collection was then organized in conjunction with the AFM.

The data collection was a one-day event gathering all the families participating in the SMA support group. It was not rare among French specialists to resent this campaign as a form of commissioned and lay-controlled research:

> The Dourdan collection was the first time AFM self-managed a research operation. Then they realized how powerful they were. . . . For the first time we got the feeling that a power alternative to the hospital and the clinicians was emerging. Medicine was debarred, eliminated. They were speaking of taking destiny back into their hands, of empowerment. Two worlds facing each other. With respect to our practice, there was a major problem: self-help and identity were forming the basis of diagnosis. No pediatrician, no semiology, no examination. Nosological analysis? We don't care. Muscle weakness or muscular diseases resembling chronic SMA? We don't care. That was the beginning of a long process which resulted in absurd public collection campaigns with 1-800 numbers. (Oral Interview, Prof. L. [pseudo.] head of a Medical Genetics Unit, February 1995)

The professional response was to reinforce the collaboration between neuropediatricians and geneticists. During the collection, letters were given to the families in order to organize a follow-up with the physicians in charge and cross-check the SMA diagnosis. The development of a French SMA research consortium thus favored the definition of "homogenized" criteria for entry into the genetic study.[17]

---

17 It was based on V. Dubowitz's review of SMA classifications: "SMA were selected on the following diagnostic criteria: i) proximal, symetrical limb and trunk weakness; ii) muscle atrophy without facial or extraocular involvment; iii) no spasticity, hyperreflexia, sensory loss, or mental retardation; iv) electromyographic studies showing denervation and diminished motor action potential amplitude with normal or slow nerve conduction velocities; and v) muscle biopsy consistent with denervation with no evidence of storage material or other structural abnormalities"

In December 1989, the research team working at the *Hôpital des enfants malades* had excluded three quarters of the human genome when the rumor circulated that an Anglo-American consortium had located an SMA gene. The French workers hurried to investigate the remaining chromosomal markers and, a few weeks later, they mapped a candidate SMA gene on chromosome 5. The first round was over. A possibility to further cash on the building of the medical network, the large pool of recruited families, and the collection of markers was to focus on the clinical diversity of spinal muscular atrophies.

One aspect of the search for SMA genes was therefore that such work required addressing at the same time nosological and technical issues. Are the different forms of SMA one or several diseases? Is each form of SMA caused by one or by several genes? Are the patient pools representative? Are the biological markers reliable? For instance, clinicians may not use the same criteria to define a "true" SMA whether they think that severe and mild forms are different pathological entities or only different "presentations" of one single disease. Similarly, laboratory geneticists could consider genetic homogeneity within a small set of SMA patients either as typical of the general population, or as an artifact reflecting classification and recruitment problems.

Some historical background is required in order to clarify the issues at stake. Classification of muscular atrophies had been a non-issue for almost fifty years following the description in the early 1890s by Guido Werdnig and Johann Hoffman of progressive muscle weakness attributed to atrophy of the anterior horn cell nerve cells that resulted in early death. Normal practice was clinical diagnosis based on symptoms including onset in infancy or early childhood, severe paralysis, areflexia, absence of pain, and electrical signs of denervation (Walton 1956). The subject was complicated in the late 1950s as neurologists followed the analysis of Ernst Kugelberg and Ludwig Welander in reclassifying cases of progressive muscular dystrophy as late onset forms of SMA, showing mental or sensory disturbances, survival into adolescence and adulthood, and motor neuron lesion under refined electro-myography. Acute (Werdnig-Hoffman) and chronic (Kugelberg-Welander) SMA then emerged as two clinical forms of a single "heredofamilial disorder." Classification was merely

(contd.)
(Dubowitz 1964, 710).

a matter of age at onset and severity which determined the clinical fate, i.e., prognosis and provision for care. Thus, as a young pediatrician in Sheffield, Victor Dubowitz argued all grades of severity existed between "severe cases with weakness present at birth, early respiratory distress" and "patients with onset of mild weakness in infancy who remain ambulant and show little subsequent deterioration" (Dubowitz 1964, 710).

Clinical heterogeneity was complemented with genetic heterogeneity in the 1960s, following the rise of genetic counseling in pediatric hospitals. Late onset SMA revealed uncertain features with X-linked, autosomal dominant, or autosomal recessive forms characterized by heterogeneous risks. In contrast, acute and autosomal recessive SMA became a stronghold for genetic counseling. Statistical surveys resulted in the attribution of 1:80 to 1:100 gene frequency and an almost predictable age of death. Heterogeneity illustrated by variable classification schemes including 3, 4 or 7 SMA classes was reinforced by the existence of two forms of genetic management. On the one hand, variable chronic or late onset SMA was of uncertain prognosis, variable risk of recurrence and difficult care. On the other hand, homogeneous chronic or acute SMA was of ascertained rapidly fatal course, 1:4 risk of recurrence and short-term care (Pearn 1982).

The emergence of a clinician's regress as a means to narrow down the uncertainties associated with the molecular diagnosis of heterogeneous SMAs became apparent following the publication of reports on the localization of genes involved in severe forms of the disease. With a handful of families as samples, the first article by the Anglo-American consortium argued for the location of a gene for chronic SMA diagnosed in childhood. The paper did not present evidence for the origins of acute SMA forms with onset in early infancy. The authors recalled the accepted norm for linkage analysis: "a maximum log likelihood ratio (lod score) of three or more is appropriate for declaring linkage." Computation based on "six families diagnosed with acute SMA," peaked at 1.6 (Brzustowicz et al. 1990). A few weeks later, four new consanguineous families had been included in the study. New computation for acute SMA cases showed that the genetic markers used to define the location of the gene for chronic SMA families peaked at 13 when two anomalous families were excluded from the pool. Linkage was emerging and the message conveyed by the paper was that there was "genetic homogeneity between acute and chronic forms of SMA" (Gilliam et al. 1990, 825). The two unlinked families, however, raised "the possibility of genetic heterogeneity or disease misclassification among the SMA family set" (ibid.). The paper informally pointed to alternative outcomes of the attempts at matching biological and medical practices, genotype and phenotype. The French consortium added considerable credibility to the unifying perspective by analyzing larger pools of chronic and acute SMA families (Melki et al. 1990). The hypothesis of genetic heterogeneity was rejected on the basis of molecular analysis conducted within a partly redefined set of SMA families.

These claims were viewed as mixed blessings by other researchers. A *Lancet* editorial accompanying the publication of the French study thus commented that "linkage analysis was compatible with gene locus homogeneity between acute and chronic SMA." However, "there is some evidence of heterogeneity in the data, which suggests that a minority of families may have a mutation that does not map to the same locus" (Anon.1990, 281). To the author of the editorial, this heterogeneity may "create practical difficulties" until the time when the families in question were clinically or molecularly distinguishable. Meanwhile, "prenatal diagnosis and carrier detection on the basis of closely linked gene markers should probably be withheld from use in acute SMA for a short while until the question of heterogeneity is cleared up. This is not just pedantry but good medicine, because to make a prediction on the basis of analysis of markers linked to the wrong gene will lead to errors, and errors in this sort of prenatal diagnosis have dire consequences" (ibid.). Thus, the research debate about nosologic schemes and genetic heterogeneity was linked to clinical evaluation and practical issues in genetic counseling.

A few weeks later, German geneticists embraced a similar scepticism. Deploying a "study of 234 cases from 206 families" which gives "evidence that the condition may be heterogeneous," they warned that the problem of true "genetic heterogeneity," illustrated by the identification of cases with dominant or X-linked inheritance or "atypical pedigrees"—i.e., unexplained by simple autosomal recessive inheritance—is made more complex because of "misdiagnosis" since cases of Duchene/Becker type progressive muscular dystrophy are often reported as chronic SMA (Serres et al. 1990, 750). The recommendation was therefore again "that prenatal diagnosis with DNA markers should be withheld until the question of heterogeneity is clarified" (ibid., 749). Over the next two years, genetic heterogeneity was not ruled out but changing nosologic standards and new counseling practices brought the debate to an end. Rather than being used to diagnose chronic SMA, which showed the clearest genetic linkage, DNA markers entered counseling sessions in the form of prenatal diagnosis for couples who had already experienced the death of an acute SMA baby. Strong demands and clinical homogeneity balanced the risk of error. Prenatal "molecular" diagnosis was advocated by the American and the French consortium (Daniels 1992; Melki 1992). Soon, DNA markers were used to supplement the clinical diagnosis of chronic SMA. Finally, they provided the basis for presymptomatic diagnosis of chronic SMA (Brahe 1993). Confidence in the use of a single set of markers was reinforced by counseling experience: prognosis of safe babies proved right except in very few instances which could be traced to "misdiagnosis".

This points to the fact that the consensus about genetic homogeneity was favored by counseling practices. Yet, it would have been much more difficult to achieve if not for collective regulation and public articulation of the clinician's regress. In December 1990, the "International SMA Collaboration" gathering both regional consortia held its second meeting to define criteria for the diagnosis and classification of SMAs. "After considerable discussion, debate and compromise, consensus was achieved" (International SMA 1991, 81). The streamlining of clinical practices aimed at producing a "framework for the molecular geneticist." On the one hand, exclusions criteria were reinforced: patients showing disfunction of the central nervous system or other neurologic systems were excluded as well as patients with "involvement of other organs, i.e. hearing, cardiac or vision." On the other hand, a simple classification was adopted to eliminate fine-tuned and skilled clinical assessment, which was then viewed as too complex for achieving consistency in large-scale studies. For example, defining courses for type II (intermediate) SMA and type III (mild) SMA were "never stand " and "stands alone" respectively.

With respect to the genetic heterogeneity question, this compromise achieved three goals. First, the international consortium focused issues on autosomal recessive inheritance, leaving out troublesome late onset dominant cases (Kausch et al. 1991). Second, the new criteria for inclusion were chosen in order to eliminate the families which did not conform to the chromosome 5 location in the French or the

Anglo-american study. These families were reassessed as showing additional features such as cardiac abnormalities or cataract which became additional criteria for exclusion (International SMA 1991, 81). Third, changes in nosological practices were complemented with a theoretical framework for handling abnormalities. In other words, the problematic links between clinical features and results of DNA analysis became usual genotype/phenotype problems viewed in terms of variable expression of the SMA gene(s). In a short time, debates over the diagnosis and nosology of spinal muscular atrophies were channeled into a reshaping of clinical practices which provided the conditions for establishing the genetic homogeneity of SMAs.

Various elements were mobilized in order to narrow down biological and clinical uncertainties: the use of exemplars, the standardization of tools, the recruitment of new families, the segmentation of medical populations, etc. Two moves, however, were of special importance. Both illustrate the new role of the genetic laboratory and contrast the dynamics of debates in the 1960s, when the norms of caring pediatricians prevailed. First, the circulation and use of a single set of DNA markers for acute and chronic SMA was facilitated by habits in genetic counseling. Second, a change of nosological categories limited the variability of the molecular analysis of SMA genes. Thus, changes of nosological categories and new prognosis procedures ended the regress. In other words, theoretical consistency was replaced by practical developments in the genetic clinic.

In France, the muscular dystrophy association contributed to this move in at least two different ways. French SMA families were very supportive of the work of bench geneticists as an alternative to the work previously done by clinicians specialising in neuromuscular disorders. Thus, they provided additional legitimacy to the molecular vision. In addition, after AFM launched the molecular project by organizing the large DNA collection, the mere existence of an AFM support group, acting as one single entity, contributed to sap the strength of dissimilar clinical experiences by families with children suffering from either chronic or acute SMA. This may explain why these families supported the use of molecular prenatal diagnosis for all forms of the disease.

CONCLUSION

When studying the 19th century controversies on cerebral local-ization, the sociologist Susan Leigh Star used the term "triangulation" to describe work processes combining biological research and clinical work in order to make local results "more robust" (Star 1986). In her account, "triangulation" means the use of "multiple methods to de-scribe the same object" and to improve validity by comparing tools. Within the context of medical genetics, debates about diagnostic cri-teria, nosologic categories or etiologic factors may be viewed as in-stances of "triangulation" between clinical and genetic methods em-ployed to describe the same disease. The cases presented in this paper however show that what is usually at stake in these debates is precisely the fact that activities at the bench and activities at the bedside deal with one single hereditary disorder. In other words, discussions over the relationship between genotypes and medical phenotypes do not only specify the correspondence between two levels of analysis, but create the relevant biomedical unit to be analyzed. This is important to understand why these debates have often been terminated by using either clinical examinations or genetic investigations as stable and clear-cut sources of knowledge in order to calibrate local practices and make uncertain results more robust.

This paper has illustrated the notions of experimenter's regress and clinician's regress by opposing the transformation of Down's syn-drome into a chromosomal abnormality in the 1960s and the estab-lishment of Spinal Muscular Atrophies as an homogenous genetic dis-order in the late 1980s. In the former case, clinical diagnosis was taken for granted and the karyotyping procedures became mere tech-niques, i.e. additional tools for the clinician occasionally used to sort out border-line cases. Although these innovations revealed chromo-somal heterogeneity, the clinical homogeneity of Down's patients was in the end reinforced. In the latter case, the clinical criteria for the classification and diagnosis of spinal muscular atrophies were chal-lenged by molecular geneticists using a single set of DNA markers for mapping the different SMAs and for making prenatal diagnosis. Molecular results were taken for granted. Finally, molecular homo-geneity practically and theoretically superseded the clinical hetero-geneity, i.e., the divide between chronic and acute SMAs.

One interpretation of these contrasted paths may be that geneticists and clinicians belong to different professions. Technical or clinical regresses may then be viewed as trials of professional standings. In the late 1950s cytogeneticists were ranked poorly within the hospital hierarchy. Consequently, the clinical practices prevailed. In the late 1980s, the development of molecular biology and genetic engineering granted a new generation of human geneticists with prestige and influence. It is therefore not surprising if their tools now dominate the skills of the clinicians. Such grant vision of the medical and the biological professions wrongly assume that geneticists and clinicians have well-defined and contrasted professional roles: examination and care of patients on the one hand, experimentation and biological measurement on the other hand. Although there is some truth in this divide, the professional approach does not do justice to the fact that medical genetics is rarely pure laboratory work done by trained molecular biologists but most often an alternation between the bench and the consulting room practiced by trained pediatricians. Fine-grained analysis of patterns of work are therefore necessary to give a better account of both the division *and* the coordination between genetics and pediatrics. Our comparison of three episodes in the development of medical genetics in France sheds some light on the local practices and historical circumstances which account for the alternation between technical and clinical closures of biomedical debates.

The work of Raymond Turpin and Jerome Lejeune shows how elements in the French medical culture and clinical activities constrained the 1960s debates about Down's syndrome. Several factors contributed to a pediatric style of work: 1) the skills of morphological analysis; 2) the participation in primary medical care; 3) the imprint of "baby-doc eugenics" upon the definition of pediatric aims—family welfare, birth rate enhancement, struggle against hereditary predispositions. In Turpin's service, Down's syndrome remained a curable constitutional disease while karyotyping became a tool of limited diagnostic use rather a source of evidence for genetic causation.

This pediatric style opposed the vision of Down's syndrome devised by human geneticists who leaned toward biological research and built on other scientific cultures. This was illustrated by the combination of Mendelian studies and biometrics, which dominated the study of Down's syndrome at Penrose's Galton Laboratory in London. In France, Lejeune's monopoly on Down's studies was not challenged.

Yet, an alternative style of practice in human genetics emerged in the medical unit headed by Maurice Lamy. Partly imported, the new approach focused on "true" hereditary diseases, meaning Mendelian disorders, and on the collection of pedigrees and reference observations which could be used to exemplify genetic causation and re-evaluate clinical categories. This genetic practice was, however, associated with routine clinical duties and with the service roles of primary-care pediatricians. Thus, it did not hinge upon the classical hierarchy between the laboratory and the clinic until the late 1960s.

The links between genetic results and clinical decisions were actually reinforced when human genetics became a form of medical practice. In other words, when the development of amniocentesis, prenatal diagnosis, and postnatal screening transformed genetic counseling into a form of "patient" management based on karyotyping techniques and biochemical testing. This paved the way for the development of medical genetics as an organized specialty centered on examination, laboratory testing, and counseling. As shown by the work of the SMA consortia, the molecular reshaping of diseases rested on changes in the clinical work done by human geneticists. The case of spinal muscular atrophies thus stresses the collective redefinition of diagnostic criteria and an homogenous use of DNA markers in prenatal diagnosis. In the French context, the links between counseling practices and molecular tools were reinforced by the mobilization of the families of patients and by AFM's articulation of a lay demand for molecular genetics.

One final comment is that the postwar history of medical genetics may be read as a shift from a clinical regime, which was dominated by experimenter's regresses toward an experimental regime in which clinical regresses prevailed. In this respect, the impact of genome research on medicine may be compared with the rise of medical bacteriology in the late nineteenth century (Fleck 1986). In other words, a terrain has been created for numerous changes in clinical diagnosis. Commenting upon this process, some authors have worried about disappearing medical skills. A more serious reason for concern is, I think, the speed with which technological irreversibility is achieved.

Genomics and Practice: a Commentary on "Whose Work Shall We Trust?"

Robert A. Bud

We may associate France with fine wine and elegant chateaus. It is also the country which has taken pride in its modernist art and technology; in its dependance on nuclear power; in the TGV; and in the *Concorde*. Perhaps because it does not rate the highest ratio of science Nobel prizes to population, the work of Ephrussi, Boris, Monod, Jacques, Jacob, François and Dausset have given molecular biology a particular status there. France is the country of the futuristic Genethon, the Paris center funded by the Association of Muscular Dystrophy whose robots scooped American teams to achieve a physical map of chromosome 21. As James told Congress in 1993, "the French have moved to super-production first" (Watson in Cook-Deegan 1994, 197). It had not been a traditional center of modern genetics. Before the Second World War, even classical Mendelism was hardly taught. There, above all, it may seem that genomics is revolutionary.

It is therefore specially striking that Jean-Paul's paper begins in a place doubly unusual for a treatment of genomics. One, the geographical location he describes is the ward of a busy hospital full of clinicians rather than the laboratory full of robots. The context is not one of Nobel prize-winning breakthroughs but of tentative diagnosis. A

95

feature of traditional evaluation of genomics is the focus on the entire population as the party affected by its cultural impact. Gaudillière is shifting attention to the clinical context where the impact is practical and immediate. Secondly, the change in this world, while certainly significant, is evolutionary rather than revolutionary. The paper shows that the new genetics has enabled a changing status and style for the clinical setting in which clinicians work with geneticists.

This introduction alerts us to this work's roots not in the history of science with its exciting tale of breakthrough and sudden discovery. Rather this is a study growing out of the histories of technology and medicine in which the uncertain relationship between model and practice are reflected in complex professional relationships between scientists and doctors or engineers. A generation ago it seemed, to use Derek Price's metaphor, that science was like a tree whose fruits would just fall. Since then it has become clear that, even in the age of technoscience, diverse skills are involved in technical innovation and science provides some but far from all. The shocking experience of penicillin in the 1940s, the *coup de foudre* experienced by scientists, doctors and patients alike has perhaps skewed our appreciation of the complex process of assimilating new drugs. The English historian Michael Worboys, Michael: on sulpha drugs has looked at the choice of anti-pneumonia therapies before World War II and the complex process by which sulphapyridine was evaluated as a therapy. Fujimura has looked at the even more complex process of diagnostics (Fujimura 1987).

The use of this lens gives Gaudillière a complex picture. He moves between the activity of a contemporary laboratory, the emergence of the chromosome interpretation of Down's syndrome in the 1960s, muscular dystrophy studies in the 1980s and back again to genetic counseling practices. While Mendelian genetics are at the heart of most histories of eugenics and of genetic diagnosis, in France as Schneider has shown, a vigourous tradition of neo-Lamarkian eugenics held sway until the Second World War. This combined heredity with infection and intoxication as explored by Schneider in his book *Quality and Quantity* (Schneider 1990). It was within this tradition that Turpin, the leading French student of Down's syndrome patients, worked in the 1930s. Even in the 1960s, pediatrics and medical genetics were closely interwoven with even the discovery of the anomalous chromosome count of patients being a minor factor in diagnostics.

The difference between the medical geneticist and the clinician was that the geneticist on the basis of evidence at his disposal could intervene before conception or even marriage to recommend a eugenically sensible course of action. By contrast to this clinically dominated view of the 1960s, the emergence of genomics in the diagnosis of muscular dystrophy in the 1980s created a quite different balance of diagnostic power. In France genomics had been promoted by the Muscular Dystrophy Association, and their scientists working with others across the world were able to develop diagnostic tests without even seeing the patients. Nosology and clinical diagnostics therefore followed behind.

Not only does Gaudillière point us towards negotiation, he even gives a language by which it may be analysed and compared. Compellingly, he uses the idea of regress: experimenter's—what is it that one doubts—taken from Harry Collins,Harry and the concept of uncertainties used by Leigh Star to indicate the process (Star 1983). The focus on the changing balance of power in the negotiations over diagnosis is refreshing. It might stimulate questions about how much further one can go. In his autobiography Cohen, Danieldescribes the conversations with political leaders such as Jacques Chirac and even François Mitterand which led to a large part of his support. Together with the enthusiasm for the prospective cure for a terrible affliction, they were excited by the opportunity for French and even Parisian glory and for technical growth. Cohen quotes Jean Devaquet, working with Chirac, then Mayor of Paris: "we absolutely must help you especially as you are working with industry, that's excellent" (Cohen 1993,140). There is an exact parallel to be found in Robert Cook-Deegan's account of the American genome project. He writes "Congressional patrons of genome research saw the genome project as a vehicle to maintain a technological advantage of Japan" (Cook-Deegan 1994, 230).

Thus, just as Gaudillière has shown how a history of medicine perspective can help us better appreciate the practical context of genomics, so a history of technology: and genomics perspective would be helpful too. Is the prestige factor to which Gaudillière so accurately points in his paper merely an output, a consequence, or is it indeed an input? To what extent is the changing balance of power in the lab a result of prestige acquired from elsewhere? I have suggested that one factor might be the association with the highest of high technologies. An explanation in terms of DNA is also, of course, statusful, as Nelkin and Lindee have recently pointed out in their book, *The DNA Mystique* (Nelkin and Lindee 1995). Students of European genomics must ask themselves to what extent the association with American practice is both a challenge and an example. The opportunity and need better to integrate macro and micro studies of genomics and diagnostics is indicated by the contrast between this work and existing literature. It is now a decade since the publication of the first edition of the book *Dangerous Diagnostics* by Nelkin and Tancredi (Nelkin and Tancredi 1989). Their account of diagnostics was at a very different level from that of Gaudillière.

In the conclusion to his paper, Gaudillière suggests that in France we have seen three professional generations: the baby-doc eugenicists, the medical geneticists and genomics. In treatments by himself and others of medical geneticists we appreciate not only the immediate clinical contexts of their work but also the ghosts of the baby-docs and the macro-cultural context. The model may inspire us to understand the ghosts as well as the uncertainty reduction that actually matter in the diagnostic lab of the emerging present.

4

A Book of Life? How a Genetic Code Became a Language

Lily E. Kay

HISTORICIZING NATURE AS TEXT

In 1967, molecular biologist Robert L. Sinsheimer, then Chairman of Caltech's Biology Division and sixteen years later the progenitor of the Human Genome Project, wrote a remarkable little book, entitled, *The Book of Life*. With the genetic code just completed—down to its last termination signal—Sinsheimer was inspired to liken the code to an ancient book: a Mayan Codex. "This"—he pointed to the horizontal array of the Codex "is a book. No man living now can fully decipher it . . . . This"—he displayed the bundles of human karyotype— "is another book—or perhaps more accurately an encyclopedia.

> It is a set of human chromosomes, the book of human life. In this book are instructions, in a curious and wonderful code, for making a human being. In one sense—on a sub-conscious level—every human being is born knowing how to read this book in every cell of his body. But on the level of conscious knowledge it is a major triumph of biology in the past two decades that we have begun to understand the content of these books and the language in which they are written. (Sinsheimer 1967, 5–6)

And just as there were many types of books—story books, cookbooks, or primers of Japanese flower arrangement—he noted, "[t]he messages our chromosomes contain are rather like instruction manuals—

they give explicit instructions on how to make some substance, when to make it, in what amounts, and under what circumstances." The language of the "Book of Life"—as any written language—was in essence a one-dimensional script, a linear sequence of symbols prescribed by convention, he pointed, proceeding to explain the "strange and wonderful devices [which] read these books of instruction and follow their programs"(ibid., 9–11).

Sinsheimer's *Book of Life* was a creative literary act, to be sure (it sold many thousands of copies) (Sinsheimer 1994, 134), but its scriptural motifs—language, text, instructions, program—were hardly novel by then. Sinsheimer recapitulated and amplified the ubiquitous articulations of the genetic code as a book, with all its attendant linguistic tropes and scriptural icons: letters, words, text, writing, reading, dictionaries, commas, sense, nonsense, and missense. Based on this potent linguistic imagery and representations of life as text, the genome could now be read and edited unambiguously by those who know. This writing technology thus laid claim to new levels of control over life. Beyond the control of matter there was now control of the word.

Where did this scriptural iconography of the genetic code come from? How had the genetic code come to be textualized? To answer these interrelated questions—and the focus of this paper—we will look back to the 1950s, to the first phase of the work on the genetic code: the theoretical and mathematical phase. For it is during that period that the discursive and semiotic apparatus of genomic textualization was set in place; primarily through the work of researchers from the physical sciences. Few biochemists ventured into that epistemic ether from their entrenchment in the world of matter. Only around 1959 did they, too, begin to recast their chemical representations of heredity as scriptural ones.[1] But to grasp the levels of significance of this textualization, its potency and authority, we must view it both diachronically and synchronically: as an age-old metaphor and as an historically specific and culturally contingent one. For ideas of nature as/and text had changed with time.

---

1 There were exceptions. Erwin Chargaff began to use scriptural and informational representations in the mid-1950s, though he later retracted them as meaningless. Sol Spiegelman began to use cybernetic and informational representations of protein synthesis already in the late 1940s. Virus researchers, notably Stanley's group at Berkeley, adopted the information discourse and its scriptural representations only in 1959. These issues are discussed in Kay 1997 and Kay 2000, chaps. One, Two, Four respectively.

As Jacques Derrida has noted, significations of the genome as a book of life in the 1960s are inextricably linked to the symbolism of "The Book" as natural, eternal, and universal writing; a metaphor pervading the entire history of the West, beginning with Plato (Derrida 1976, 15–16). Nature—animate and inanimate—had since then been continuously textualized. Pondering Nature's *logos* and diversity of life, Lucretius (50 B.C.) postulated that

> Each thing, then, has its own engendering matter.
> If not, how could the *mother* stay the same?
> But since all creatures rise from their own seeds
> They need a place stocked with their stuff, their atoms,
> Whence they are born and emerge to the shores of light.
> Everything can't just spring from anything;
> Inner capacities make things what they are.

And given the manifestations of this *logos* he concluded,

> Believe, then, As all words share one alphabet, so too
> Many things may be made from the same atoms—.
> (Lucretius 1995, Bk. 1, lns. 170, 195)[2]

Such textualizations of nature were revived and reshaped in the twelfth century.

As Brian Stock has shown, Naturalism in the high middle-ages bore its own historicity and culture. With the growth of a more literate society, the science of nature had to accord with the inner logic of contemporary texts. Nature was constituted through interconnections between words, thoughts, and things; Medieval structures of knowledge: logic, grammar, rhetoric, and theology, were mobilized to unravel the "secrets of nature." For William of Conches, Nature consisted of many books, corresponding to the number of controlled interpretations. For Alan of Lille, deviance was a grammatical error. Hugh of St. Victor saw knowledge of Nature as inextricable from logic. Ideally, signification through things was preferable, but the philosopher limited to *scientia*, knew only meaning of words. Words, texts, reason, and nature formed the seamless fabric of medieval natural knowledge (Stock 1983, esp. 315–25).[3]

In that history of Nature as text, a decisive moment came in the

---

2 I am grateful to Matthew Meselson for calling my attention to this source.
3 I am grateful to Lorraine Daston for suggesting this book.

seventeenth century with the birth of modern science. With the rise of an autonomous experimental tradition, representations of nature became intertwined with interventions, or to use Derrida's terms: (reading) the book of life/nature has become inseparable from its writing. The "book of nature" thus awaited "decoding" by the experimental investigator equipped with an "ideal language." Bacon tells us that we read God's natural truth with the "alphabet of Nature." Both Descartes and Galileo spoke of the writing and reading of the great book of Nature; Leibnitz searched for the ideal language, *"characteristica universalis,"* which would correspond exactly to Nature's written language; Bonnet presumed that "our earth is a book that God has given to intelligences far superior to ours to read . . . ." And for Karl Jaspers, "The world is the manuscript of an other, inaccessible to a universal reading, which existence deciphers" (Derrida 1976, 15).[4]

The advent of telegraphy in the second half of the nineteenth century added new dimensions to the meaning of the body as a system of communication. The metaphor of the human nervous system as an electrical (telegraph) network conducting "nervous energy" was ubiquitous in moralizing the body in late nineteenth-century America. And as Evelyn Keller has pointed out, in pondering the enormous differences between the hereditary particles in the germ cell and the adult form in 1904, August Weismann thought that the variation was "very like supposing that an English telegram to China is there received in the Chinese language" (Rosenberg 1976, 4–6; Edge 1974; Keller 1995a, 79–81). Yet, such metaphors of telegraphic communications were absent in genetics in the first half of this century.

In the 1950s, with the spread of information theory, electronic communication technologies, and computers, the very notions of text and language were profoundly transformed, once again. It now seemed technically legitimate to speak of molecules and organisms as texts, namely, as information storage and retrieval systems. Heredity became a communication system which transferred information through the cell and the cycles of life. Except that in information theory language was purely syntactic and information had no semantic value (Kay 1994; 1995; 1997). Thus, ironically, the textualization of

---

4 Hacking 1982, esp. 130–46. On the scriptural representations of nature and the "book of nature" metaphor in the seventeenth century, see Arbib and Hesse 1986, chap. Eight, esp. 149; Bacon, "Natural and Experimental History," in Bacon 1937.

the genetic code, while deriving much authority from the age-old metaphor of "the book," also grounded its scientific legitimacy in this purely syntactic notion of language which had little to do with human communications, and which was deeply embedded in Missile Age technologies and Cold War culture. Histories of the genetic code have tended to emphasize the roles of Erwin Schrödinger in the 1940s and Francis Crick and his Cambridge circle in the 1950s, though George Gamow's role—limited to articulating the coding problem—has been acknowledged (Olby 1974; Stent and Calendar 1978; Doyle 1993; Judson 1996). But, as we shall see, those contributions form only part of the picture. There are other narratives; other histories of the genetic code. Eminent physicists, biophysicists, chemists, mathematicians, communication engineers, and computer analysts—whose own projects situated them at the hub of weapons design, operations research, and computerized cryptology—joined in the effort to "crack the code of life." And they transported the tropes and icons of the new communication sciences into molecular biology, thus extending and amplifying the discursive space which emerged in the late 1940s, in which organisms and genes were textualized: represented as electronic messages, words, letters, instructions, and texts. These semiotic tools were soon followed by other linguistic tropes: commas, dictionaries, sense, nonsense, and missense.

We shall examine first the emergence of the information discourse in biology at the beginning of the 1950s decade, and then follow its instantiation in the later analyses of genetic codes, where it resulted in a deconstruction: a book of life written in a language devoid of meaning.

ORGANISMS, TEXTS, INFORMATION

First, it is important to underscore that the information discourse was not an outcome of the internal momentum of molecular biology, nor of the logic of DNA genetics. The shift occurred earlier (still within the protein paradigm of heredity) and impacted on every discipline in the life and social sciences in the 1950s. The impact of digital control systems on representations of life processes was both anticipated and promoted by MIT mathematician Norbert Wiener and his circle already in the early 1940s, while working on computational problems of ballistics in Warren Weaver's mathematical division within the OSRD (Office of Scientific Research and Development). These interests were first articulated in 1943 in a noted joint article, "Behavior, Purpose, and Teleology," by MIT-trained engineer Julian Bigelow, Harvard physiologist Arturo Rosenbleuth, and Wiener—the first link between servo-mechanisms, physiological homeostasis, and behavioral processes (Rosenblueth et al. 1943, 18–24; Kay 1994; 1995; Galison 1994; Heims 1980). Negative feedback as a paradigm of thought and action became Wiener's disciplinary mission. Even before the war ended, Wiener, Hungarian émigré and mathematician, John von Neumann, and their circle, began to campaign vigorously for the field of automated digital control, yet unnamed (Wiener to von Neumann, 17 October 1944, Wiener Papers MC22, Box 2.66 MIT archives).

Von Neumann—by 1945 a key figure in strategic military planning—was becoming interested in biology. His interest in biology in general and genetics in particular became closely linked to his mission of developing self-reproducing machines. In fact, he became interested in viruses and phage as the simplest models of reproduction. But in linking viruses to information processing, von Neumann, like most researchers (especially in the United States) operated within the dominant paradigm in life science. He conceptualized reproduction within the protein view of heredity, in which autocatalytic mechanisms of enzymes served as explanations of gene action and virus replication, and genetic specificity arose from the combinatorial properties of amino acid sequences. Von Neumann made contacts with the biomedical community, participating in meetings and communicating with life scientists (e.g., Max Delbrück, Sol Spiegelman, Joshua Lederberg, and John Edsall). Through these exchanges he was encouraged to develop his self-duplicating machine as a possible heuristic model of gene action. Although von Neumann's project in biology did not go far (he died in 1957), its discursivities began to permeate genetic representations.[5]

---

5 See von Neumann to Wiener, 29 November 1946 (Wiener Papers, Box 2.72); Spiegelman to von Neumann, 3 December 1949, and von Neumann to Spiegelman, 10 December 1946 (von Neumann Papers, Box 7.1, Library of Congress (hereafter, LC); letters between Joshua Lederberg and von Neumann (1955) (von Neumann Papers, Box 5.8, LC); I am grateful to Joshua Lederberg for providing me with the missing letters. See also von Neumann 1951a; and Aspray 1990.

The cybernetic view of life gained momentum after 1948. That year Wiener's book, *Cybernetics: or Control and Communication in the Animal and Machine* was published simultaneously in France and the United States. In this remarkably influential book, Wiener expounded two central notions: that problems of control and communication engineering were inseparable—communication and control were two sides of the same coin—and that they centered on the fundamental notion of the message—discreet or continuous sequence of measurable events distributed in time. There he expounded on what Michel Foucault would later call a new episteme: the coming of the information epoch: "If the 17th and early 18th centuries are the age of clocks, and the 18th and 19th centuries constitute the age of steam engines, the present time is the age of communication and control" (Wiener 1965, 39; see also ibid., "Introduction;" Foucault 1970).

His historicization applied also to animate phenomenology. The twentieth-century body was sharply demarcated from that of the nineteenth century, having moved from material and energetic representations to informational ones. "In the nineteenth century," Wiener observed (supplying the discursive framework of information for Crick's Central Dogma a decade later),

> [T]he automata which are humanly constructed and those other natural automata, the animals and plants of the materialist, are studied from a very different aspect. The conservation and degradation of energy are the ruling principles of the day. The living organism is above all a heat engine. . . .The engineering of the body is a branch of power engineering. . . .[T]he newer study of automata, whether in the metal or in the flesh, is a branch of communication engineering, and its cardinal notions are those of the message, amount of disturbance or "noise". . . . quantity of information, coding techniques, and so on. (Weiner, 1965, 41–42)

The same conclusions held true both for inanimate and animate systems: for enzymes, hormones, neurons, and chromosomes, he stressed. He was in regular contact with his old friend, John B. S. Haldane, one of his many links to life science. And through these dialogues and visits, Wiener's cybernetic view of heredity, like Haldane's, became grounded in the primacy of proteins. Like von Neumann, Wiener forecast in 1948 that combinatorial mechanisms by which amino acids organized into protein chains, which in turn formed stable associations with their likes, could well be the mechanisms of reproduction of viruses and genes (ibid., 41–42; 93–94).

Also in 1948 the MIT-trained engineer Claude E. Shannon at Bell Labs published a major article, "The Mathematical Theory of Communication," where he developed the salient features of information theory. His work implied that information can be thought of in a manner entirely divorced from content and subject matter—devoid of semantic value—and established the basic unit of information as the binary digit, or bit. His highly technical article gained wide exposure through a joint effort with Warren Weaver (director of the Rockefeller Foundation's Natural Science division and the molecular biology program), who explained the mathematical concepts of information in his characteristically lucid and eloquent manner (Shannon 1948; Shannon and Weaver 1949).

Simultaneously, while developing the mathematical theory of communication, Shannon also worked on secrecy systems (indeed, the two were intertwined). His work transformed theories of coding, cryptanalysis, and linguistics by introducing concepts such as (mathematical) redundancy and binary coding. But these linguistic analyses had little to do with human language: they were designed for machine communications. Shannon was mindful of these differences. Though he pondered the relation between information theory and human communication, he scrupulously circumvented semantic questions. As we shall see, the new cryptology, coupled with electronic computing, was applied to analyses of genetic codes (Shannon 1949; Kahn 1967, 743–56; see also Edwards 1995).

The lessons of the Wiener-Shannon theory found an even wider circulation through Wiener's book, *The Human Use of Human Beings*, published in 1950. His thesis was that contemporary society could only be understood through a study of messages and communication facilities. The individual and the organism must be recast in terms of information. In a key chapter entitled, "The Individual as the Word," Wiener elaborated this concept and its corollary; the technical possibility of writing the book of life.

> The earlier accounts of individuality were associated with some sort of identity of matter, whether of the material substance of the animal or the spiritual substance of the human soul. We are forced nowadays to recognize individuality as something which has to do with continuity of pattern, and consequently with something that shares the nature of communication. (Wiener 1950, 103)

It is not matter, but the memory of the form, that is perpetuated during cell division and genetic transmission, he insisted. He prophesied that in the future it will be possible to transmit the coded messages that comprise organisms and human beings (ibid., 103–09).

We witness here the opening of a new discursive space in which the word, or the message, became configured together with concepts of the gene as the locus of scriptural-technological control, and with notions of control over bodies, in ways which bypassed their physicality. Their three-dimensionality was flattened into a one-dimensional magnetic tape; their material density symbolically represented as a digital code.

The impact of Wiener, Shannon, and von Neumann was enormous, not only affecting almost every academic field but also cultural sensibilities, especially in the United States but in England, France, and the Soviet Union as well. Even in those disciplines where the technical promises of information theory did not materialize, the information discourse, with its icons and linguistic tropes—information transfer, messages, words, codes, texts—shaped the thinking, imagery, and representations of phenomena. Such was the case with the genetic code, where the discursive framework of information outlasted the authentic technical attempts to apply the theory to genes and the problem of protein synthesis.

The task of building a new discipline—an information-based biology—became the mission of the Viennese émigré radiologist Henry Quastler, who in the mid-1950s moved from the University of Illinois, Laboratory of Control Systems to Brookhaven Laboratories. From 1949 until his untimely death in 1963, Quastler, inspired by Wiener and Shannon, and funded through military sources, channeled relentless energy to rewrite biology as an information science. His output—articles, reports, symposia, and books—was prolific. He achieved a measure of acclaim in the 1950s, with hundreds of references to his work in *Science Citation Index*. In 1952 he organized the first major symposium at the Control Systems Laboratory on *Information Theory in Biology* (Quastler 1953).[6] Two things are striking about the proceedings: the discursive shift to scriptural representations of life, and the conceptualization of information storage and transfer within the protein paradigm of heredity.

---

6 For a detailed account see, Kay 2000, chap. Two.

Working out the measure of biological specificity in terms of information content, Quastler provided a quantitative formulation of what five years later Crick would articulate qualitatively as the Sequence Hypothesis. Similarly, several symposium participants investigated the informational specificity inherent in the combinatorial properties of proteins. Proteins were especially attractive for information theory, they observed: "[t]hey are constructed much as a message . . . the protein molecule could be looked upon as the message and the amino acids residues as the alphabet" (Branson 1953, 84–85). They proceeded to study the intersymbol correlations in the protein text. This was not the medieval scribal text of nature, nor the printed text of seventeenth-century natural imagination, but the postwar text of communication and control systems. Just two years later such textual analyses would become the prime resource in the attempts to break the genetic code.

In a 1949 paper by Quastler and physicist Sydney Dancoff, the information content of a human organism was shown to be 5×10^{25} bits. The information content of single printed page is about 10^4 bits, they noted; thus the genetic description of a human being would entail 5×10^{21} pages—on the order of an enormous library. Based on this logic, the information content of a germ cell was set at about 10^{11} bits; and what they called the "genome catalogue" at the order of about million bits (Dancoff and Quastler 1953; also Quastler 1953b; Branson 1953; Augustine, Branson, and Carver 1953). These scriptural representations of life and heredity—texts, alphabets, letters, words, messages—were then new to biology. They were grounded in the discursive framework of information theory and occurred well before DNA replaced proteins as the source of hereditary information.

Information theory seemed to promise a great deal to the new biology in the early 1950s. A week before Watson and Crick's 1953 famous report on the double-helical structure of DNA appeared in *Nature* (Watson and Crick 1953a), Watson, with geneticist Boris Ephrussi and physicists Urs Leopold and J. J. Weigle, sent a note to *Nature*, suggesting new terminology in bacterial genetics (Kay 2000). Arguing for imposing rhetorical order on the proliferating semantic confusion in bacterial genetics (transformation, recombination, induction, transduction, etc.), they proposed to replace these uses with the term"inter-bacterial information" explaining that "[i]t does not imply necessarily the transfer of material substances, and recognizes

the possible future importance of cybernetics at the bacterial level"
(Ephrussi, Leopold, Watson, and Weigle 1953).

This expectation of the relevance of the information concept to the
study of the combinatorial properties of DNA clearly guided the
thinking of Watson and Crick in their *Nature* report:

> It follows that in a long molecule many different permutations are possi-
> ble, and it therefore seems likely that the precise sequence of bases is the
> code which carries the genetic information. (Watson and Crick 1953b,
> 966)

To emphasize: this was no longer Schrödinger's quaint code-script,
which, as I have argued elsewhere, had drawn on a fifty-year old
tradition (e.g., Emil Fischer, Charles Sherrington) of embedding biological specificity in the combinatorial elements of protein sequences
(Yoxen 1979; Kay 2000, chap. One). And it did not merely reflect a
paradigmatic shift from protein to nucleic acid sequences as the origin
of that specificity. The very idea of biological specificity was being
displaced by the notion of information, as were representations of
texts and codes. Watson and Crick's code bore its own historicity, that
of the postwar era: it governed the transfer of information of heredity, now envisioned as an electronically–programmed communication
system.

THE RISE AND FALL OF OVERLAPPING CODES

George Gamow, the Russian-émigré physicist (professor at George
Washington University), cartoonist, science popularizer, and military
strategist, had been enthusiastic about biology for more than a decade.
But Watson and Crick's 1953 publication in *Nature* excited Gamow
even beyond his normal exuberance.

> I remember very well this day, I was for some reason visiting Berkeley
> and I was walking through the corridor in Radiation Lab, and there was
> Luis Alvarez going with *Nature* in his hand (Luis Alvarez was interested
> at this time in biology) and he said, "Look, what a wonderful article
> Watson and Crick have written." This was the first time I saw it. And

then I returned to Washington and started thinking about it . . . . (Gamow to Delbrück, 13–22 April 1941, Delbrück Papers, Cal Tech archives, Box 8.21)[7]

Gamow immediately perceived the relationship between DNA structure and protein synthesis as a cryptanalytic problem: how can a long sequence of four nucleotides determine the assignment of long protein sequences comprised of twenty amino acids. "And the question was to find, is it possible?"

> And at that time I was consultant in the Navy and I knew some people in this top secret business in the Navy basement who were deciphering and broke the Japanese code and so on. So I talked to the admiral, the head of the Bureau of Ordnance. . . . So I told them the problem, gave them the protein things [list of amino acids], and they put it in the machine [computer] and after two weeks they informed me there is no solution. Ha! (Weiner-Gamow interview 4 April 1968, 80)

Cryptanalysis too was being transformed by information theory and computing. Although it had utilized letter frequency analysis and their neighbor relations ever since the Renaissance, these methods had become mechanized and mathematized during World War II, when cryptology became a prime source of intelligence. But by the time Gamow consulted with the Navy cryptologists, even those recent practices were already being transformed by Shannon's statistical theory of communication which had set up a measure of redundancy in codes (redundancy equals one minus the relative entropy which in turn is a measure of information). Shannon argued that redundancy arose from an excess of linguistic rules, and that—backed by letter frequency counts—it could furnish the ground for cryptanalysis. Information theory showed how to raise the difficulty of cryptanalysis and how much ciphertext was needed to reach a valid solution. These approaches were now coupled with electronic computers. As such they did not merely raise the sophistication and efficiency of cryptanalysis but reconstituted it within the new sciences of electronic communication, symbolic logic, linguistics as logical syntax, and me-

---

7 See also George Gamow, interview by Charles Wiener, 4 April 1968, American Institute of Physics Oral History archives, 76. In fact, back in the early 1940s, Gamow planned to coauthor a book on biology with Max Delbrück (Delbrück Papers, CIT Box 8.21, Gamow to Delbrück 13–22 April).

chanical translation; cryptanalysis was resignified within the compelling logic of guidance and control (Kahn 1967; Pratt 1942; Welchman 1982; Hinsely and Stripp 1993; Bamford 1982; Richelson 1989; O'Toole 1991).

Gamow's initial solution to the cryptanalytic problem (how sequences of four nucleotide bases specified sequences comprised of twenty amino acids) became known as the "diamond code" (Gamow 1954a; 1954b). It was an overlapping triplet code (but not called so until a year later), based on a combinatorial scheme where four nucleotides, arranged three at a time (4x4x4=64) were more than sufficient to specify twenty amino acids (whereas nucleotide doublets, 4x4=16 clearly fell short). The scheme was based on DNA-protein "translation": The amino acids would fit into the rhomb-shaped "holes" (diamonds) formed by the nucleotides in the DNA chain. Drawing on Quastler's work, Gamow argued that there was a mathematical correspondence between the "expected intersymbol correlation" of the overlapping nucleic acids of the diamonds and the "observed intersymbol correlation" between the amino acids in polypeptide chains. It so happened that there were exactly twenty such diamond-shaped configurations, corresponding exactly to the twenty amino acids in the alphabet; "magic twenty" as it soon came to be called.

Several researchers—Linus Pauling, Erwin Chargaff, Francis Crick, and Martynas Ycăs—raised various objections to Gamow's model as soon as it was published. Both Crick and Ycăs pointed out that his diamond code contradicted the small amount of sequence data then available (Fred Sanger recently sequenced insulin). Ycăs, a Russian émigré and Caltech-trained biologist working at the Pioneering Research Division, Quartermaster Research and Development Command, U.S. Army, became Gamow's decoding partner. As Gamow explained, "Dr. Ycăs takes care of the biological part of the problems we are trying to attack, while I handle the more mathematical part of the picture" (Pauling to Gamow 2 March 1954, Pauling Papers Box 263.39).

In the fall of 1954—then a visiting professor at Berkeley—Gamow engaged several Caltech scientists in the coding problem: Max Delbrück, molecular biophysicist Alexander Rich, and physicist Richard Feynman. They too were skeptical of his diamond code and proposed a coding scheme based on RNA; the role of RNA as an in-

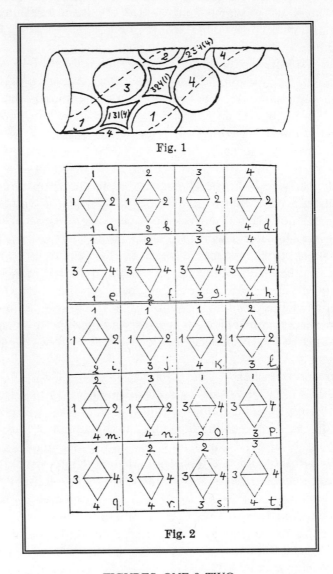

Fig. 1

Fig. 2

FIGURES ONE & TWO
George Gamow's Diamond Code model of protein synthesis illustrating (1) spaces in DNA helix into which
amino acids fit as rhomboids; and (2) table of the twenty
amino acids forming the different diamonds. Source:
Gamow 1954a. Reprinted with permission from *Nature*,
173: 318. Copyright 1954. Macmillan Magazines Limited.

termediary in protein synthesis was then increasingly accepted. But decoding along these new lines would require a computer, Gamow argued,

> As in the breaking of enemy messages during the war (hot, or cold!), the success depends on the available *length* of the coded text. As every intelligence officer will tell you, the work is very hard, and the success depends mostly on luck. There are 20! $=10^{17}$ possible assignments of the aa's [amino acids] to base triplets! . . . I am afraid that the problem cannot be solved without the help of electronic computer. (Gamow to Chargaff, 6 May 1954, Gamow Papers, Box 7,Ycǎs folder, 2)

Gamow planned to put the problem on Los Alamos' Maniac in July.

With the growing enthusiasm for the coding problem, Gamow formalized the coding network by establishing the RNA Tie Club, a select group of twenty members (selected by Gamow) corresponding to the twenty amino acids (another "magic twenty"?). Each member was to have a diagrammed tie, made to Gamow's design, and a tie pin bearing the abbreviation of the member's amino acid. Thirteen members were physical scientists—physicists, chemists, mathematicians, and computer scientists. The club's aim was to foster communications and camaraderie by circulating notes, manuscripts, and articles pertaining to the coding problem; its hope was to locate funding for regular meetings. And indeed for a while it seemed certain that the Army's Quartermaster Research and Development Command would sponsor the RNA Tie Club. But the scheme fell through. Delbrück's motto: "Do or die, or don't try," graced the club's stationary (Gamow Papers, Box y.30, RNA Tie Club; Judson 1996, 269; Crick 1988, 95). Though it fell short of Gamow's high expectations, the club served to promote the coding problem, with all its discursive and material resources, among geneticists and later among biochemists.

Several coding schemes were soon proposed by members of the RNA Tie Club: major-minor code, compact and loose triangular codes, and Edward Teller's sequential code. All were quite restrictive—either fully overlapping (sharing two out of three nucleotides in a triplet) or partially overlapping (sharing just one). They were all based on the principle of matching polypeptide fragments with overlapping nucleotide triplets along a single–stranded nucleic acid; and all were extraordinarily complicated, necessitating the use of computerized cryptanalysis (Gamow Papers, Box 7, Ycăs folder, 1954 and 1955 passim).[8]

That summer (1954), Gamow and Nicholas Metropolis—a theoretical physicist and expert in electronic computing and logic design—tested the various overlapping (restrictive) codes on the Los Alamos' Maniac, using the so-called Monte Carlo method. Applying statistical frequency analysis, they compared artificial amino acid sequences constructed on the basis of various proposed codes with naturally occurring known sequences; the stronger the restrictions on the association between amino acids in the sequence, the smaller the number of different neighbors as compared to random distribution.

The results were negative. In spite of the strong intersymbol restrictions in the proposed codes, both the artificially constructed sequences and the naturally–occurring sequences produced a random amino acid distribution. Rather than question their guiding premise of an overlapping code, or more fundamentally, that the scheme was, in fact, a language-like code, the team inferred that the method employed was not sensitive enough (Gamow and Metropolis 1954).

Another member of the RNA Tie Club offered a different angle on coding: symbolic logic. Mathematical biophysicist Robert Ledley from Johns Hopkins' Office of Operations Research applied his expertise in digital computational methods of symbolic logic. He showed how this method could work for the limited case of $3!=6$ (three amino acids); this took only a hundred hours on the Maniac. But to solve the coding problem for the case of $20!=2.3 \times 10^{17}$ (twenty amino acids) he estimated that, "a computer put to work in the days of the Roman Empire, at a rate of one million solutions per second, 24 hours a day, all year round, would not yet be close to finishing the job" (Ledley 1955, 498).

Two other decoding systems were devised in 1955: Gamow's and Ycăs' statistical analysis of the correlation between nucleic acid and protein composition in two viruses (TMV, tobacco mosaic virus, and TYV, turnip yellow virus), which led to contradictory results; and Rich's and Gamow's neighbor distribution plots, based on a table of numerous dipeptides meticulously compiled by the South African

---

8 For a detailed review, see Gamow, Rich, and Ycăs 1956. For a technical analysis of these codes, see Sarkar 1989.

molecular biologist, Sydney Brenner. The table showed that amino acid distribution in peptides followed a Poisson (random) distribution, and thus negated the restrictiveness (nonrandomness) of overlapping codes. Two years later, Brenner would provide an elegant and decisive proof that fully overlapping codes were logically impossible (Gamow, Rich, and Ycăs 1956; Brenner 1956; 1957).

By the end of 1955, when Gamow, Rich, and Ycăs gathered all their findings in a major review, "The Problem of Information Transfer from Nucleic Acids to Proteins," it was clear that the code was nonoverlapping. Yet no one confronted the broader epistemic implications of these findings: the problematic linguistic attributes of nucleic acids and proteins; and the questionable validity of the code idiom. The lure of information veiled these problem.

Indeed, Gamow, Rich, and Ycăs did not deploy the notion of information unconsciously or casually, as did many life scientists in subsequent years. They used information—with all its valencies to mathematics, logic, cryptanalysis, linguistics, computers, operations research, and weapons systems—as a means of framing the coding problem. The trope of information served to integrate mechanisms of molecular specificity, structural considerations, mathematical relations, linguistic attributes, and coding, within a single explanatory framework.

Even if one grants for a second the problematic analogy between combinatorial elements in a molecule and the alphabetical elements of a language, the "translation" between the four nucleic acid bases and the twenty amino acids would obey the rules of cipher, not code. Codes operate on linguistic entities, dividing raw material into meaningful elements like words and syllables (while ciphers will split the t from the h in "the", for example). More fundamentally, whether code or cipher, the tacit operating assumption behind all the so-called deciphering or decoding attempts was that the code operated on language-like entities. And once the analogy between combinatorial elements in nucleic acids or proteins and the letters of the alphabet took root, the comparison with language took on a life of its own, until the metaphor of language, with all its ambiguities, aporias, and tautologies, began to be taken literally. "The Code"—that logocentric icon preceding its own representation—demanded the existence of language as its predicate. No language, no code.

The commitment to "The Code" was unwavering, even in the face of empirical obstacles. The authors even compared cryptoanalytically the neighbor frequency in the English language (using Milton's *Paradise Lost*) with distributions of artificially constructed amino acid sequences based on the various codes. The English language sample (being highly redundant, or restrictive), deviated markedly from Poisson (randomness), while known amino acids followed random distribution. They even used Shannon's cryptanalytic innovations to determine the information content of these sequences; only to learn that unfortunately, once again, their theoretical sequences deviated markedly from the Poisson distribution and from their experimental sample (Gamow, Rich, and Ycăs 1956, 64–66). Herein lies another aporia. Clearly the amino acid "text" did not obey the rules of any known language (as Quastler's colleagues had already demonstrated); either the code overlaps and thus contradicts the experimental evidence, or it does not overlap and contradicts the meaning of language. But precisely because language and code necessitated each other in the author's representations, the tautology survived intact.

SYNTAX TO SEMANTICS? COMMA-FREE AND OTHER CODES

Once the empirical evidence persuaded the code workers that the code was non–overlapping—namely, having no intersymbol restrictions—the analogies with human communication had lost their validity. Unlike human languages, possessing semantic attributes, the code's language—like machine languages—would have to be purely syntactic. When their article went to press, Gamow and Ycăs had already constructed a non–overlapping code; the so-called "combination code," where only the combination of the bases in a triplet mattered. The number of these combinations added up exactly to what became known as the "magic twenty"—the twenty amino acids (Gamow and Ycăs 1955).

Nicholas Metropolis and Stanislaw Ulam at Los Alamos were already testing that code to address the problem of randomness of amino acids. Gamow called upon his friend, John von Neumann, to supply an analytical solution to the thorny problem of explaining the non–random distribution of amino acids, predicted by the combination code, in relation to the arrangements of nucleotide triplets in RNA. "[W]hat do *you* think about this 'new' trouble?" Gamow prodded Ycăs. But after weighing the pros and cons of randomness, Ycăs concluded that "it would be best to say that when such eminent gray matter as von Neumann and Ulam are working at high speed it is best for myself to remain blank."[9] By the time they submitted their paper for publication (1955) they managed to make some sense of the contradictions and were quite satisfied with their results.

But their code had drawbacks, Ycăs admitted to Crick,

> Chemically, of course, this [combination code] makes no *obvious* sense. Since the triplets are non-overlapping, we have a "punctuation mark" problem. Also the "degeneracy" to get magic 20 raises stereochemical problems. (Ycăs to Crick, 15 February 1955, Gamow Papers, Box 7, Ycăs folder)

With the problem of "punctuation marks" Ycăs articulated what would soon become a key preoccupation for analysts of nonoverlapping codes: how to distinguish between consecutive triplets along the nucleic acid sequence.

Crick concurred. Having just vetoed all extant codes, he pointed out that in the DNA double helix one chain runs up while the other runs down, as Delbrück had already emphasized, implying that "a base sequence read one way makes sense, and read the other way makes nonsense" (Crick 1955).[10] We see how molecular directionality was no longer merely analogized to the directionality of language, it was conflated with it through the act of reading and sense-making. Once again we witness how the tropes of language—reading and sense—became constitutive of the decoder's modes of reasoning, and how the information discourse (erroneously) bestowed semantic attributes on the (syntactic) arrangement of molecular symbols. We see the process by which Nature was continuously being textualized.

There was much discussion about the "punctuation mark problem," namely how the "reading mechanism" recognized the start and end of a triplet, Ycăs recalled. Without a marker ("punctuation"), a sequence of bases could be read ambiguously, for example, ABC, DCC, BDA,... or, A, BCD, CCB, DA... . There were three ways out of that

---

9 Gamow to Ycăs, 1 August 1955 and Ycăs to Gamow, 5 August 1955. On the nearly daily communications about their study see Gamow Papers, Box 7, Ycăs folder (1955), passim.

10 This is undated but certainly from mid-January. I am grateful to Crick for sending me a copy of this paper.

conundrum: there could be an initial configuration, marking the beginning of the count (as it was later shown); the "coding" triplets could be separated by some kind of "noncoding" (nonspecific) arrangement of bases; or, the nature of what soon became known as the "dictionary," might itself prevent ambiguities. Namely, only those particular configurations of triplets which specified amino acids would have "meaning" and count as "words." It was the third option which Crick, with chemist Leslie Orgel and biophysicist John Griffith, chose to elaborate in their ingenious comma-free code in 1956.

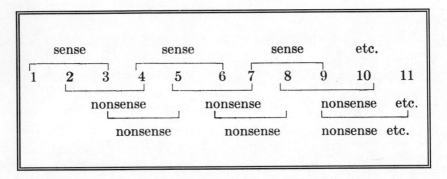

FIGURE THREE

The Crick, Orgel and Griffith model of the "Comma-Free" triplet code. Source: Crick, Orgel and Griffith 1957. Reprinted with Permission of the National Academy of Sciences.

And as Ycăs observed, this type of code was already well established in communication engineering (Ycăs 1969, 31).

Almost miraculously it turned out to be possible to select twenty "sense" triplets in such a way that juxtaposition of any two produced overlapping "nonsense" triplets, thus preserving the "magic twenty." However the solution was not unique for there were two hundred and eighty eight solutions. The authors also offered a mechanism for the assembly of polypeptides. Although the authors were rather cautious about their scheme, the comma-free code generated much excitement, "It seemed so pretty, almost elegant," Crick recalled (Crick 1988, 99). Parenthetically, physicist Leo Szilard, then deeply immersed in molecular biology, came up independently with a similar model (Szilard 1957; Kay 2000, chap. Five).

Comma-free codes attracted considerable attention. While their novelty, pristine logic, and seductive elegance intrigued molecular biologists, this class of codes were being used in advanced communications systems. It seemed natural, therefore, to draw on the expertise of mathematicians and communication engineers to analyze and elaborate the general properties of comma-free codes. Such expertise was readily available at Caltech's Jet Propulsion Laboratory (JPL)—a stone's throw from Delbrück's group—which since the end of World War II had been sponsored by the United States Army, and served as a major resource for the development of the ICBM program (intercontinental ballistic missiles) and the militarized aerospace technologies (Leslie 1993).

Three JPL scientists responded to the challenge of comma-free codes: Solomon W. Golomb, a young mathematician and communication engineer; mathematician Basil Gordon, specialist in combinatorial analysis, and mathematician and electrical engineer Lloyd R. Welch, who two years later joined the Institute for Defense Analysis. In the summer of 1957 they tackled the task of providing a mathematical generalization of the coding problem: What was the maximum size of a comma-free dictionary in the case of an arbitrary number of symbols and arbitrary length of words? They defined the set D of k-letter words ("sense") as a "comma-free dictionary" in such a way as to exclude all overlapping triplets ("nonsense"). Their solution, though partial, demonstrated that it was possible to construct such codes (Golomb, Gordon, and Welch 1958).

The term "dictionary" added yet another linguistic trope to the discursive and semiotic space configured by information theory, electronic computing, and linguistics, within which the genetic code was being constituted as a scriptural technology. Their usage of that idiom seemed reasonable from the standpoint of cryptanalysis based on information theory, where a dictionary simply signified a specific set of symbols. But technically, these symbols could never transcend their syntactic bonds to attain the semantic value—or "sense"—of human language.

Rather unexpectedly, comma-free codes could even contribute to the mathematical theory of coding, since they were a subset of a larger class of related codes, the authors argued. In their search for optimum coding techniques, Claude Shannon and Brockway McMillan had studied theoretical codes where the entire message was available.

But in actual communication application—as with the problem of the genetic code—only disjointed portions of a message were likely to be received. This is where comma-free codes could indeed be useful, they concluded. Genetic codes were now becoming "boundary objects," migrating along the two-way traffic between molecular genetics on the one hand, and the militarized world of mathematics and communication engineering on the other (ibid., 209; McMillan 1956).[11]

The treatment of Golomb, Gordon, and Welch was purely mathematical. But soon after, Delbrück teamed up with Golomb and Welch to address comma-free codes as both a mathematical and genetic problem. They showed that there were five classes of such codes (a total of four hundred and eight codes). They also expanded the range of possible coding errors—"misprints" as they called them—to include what they termed "missense."

> Certain misprints in the coded message will produce nonsense (the resulting triplet does not code for any amino acid), other misprints will produce missense (the resulting triplet codes for a different amino acid). The codes were studied with respect to missense/nonsense ratios produced by various classes of misprints. (Golomb, Welch, and Delbrück, quote on unpaginated abstract)

But the most thoughtful feature of their model—from a biological standpoint—underscored that such genetic codes had to meet another requirement: transposability (directionality relative to a DNA strand). As Delbrück had stressed for years, DNA had no intrinsic sense of direction; its two helices run in opposite direction. And since there were no known cases of one genetic locus coding for two different proteins, it seemed likely that only one DNA strand was read. Thus, Delbrück argued, the dictionary had to be not just comma-free but transposable as well. Namely its reversed complement had to be all nonsense.

> We wish to emphasize that we consider the postulate of comma-freedom and the postulate of transposability to be almost on the same footing. Indeed the principal virtue of comma-freedom is that any message can be

---

11   On this two-way traffic see Golomb 1962 where he refers to Jayne 1959 and to Cal Tech, JPL 1961. On the meaning and significance of boundary objects, see Fujimura 1988; Star and Griesemer 1989.

read unambiguously starting at any point, with the proviso, however, that one must know in advance *in which direction to proceed.* Since the equivalence of the two opposite directions in a structural sense seems to be one of the more firmly established features regarding the DNA molecule the advance knowledge as to the direction in which to read cannot come from the basic structure. Comma-freedom would therefore seem to be worthless virtue unless it is coupled with transposability. (Ibid., 11; also Ycǎs 1969, 33)

And as he had done for a couple of years, Delbrück framed his arguments within the discourse of information. Given the mounting evidence that DNA functioned through the intermediary RNA, he felt that "it would be unwise not to give some currency to 'information transfer' as a possible replication mechanism" (Delbrück and Stent 1957, 730).

Delbrück did not invoke the information idiom lightly. Though, like most molecular biologists, he tended to use the term generically (rather than in its mathematical sense), unlike them he was well-versed in information theory. "I am teaching information theory this term," he wrote in the fall of 1955 to his colleague Robert Sinsheimer, then at Iowa State College, "and we have been using your DNA data to illustrate the notions of intersymbol influence and statistical properties of information sources" (Delbrück to Sinsheimer 21 November 1955, Delbrück Papers, Box 20.3). Clearly, his close interactions with the mathematicians and communication engineers at JPL shaped his thinking, thus reinforcing the growing trend of representing heredity as a communication system of guidance and control; even if by the conflation of technical and genetic uses of information he actually undermined the validity of DNA, RNA, and proteins as cryptanalytic texts possessing sense.

Robert Sinsheimer too tried his hand at decoding in 1958 (he had by then moved to Caltech). He took a bold leap: a binary code. As a graduate student, Sinsheimer had become familiar with feedback controls, cybernetics, and communication theory, when he worked at MIT's Radiation Laboratory during the war years. "It therefore seemed very natural to apply the concepts of information content, stability to thermal and other interference (noise), etc. to the issue of genetic inheritance, mutation, et al," he recalled. At Iowa State he gave a series of lectures on these subjects, in John W. Gowen's genetics seminar, lectures based on Wiener's *Cybernetics,* Shannon and Weaver's *The Mathematical Theory of Communication,* and Quastler's

Essays on the Use of Information Theory in Biology (Sinsheimer to Kay, 19 July 1994).[12]

Given the recent finding that base composition in nucleic acids varied widely in different organisms and that the pattern of variation followed the equivalence of bases A to C (Adenine and Cytosine, with amino groups) and T to G (Thymine and Guanine, with keto groups), only two symbols mattered, he reasoned:

> There is an elementary theorem in information theory (Shannon & Weaver, 1949) that if a message is to be written in a code of T symbols, the message can be written most efficiently (i.e., make use of the least quantity of symbols) if each symbol is used to an equal extent. In our case the message would be the protein content of a cell; this is to be expressed in a two-symbol (N, K) RNA code. (Sinsheimer 1959, 219)

In his informationally efficient two-letter code the quantity of the letters (N, K) was always equal—thus satisfying Shannon's theorem—and the effective composition of DNA was invariant.

With such a proliferation of codes—all several hundreds of them—combination code, comma-free codes, transposable comma-free codes, binary codes, quadruplet, and sextuplet codes—the optimism surrounding the coding problem had dwindled markedly by the closing of the 1950s decade. The sedentary life of the RNA Tie club came to an end; its president, Gamow—preoccupied with his new life at the University of Colorado and beleaguered by alcohol-related ills—had become primarily a spectator and commentator. And it was not much of a spectacle. Assessing the "Present Position of the Coding Problem" at the 1959 summer symposium at Brookhaven National Laboratory, Crick was glum. He now was ready to reassess some of the fundamental premises of the coding problem (Crick 1959).

There was little to hold on to but the cardinal belief in the existence of "The Code"—that object prefixed in the mind and preceding its experimental warrant. For five years now—despite the contradictory data and dubious results, and with some of the most "eminent gray matter working at high speed," as Ycǎs had put it—there were intriguing theories but few concrete returns. Yet the champions of the genetic code clung tenaciously to their logos. And in the next two

---

12 Personal communication (letter). I am grateful to Sinsheimer for providing me with this information. See also Sinsheimer 1994, 56.

years there were several other attempts to "crack the code of life," relying primarily on the recently sequenced TMV as a "Rosetta stone."

As late as the spring of 1961 Delbrück's colleague, Golomb, was still enthusiastically refining his promising comma–less code. "New Way to Read Life's Code Found," announced the *New York Times*. "A 'dictionary' of 24 Words Appears Able to Describe Inheritance Mechanism."

> The scientist, 29-year-old Dr. Solomon W. Golomb of the Jet Propulsion Laboratory of Pasadena Calif., told a meeting of American Mathematical Society at the New Yorker Hotel of building what might be called a "dictionary of life" .... Biologists who have examined the new theory say it fits all of the facts now known about the way nucleic acids and proteins work in transmitting hereditary information .... Scientists from several fields have been trying to crack the code of life since the middle Nineteen–Fifties .... Dr. Golomb decided that this [nature's alphabetic] extravagance probably produced the redundancy in the coded messages that would be necessary to minimize mistakes and assure a high degree of reliability in transmitting genetic information .... [H]e is satisfied that his code is a good and workable one and could be found in nature—if only on another inhabited planet. (Osmundsen 1961)[13]

But within a few weeks of this announcement something which no one expected happened. The genetic code was broken by two young obscure biochemists, Marshall Nirenberg and Heinrich Matthaei at NIH, who broached the problem from a radically different vantage point. And within a few years the problem of correlating amino acids with nucleotide triplets was solved biochemically.

CONCLUSION

Scientists have generally tended to regard the theoretical work on the genetic code in the 1950s as naively optimistic at best, or erroneous and unfruitful at worst: the demise of the Pythagorean ideal by the world of matter. But as one of the code breakers Carl E. Woese (physicist, microbiologist, and author of the important text, *The Genetic Code*) reflected,

---

13 The report refers to Golomb 1962.

What has not been generally appreciated is that the subsequent spectacular advances in the field, occurring in the second period [1961–
1967], were interpreted and assimilated with ease, their value appreciated and new experiments readily designed, precisely because of the
conceptual framework that had already been laid. (Woese 1967, 17)

But it is not merely the conceptual framework which had been laid
down. Rather a knowledge-power nexus had formed within which
molecular biology reconfigured itself as a pseudo-information science
and represented its objects in terms of electronic communication systems (including linguistics).

Several scholars have noted the cognitive and disciplinary impact
of the physical sciences on molecular biology, as well as the impact of
Cold War technosciences on the life and social sciences. As Evelyn
Fox Keller observed, physics was a social resource for biology in
borrowing physics-like agendas, language and attitudes, thus eventually reframing the character and goals of biology (Keller 1990).[14] But
not only physics. By examining the work on the genetic code in the
1950s—cognitive strategies, discursive practices, semiotic tools, and
institutional resources—we have seen how the old problem of genetic
specificity was reframed. It was recast through scriptural technologies
poised at the interface of several interlocking postwar discourses:
physics, mathematics, information theory, cryptanalysis, electronic
computing, and linguistics.

We have seen that the textualization of Nature has been a diachronic process, pervading the entire history of the West. But it also
must be viewed synchronically. The "book of nature/life" bears its
own historicity, for the very meaning of "text" had changed over the
centuries: from scribal, to printed, to electronic. Thus to extract the
levels of significance of the genetic code as a book of life, and understand its tropes—language, dictionary, message, instructions—we
must understand the age-old authority invested in that logocentric
book, as well as its scientific legitimation through information theory
and within the regimes of signification of Cold War Military Culture.
Ironically, these modes of signification led to a deconstruction: since
the language of information theory is purely syntactic the book of life
conveys no semantic meaning.

---

14 See for example Fleming 1968; Olby 1974, section Four; Kohler 1976;
Abir-Am 1982; Kay 1985; Haraway 1979.

Symbols, Icons, Indices:
A Commentary on Lily Kay's
"A Book of Life"

Hans-Jörg Rheinberger

Throughout her paper, Lily Kay confronts us with a fascinating account of how, as she puts it, "linguistic tropes and scriptural icons" pervaded the growing space of molecular biology during the 1950s, culminating in the deciphering, at the beginning of the 1960s, of what by then had come to be called the genetic code. I think Kay convincingly shows that the "language of information theory: and molecular biology" which leaked into and finally filled molecular biology's capillary system, had its roots to a lesser extent in the development of that new biology itself, but rather in the broader context of a dawning age of cybernetics, of new communication media, and of servomechanisms.

I will organize my commentary, not around the central argument of Lily Kay's paper, but around a suggestion that came to my mind when reading her text. It is a semiotic suggestion, and whether it will survive criticism remains to be seen. It is for the sake of such critique that I present it to the reader.

If it holds that molecular genetics can be seen as the result of a linguistic and communicational conquest of a new continent, that is, the continent of the organic, then we need an understanding of how such representational conquest works. That is, in other words, we need an assessment of scientific representation or, as Hans Blumenberg put it, an investigation into the "lisibility of the world" (Blumenberg 1986). To put it somewhat less traditionally, we need what I would like to call—borrowing terms from Elisabeth Stengers—some idea about "operations of propagation," or "operations of passage" (Stengers 1987). The question is, how do representations travel and become inseminated in a particular research field, and how do they acquire their particular dynamics and capacity for signification in the new surroundings for which they were not created?

So, first, what are we talking about when we speak of representation? Intuitively, we connect, 'representation' to the existence of some entity that is represented. In the words of Bas Van Fraassen, Bas: on representation: "In sum, representation of an object involves producing another object which is intentionally related to the first by a certain coding convention which determines what counts as similar in the right way" (Van Fraasen and Sigman 1993, 74). Upon closer inspection, however, the term 'representation' reveals itself to be polysemic and locatable only in the context of a general metaphorology. If we speak about the "representation of something given," the common sense of the notion appears to be plain: we speak about the representation, of something. If, however, we claim that we have seen an actor yesterday evening representing Hamlet, we speak of a representation *as*. In this case, representation takes on a double meaning: that of vicarship and that of embodiment. Every play is governed by this tension, this "paradoxical trick of consciousness, an ability to see something as 'there' and 'not there' at the same time" (Mitchell 1986, 17). If, finally, a chemist tells us he has represented (in German, *dargestellt*) a particular substance in his laboratory, the meaning of 'representation of' has disappeared, and instantiation, in the sense of the production of a particular substance, has taken over. In this latter case, we deal with the realization of a thing. These instances need not be sharply separated. There is a continuum from one extreme to the other: from vicarship, to embodiment, to instantiation.

The three connotations of 'representation' have their equivalent in the realm of scientific representation. Roughly speaking and without unnecessarily stretching the parallel, we are, in the first case, accustomed to speak of analogies,[1] that is, of hypothetical, more or less

---

[1] I will not deal with the vast body of literature on the role of analogies in science in this sketch.

arbitrary constructs, symbols in the sense of Charles Sanders Peirce, Charles Sanders: on symbols ([1893/1910] 1955, 102–03). In the second, we may speak of models or simulations, which amount to Peirce's icons. In the third, we deal with an experimental realization, comparable to an index in Peirce's semiotic system, that is, a primary trace within an experimental arrangement. François Jacob, François: on models and analogies argues in a similar vein when he casts this triplet in an hierarchical order and claims that experimental biology proceeds from "analogies" to "models" to "concrete models" (Jacob 1974, 203–04). Yet it appears historically contingent and a matter of case by case evaluation as to which of them figures prominently within a given historical context.

Let me make the following proposition with respect to Lily Kay's example, that is, the passage and dissemination of reasoning in terms of information transfer in molecular biology. I claim that there is a first phase in this historical process, in which talk of information in molecular genetics takes on a predominantly *symbolic* meaning. The connection between molecular entities and metaphors of information, of communication, of code, of language and of text remains largely arbitrary, conventional and a matter of *gusto* at this stage, due mainly to diffusion, and remaining diffuse for that very reason. Indeed, many researchers whose work was embedded in thoroughly biochemical research traditions, for instance, did not adopt, or even explicitly reject, the emerging terminology without damaging their research agendas for that reason. They carried on their work in the framework of energy fluxes, metabolic intermediates, and molecular structures. This phase roughly covers the period Lily Kay has been dealing with in her paper.

During the same time, however, the ground was prepared for a set of possibilities that enabled information talk to take on an *iconological* meaning. Most interesting, this preparation of new ground for conversing about information in biology did not directly flow from the symbolic efforts of the first generation of molecular biologists. It largely resulted from the efforts to establish biochemical *in vitro* systems of protein synthesis, whose promoters at first resisted the notion of information, but in whose hands the general concept of RNA as an intermediate in the synthesis of proteins gradually differentiated into ribosomal RNA, transfer RNA, and messenger RNA during the years between 1955 and 1961. The result of this iconological period is that people could begin to speak of genetic transmission and expression as being a model, that is, a particular embodiment of the general notion of information transfer. Concomittantly, the terms that first had been borrowed from another field, such as code, information, and translation, began to take on a meaning of their own within the new context. At this point of passage, to speak with Stengers, a transition occurred with the new form of reasoning from a process of diffusion to a process of infection. Concomittantly, the new research entities took on an endemic configuration. We could also speak, by manipulating the notions of Jacques Derrida, Jacques, of a slippage from operations of grafting to operations of insemination. This period resulted in the deciphering of the genetic code. Towards the end of the 1960s, the iconological vocabulary of molecular biology was entirely in place.

It is, at this point, absolutely unnecessary to worry about the fact that molecular biologists no longer used the term 'information' in the technical sense of contemporary information theory—i.e. as a measure devoid of semantic connotations—,but rather used it generically. By that time, every worker in the field could point to what he was speaking about when he was speaking about information, transcription, and translation. People deliberately dealt with messenger RNA sequences that were read in triplets with the help of transfer RNA's anti-codons that in turn allowed to specify the sequential order of amino acids in a nascent protein.

Here is the place to indicate that I miss one point in Lily Kay's account of the early coding debate between 1955 and 1960 which appears to me to be of central importance. This is the visionary postulation by Francis Crick, Francis: on code and protein synthesis, around 1955, of the concept of "adaptation" between nucleic acid code and protein product (Crick 1957). With that proposal, all the constraints posed by the assumption of a direct physico-chemical interaction between nucleic acids and amino acids were sidestepped. With the idea of RNA adaptor molecules accepted, the decoding process could be assumed to obey the same rules as the replication process: base-pairing. The problem of translation proper, that is, the attachment of amino acids to their adaptors, became dissociated from the problem of coding proper—the combinatorics of triplet identification. Indeed, it was probably not by chance that the code became solved *via* base-pairing experiments.

This was followed by a third period where talk of information in

molecular biology assumed an *indexical* meaning. The beginning of
this period is roughly identical with the establishment of the molecular tools for effectively practicing genetic engineering on a molecular
level during the 1970s: the characterization of restriction enzymes,
polymerases, plasmids and vectors on the one hand, and the development of rapid nucleic acid sequencing methods on the other. The
"indexical" period was characterized by a set of techniques that made
DNA directly readable as a veritable text from material traces that
are causally linked to the molecule itself. The exemplar of such a
primary indexical device is a sequencing gel, where DNA can be con-
secutively and automatically read off as a string of Gs, As, Ts and Cs.
The very technical possibility of the Human Genome Project: and
indexical meaning rests on this indexical transformation. The
laboratory reality of an organism now is no longer represented by, or
embodied in, texts, but rather is an instantiation of a text.

I wonder how this semiotically-grounded periodization sounds to
the ears of molecular biologists and to historians and philosophers of
science? If one speaks of molecular biology as a continent conquered
by information, communication, and control, then the appropriate
analytical tools to analyze this conquest should, or at the very least
could be, semiotic ones. By this, of course, I do not contend that other
narratives of the history of molecular biology become obsolete or
should be abandoned. What I do think, however, is that such conceptual tools will help us to understand what it means in different scientific and historical contexts to engender and to disseminate inscrip-
tions.

Let me end with a note on the distinction between "information"
and "language," and the correlated distinction between syntax and se-
mantics in the realm of the informatization of molecular biology, on
which Lily Kay puts so much emphasis, and which also lies behind the
title of her paper. This distinction is in need of explication, and perhaps clarification. What is the difficulty here? Does Kay assume that
talk about information excludes meaning, and therefore reduces the
view of life to mere bits? Is hers a "romantic" sentiment? Or does she
assume that talk about information and talk about language at the
same time exclude each other, because language cannot be dissociated
from meaning, whereas information can? Thus, is her unease motivated by horror of conflation between language and information? Of
course, I should say, the genetic book of life *per se* has no meaning. I

assume, leaving for a moment McLuhan, Marshall's slogan in suspense, that the message is not the meaning. Let us posit that meaning is something transiently generated at the intersection between two chains of signifiers. Neither natural languages nor technical information processing are isomorphic with the generation and transmission of information in living systems. In the last resort it is the differences that count. If we look for meaning in the organism, we must not look at its genes, we must look at the interface(s) between the genome and the *body*.

Coping with this basic interface and all the others derived from it, and reaching out into the facets of the social body, is what the Genome Project leaves us with. That this task urgently needs to be taken up, is, for me, one of the shocking experiences of this conference.

5

Origins of the U.S. Human Genome Project: Changing Relationships between Genetics and National Security

John Beatty

Friend or Foe? The FSX becomes a symbol of mounting strain between the U.S. and Japan. . . .

American television manufacturers were the first to fall. Then Japanese firms rolled through markets ranging from autos to semiconductors. Now many Washington politicians fear that U.S. plans to develop the FSX fighter jet with Japan could give Tokyo a vital jump start in the aerospace industry, one of the few high-technology fields in which American companies still dominate. The growing outcry has transformed the proposed jet, an advanced version of the F-16, into a powerful symbol of the rising tensions between two countries that are close military and diplomatic allies but also archrivals for the economic leadership of the world. "What we're seeing is the emergence of an entirely new concept of national security," says Wisconsin Democrat Les Aspin, chairman of the House Armed Services Committee. "It embraces economics and competitive, commercial relations." (Greenwald 1989, 44)

INTRODUCTION

During the Cold War, American aeronautics—in the concrete form of fighter jets like the F-4 and the F-15—was viewed as an instrument of the U.S.-Japan alliance against Communism. The recent

131

F-16 variant FSX, on the other hand, has come to symbolize the end-of-the-Cold War, economic conflict between the U.S. and Japan, and changing views of national security. As the *Time* article proceeds to report, "Mindful of polls showing that many Americans are more fearful of Japan's economy than of the Soviet Union's military strength, President Bush has made the FSX an example of U.S. willingness to get tough with Japan. . . ."(ibid., 44).

Changing conceptions of national security from the Cold War to the post-Cold War period, and changing U.S.-Japan relations, are reflected in the histories of many areas of science and technology—including genetics. During the Cold War, genetics was seen as a means of allying Japan against Communism—hardly as specialized an instrument of international relations as a fighter jet, but an instrument of diplomacy nonetheless. More recently, genetics has become embroiled in the post-Cold War, economic conflict between the U.S. and Japan. This change can be seen in the history of the Human Genome Project.

The U.S. Human Genome Project has a long history and a short history, corresponding to its ancestry, and its conception and development. The particular line of ancestry that I will emphasize dates back from the Human Genome Project to the Atomic Bomb Casualty Commission, which was established in 1947, and which included a large genetics component. The ABCC was funded by the Atomic Energy Commission, which was succeeded by the Department of Energy, which is one of two major sponsors of the Human Genome Project.

The Human Genome Project, which was conceived in the mid eighties, has more than its sponsorship in common with the ABCC; there are also important intellectual connections. Moreover, the long history and the short history of the Human Genome Project both involve issues of national security and U.S.-Japan relations. But the Human Genome Project plays a very different role in national security and international relations than was played by the ABCC, in large part because of global political and economic changes, and corresponding changes in national security and foreign policy agendas. The ABCC was a Cold War project; the Human Genome Project is a post-Cold War project; and I do not just mean chronologically speaking.

THE LONG HISTORY

When Department of Energy (DOE) representatives conceived the Human Genome Project, they realized that it would not be an easy sell. What, after all, is the connection to DOE's mission and/or expertise? DOE representatives stressed two main reasons why the agency should play a major role in the Project: 1) continuity with previous large-scale and largely successful efforts in human genetics funded by the agency and its predecessors, and 2) the agency's success in developing novel technologies, together with its commitment to facilitating commercialization of new technologies through closer relations between its labs and U.S. firms (Galas 1990, 28–29; Wood 1989, 30–38). These two reasons correspond to the long and short histories of DOE's involvement in the Human Genome Project.

The long history of DOE's involvement in human genetics dates back to the Atomic Bomb Casualty Commission (ABCC), a joint Japanese-American investigation of the long-term radiation-biological effects of the bombings of Hiroshima and Nagasaki.[1]

Of all the ABCC's component studies, the one that generated the most interest was the study of genetic effects. The mutagenic effects of radiation had been discovered by H. J. Muller in 1927, and studied extensively in the interim. Coincidentally enough, just a year after the bombings, amid considerable speculation about the genetic consequences, Muller was awarded the Nobel Prize. He seized upon his newfound eminence to drive home to the public the genetic hazards of

---

1 The biological effects of radiation exposure had been studied intensely during the war years by Manhattan Project investigators. This research was undertaken not out of concern or interest in the biomedical effects of the explosions, but rather out of concern for workers who might be exposed to radioactive materials. The bomb's designers did not anticipate significant radiation-biological effects at the target sites; they believed instead that most everyone within range of significant radiation exposure would be killed by the blasts. Indeed, while Manhattan Project officials planned ahead of time to survey the physical effects of the explosions, they made no special provisions for investigating the biological effects. Immediately after the bombings, however, it became clear that radiation exposure among survivors was a real problem.

In 1947, after preliminary studies on the more immediate radiation-biological effects of the bombs, President Truman directed the National Academy of Sciences / National Research Council to proceed with an investigation of the long-term effects. To this end, the Academy (through the Council) established the ABCC, with funding secured from the AEC. See Beatty 1991; Lindee 1994; Neel 1994; Schull 1990; 1995.

radiation. The survivors in Hiroshima and Nagasaki served him well in this regard: as he said of them, "if they could foresee the results [mutations among their descendants] 1,000 years from now. . . , they might consider themselves more fortunate if the bomb had killed them" (Muller in *New York Times* 1946).

The leaders of the ABCC genetics study were James Neel and William Schull, both of whom subsequently became leaders in the newly emerging field of human genetics. The results of their one-generation study of more than 75,000 pregnancies were released in 1954. The results were "negative" in the technical sense, which does not mean that no increase in mutations occurred, but rather that there was no detectable increase in mutations, or in other words there was no very large increase (Neel and Schull 1956; see also Wood 1989; Galas 1990).

Neel and Schull's somewhat reassuring results, in light of Muller's speculations about possibly very unconventional aspects of the bombings, are suggestive of the significance of the ABCC for U.S.-Japan relations in the postwar period. But the full significance can only be understood in the broader context of U.S.-Japan relations, and in the context of views about the role of science in American foreign policy (Beatty 1991; 1993; Lindee 1994).

U.S.-Japan relations in the postwar period begin, of course, with the Occupation, the "ultimate objective" of which was "to insure that Japan will not again become a menace to the United States or to the peace and security of the world." But the intensification of the Cold War occasioned a revision of the goals of the Occupation. What planners and policy makers came to fear about the post-Occupation period was not Japan unguarded, but rather the possibility that Japan unguarded might fall under Soviet influence. On the other hand, they hoped that if the U.S. prepared appropriately for the end of the Occupation, then Japan could play an important role in the containment of Communism. For example, George Kennan, who was so instrumental in the elaboration of containment policy during the Occupation years, came to regard Japan and Germany as at once the most strategic battlefields and the most strategic instruments of the Cold War: "the theaters of our greatest dangers, our greatest responsibilities, and our greatest possibilities," and "two of our most important pawns on the chessboard of world politics"(Kennan 1983, 368, 369).

Accordingly, the emphasis of the Occupation shifted from disarming and reforming, to allying Japan and making Japan resistant to Soviet influence: "Our essential objectives with respect to Japan are its denial to the Soviet Union and the maintenance of Japan's orientation toward Western powers" (U. S. National Security Council 1949, 49/1, 871). U.S. alliance and Soviet resistance were only partly military matters; there was general agreement that the Soviet intellectual threat to Japan was as much or more subversive as the military. Whether Japan would resist or fall depended on "the attitude—the orientation—of the Japanese people," this "attitude" or "orientation" being "a subjective political-psychological condition" (ibid.).

Optimally, the U.S. should aim to inspire among the Japanese a pro-western attitude; minimally, the U.S. should aim for an anti-Soviet attitude (ibid.). The Soviet orientation to be resisted was seen to be not so much a doctrine as a way of reasoning, or refusing to reason; it was, above all, adherence to authority: "Like the white dog before the phonograph, they (the Soviets) hear only the 'master's voice'" ([Kennan] 1947a, 574).

Resistance depended not only on promoting a democratic, anti-authoritarian mindset, but on ensuring Japan's economic security, on the grounds that economic insecurity breeds political discontent. Japan's economic recovery became the highest priority for the successful completion of the Occupation. This in turn was seen to depend on reviving and further developing Japan's exports ([Kennan] 1947b, esp. 541; and U. S. National Security Council 1948, 858–62).

Containment strategy loomed large in discussions of the role of science in diplomacy in the postwar years, and in discussions of the role of science in U.S.-Japan relations in particular. A prime example is Lloyd Berkner's influential 1950 report to the State Department, *Science and Foreign Relations.* Acknowledging "the stated present policy of this country to buttress the strength of free peoples throughout the world against the encroachment of Communism," Berkner proposed that

> The international science policy of the United States must be directed to the furtherance of understanding and cooperation among the nations of the world, to the promotion of scientific progress and the benefits to be derived therefrom, *and to the maintenance of that measure of security of the free peoples of the world required for the continuance of their*

intellectual, material, and political freedom. (Berkner 1950, 2 emphasis added)

Science had, according to Berkner, two important diplomatic roles to play, the first having to do with the norms of science, and the second having to do with the economic benefits of science. "Since science is essentially international in character," Berkner argued, it can open channels of communication between countries. And when two countries are linked by scientific lines of communication, then the "scientific temper" can contribute to the resolution of political differences between them. The relevant aspects of the "scientific temper" in this regard are "openmindedness" and the fact that "science is not based on authority" but instead "owes its acceptance and universality to an appeal to intelligible, communicable evidence that any man can evaluate" (ibid., 20, 22). In other words, the scientific temper is antithetical to the Soviet temper, as elaborated by Kennan and others.

Another important role for science in diplomacy, Berkner argued, has to do with the contribution of science to economic welfare, which in turn contributes to political stability. Science can in this way help to guard against "economic depression and thus to offset the threat of Communist infiltration" (ibid., 3).

Considering the important role of Japan and Germany in the containment of Communism, Berkner was especially concerned to promote science in those countries:

> There is a reservoir of scientific talent in Germany and Japan, much of it now adhering to the West.
> Have we intelligently joined in the formulation and application of the occupation control statutes as they relate to science and technology?
> Have we encouraged research that can be beneficial to ourselves and our Allies or have we neglected and dissipated scientific and technological resources in occupied countries that are sorely needed for the defense of freedom?
> Have we exercised scientific leadership that will generate the desire by scientists in occupied countries to recognize and work for our western democratic ideals? (Ibid., 23)

To carry out this science-for-containment policy, Berkner proposed that the State Department establish "science staffs" at selected foreign embassies, and in conjunction with occupation efforts in Japan and Germany (and notably not in the Soviet Union). With respect to the importance of a scientific diplomatic mission in Japan, he argued that, "because of its geographical position and its affinity with other Asiatic peoples, it [Japan] is potentially a strategic channel for the spread of United States scientific information and of *United States ways of thinking* in the Far East (ibid., 111–12, emphasis added). The State Department began to assign science officers to foreign embassies in the mid fifties; the U.S. Embassy in Japan was among the first recipients (Greenwood 1971a, 15; 1971b, 21–25).

Genetic diplomacy during the Occupation took a variety of forms. Consider two related, small-scale, but not insignificant episodes. In 1948, Warren Leonard, Chief of the Agriculture Division of the Natural Resources Section of the Occupation, wrote to H. J. Muller to solicit his advice on how to counter Soviet influence on Japanese genetics. In particular, Leonard was concerned about Japanese interest in Lysenkoism. By this time, Muller was probably the most outspoken critic—worldwide—of the Soviet suppression of Mendelian genetics in favor of Lysenko's views. In asking for Muller's help, Leonard explained that,

> Several things have happened in Japan to give me great concern about the genetics viewpoint in this country. I suspicion Lysenkoism in several places. In fact, I was told by one so-called Geneticist that Mendelian genetics was obsolete and that Japanese geneticists should be interested in the "new genetics." I immediately labeled it Lysenkoism. It is quite possible that some Japanese geneticists are intimidated with the belief that they must have the right "party line" on account of their close proximity to the U.S.S.R . . . .
> Please do not give out my statements on suspicioned Lysenkoism in Japan because of the possible embarrassment such action might cause us in taking counter-measures. Many press statements in the United States come back to Japan in the Japanese press. (Leonard to Muller 17 May 1949, Muller Papers Box 10, Folder "1949, April-December")

Several years later, Muller travelled to Japan on a related diplomatic mission. William Schull recounts how, in 1951, he and fellow ABCC geneticist, Masuo Kodani, were "summoned" by Occupation authorities to travel to Tokyo in order to greet Muller and accompany him on a lecture tour. To Schull, the purpose of the tour was transparent: Muller was not only one of the world's leading geneticists, but was also by this time an ardent anti-Communist, who could speak from his own experience in the Soviet Union about the repression of thought there. Muller took his message to one of the most radical student bodies in Japan, Kyoto University, where he urged the students to pursue the truth objectively and doubt all authority, and in this way become "the most active workers for that freedom, both material, intellectual, and spiritual, which can be won by the use of science in the service of man, but which will be lost if men allow their civilization to be destroyed by totalitarianism" (Muller 1951, 28–29). As Schull recollects, Muller's presentation "projected an experience and authority that brooked no opposition. Student objections, if they existed, were so feebly and ineffectually advanced as to be embarrassing" (Schull 1990, 98–105, esp. 104). Muller was also visiting Japan as a representative of the Committee for Cultural Freedom (Carlson 1981, 370–71).

Among the various scientific/diplomatic projects more or less clearly associated with the Occupation, one of the most enterprising was the ABCC, which was originally associated with the Public Health and Welfare Section of the General Headquarters of the Supreme Commander Allied Powers. After the Occupation, the ABCC was assigned formal diplomatic status by way of a "note verbale"—basically, an understanding—between the Japanese Ministry of Foreign Affairs and the U.S. Embassy in Tokyo (NAS Archives, ABCC Records 1952). From the beginning, the ABCC was viewed as a diplomatic as well as a scientific enterprise. For instance, it was noted in a preliminary ABCC report that,

> a long-term study of atomic bomb casualties in collaboration with the Japanese, affords a most remarkable opportunity for cultivating international relations of the highest type. . . . Japan at this moment is extremely plastic and has great respect for the Occupation. If we continue to handle Japan intelligently during the next few years while the new policies are being established, she will be our friend and ally for many years to come; if we handle her unwisely, she will drift to other ideologies. The ABCC or its successor may be able to play a role in this. (Henshaw and Brues 1947, 4–5)

The diplomatic significance of the ABCC lay very generally in the establishment of scientific lines of communication for purposes of alliance and containment, à la Berkner. It lay more specifically in the prospect of U.S.-Japan cooperation to resolve a lingering diplomatic problem, namely the possibly unconventional effects of the bombings of Hiroshima and Nagasaki, especially the genetic effects (Beatty 1993). If the ABCC did not pursue the issue in cooperation with the Japanese, the Soviets would surely bid to do so, with the possibility of results deliberately tilted to undermine the U.S.-Japanese alliance. This belief was expressed over and again. According to ABCC geneticist, James Neel, General MacArthur held this view during the Occupation (Neel 1994, 88). It was expressed after the Occupation, in the mid fifties, by the Director of AEC's Division of Biology and Medicine, Charles Dunham, to the President of the National Academy of Sciences, Detlev Bronk:

> The AEC has a two-fold interest in seeing that the program is not interrupted: the necessity for making the most scientifically of all the available material on the effects of ionizing radiations on humans coupled with a need for assuring that misleading and unsound reports of the effects of radiation on man emanating from Nagasaki and Hiroshima are kept to a minimum. Were the United States to pull out, the vacuum created would assuredly be filled with something, and it might well be something bad, even flavored with occasional tinges of red. Especially might this be the case at Hiroshima. In such an event, both the world scientific community and the United States as a country would be the losers. (Dunham to Bronk, 20 December 1955, NAS Archives, ABCC Records 1955)

The view was expressed again, over a decade later, by Richard Petree, the AEC's Acting Country Director for Japan:

> The ABCC was one of the first institutionalized joint U.S.-Japan scientific projects and, as such, it has a definite value and status in our relations, particularly in the scientific field, with Japan. It has also done much to demonstrate American concern for the victims of the atomic bombings, thereby forestalling the development of adverse attitudes that would have seriously hampered our postwar relations with Japan. The scientifically documentable information obtained through the ABCC program has also provided us with irrefutable grounds to counter hostile charges which might have been made concerning the effects of the bombings. In many ways, therefore, the ABCC has served to turn a liability in our relations with Japan into an asset, and it has become an accepted and reputable U.S.-Japan scientific program. (Petree to Clark 28 December 1966, NAS Archives, ABCC Records 1967)

The diplomatic success of the ABCC was sometimes judged more circumspectly than the preceding quote suggests. That is, while the ABCC did serve that positive role, it also served the negative role of being a painful reminder of the possibility of unconventional warfare—the very name "Atomic Bomb Casualty Commission" was of questionable public relations value to say the least. The fact that the ABCC was sponsored by the AEC, the successor agency to the Manhattan Project, only made matters worse in this regard. As was acknowledged by Keith Cannan, the NAS-NRC officer most reponsible for overseeing ABCC operations,

> the dilemma persists that ABCC is, at once, a scientific project and a diplomatic front. In the latter respect it has established itself as a significant and sympathetic component of the community in which it operates. On the other hand, it is a popular target for anti-American sentiment and will remain so as long as the project is known to be operated from the U.S.A. and to be sponsored exclusively by the A.E.C. (Keith Cannan to Detley W. Bronk, 9 February 1955, NAS Archives, ABCC Records 1955)

Even in acknowledging the negative diplomatic aspects of the ABCC, though, this NAS official did acknowledge that the ABCC served as a "diplomatic front" as well as a "scientific project."

Again, it was the ABCC's genetics project that served the most positive diplomatic function, if only because extensive genetic effects would have represented the most unconventional aspects of the bombings. As Director of the ABCC, George Darling, described the negative results of the genetic investigations,

> This study unquestionably is the single most important work so far reported by ABCC. . . .
> From the scientific point of view this answer is incomplete because of the complexity of the problems of human genetics and because too little is known about the subject. From the human point of view, however, results of this study are reassurance of high degree. . . . [I]t is difficult to imagine what service ABCC could have performed for the survivors as individuals or the community as a whole which would in any way compare with the reassurance parents are able to give their teenage children today as a result of this six year study. The discovery that the best scientific evidence now available indicates that they need not be afraid to marry if their parents were exposed nor fear marriage with a partner whose parents were exposed is a service of incalculable value. (Darling 1958, 10)

THE SHORT HISTORY

Just as the long history of DOE's involvement in the Human Genome Project has to do with issues of national security and U.S.-Japan relationships, so too does the short history. But by the time that the Human Genome Project had been conceived and was being promoted, there had occurred considerable changes in the national security context, in U.S.-Japan relations, and in discussions about the role of science in foreign relations. The thawing of the Cold War during the mid to late eighties, combined with the rapidly increasing budget deficit of the U.S., an immense trade deficit with Japan that could not be substantially reduced even by substantial devaluations of the yen relative to the dollar, and the continued erosion of America's position in world markets, especially relative to Japan, led many analysts to argue for a broader notion of national security—one that emphasized economic as well as military security (Prestowitz 1988; Friedman and Lebord 1991; Okimoto and Raphael, 1993; Tolchin and Tolchin, 1992; Romm 1993; Inman and Burton 1990; Huntington 1991; Sorenson 1990). This change in attitude is well exemplified by the quotation from Les Aspin in the opening passage from the 1989 *Time* article on the FSX: "What we're seeing is the emergence of an entirely new concept of national security. It embraces economics and competitive, commercial relations"(Aspin in Greenwald 1989, 44). And as Clyde Prestowitz, former U.S. trade negotiator and advisor to the Secretary of Commerce, was quoted in the same article: "Trade *is* defense"(ibid., 45). More specifically, this change involved a shift in goals from containing Communism to "containing Japan" (Fallows 1989). In words and pictures, Japan was represented as the enemy of old.

The trade imbalance with Japan, and Japan's capture of so much of the market share lost by the U.S., was widely attributed to Japan's ability to quickly commercialize new scientific and technological discoveries and inventions. There was also growing concern that a major source of the ideas commercialized in Japan was American science and technology, and that the direction of influence was unfairly one-way. Statistics were cited to show that Japan invests much less in basic research than does the U.S., with the implication that the U.S. continues to pay for Japan's ideas. Statistics were also cited to show that a much greater proportion of the basic research carried out in Japan is conducted in the private sphere, with the implication that discoveries and inventions in Japan are not so freely exchanged as in the U.S. Thus, the free exchange of ideas that supposedly made science such an effective instrument of diplomacy—à la Berkner and others—came to be scrutinized more carefully.

Indicative of the overall difference in climate was the debate leading up to the U.S.-Japan Cooperative Science and Technology Agreement of 1988. The first U.S.-Japan cooperative science agreement was signed in 1961. It was part of a broad effort to repair relations that had deteriorated during renegotiation of the U.S.-Japan Security Treaty. The Security Treaty had originally been signed in 1951 in conjunction with the Treaty of Peace—the latter ended the Occupation, while the former guaranteed continued American military presence in Japan. The treaty was revised in various respects in 1960—e.g., to emphasize the importance of peaceful diplomacy—although it still guaranteed American military bases in Japan. A coalition of left-wing groups in Japan, with broad student support, vigorously and at times violently opposed such allowances. On the eve of the ratification of the 1960 revisions, there were massive student demonstrations, leading to the cancellation of a scheduled meeting between President Eisenhower and Prime Minister Kishi in Tokyo (Packard 1966).

The new administrations of President Kennedy and Prime Minister Ikeda attempted to improve relations by shifting the emphasis from military connections to broadly cultural and educational connections between the U.S. and Japan, one manifestation of which was the U.S.-Japan Cooperative Science Program, established in 1961. Among the motives for the broader agreement, and for the science agreement in particular, was the concern, reinforced by the recent demonstrations, that Japanese academics tended too far to the left. In particular, they viewed American science with suspicion since it was heavily funded by the military and the AEC. It was believed that more exposure to American scientists would show Japanese scientists the nonmilitaristic aims of their U.S. counterparts, and would thereby serve an allying role. As Eugene Skolnikoff, a science advisor to Kennedy at the time, later explained the rationale:

An excellent example of. . . a scientific/political initiative exists in the current U.S.-Japan Cooperative Science Program.... Politically, the Japanese scientists have become an important influence on the public and youth of Japan. Therefore, the political attitudes of many of the scientists is a source of concern to the United States with regard to the future development of Japanese foreign and domestic policies. In recent years, these scientists have had poor relations with their own government and few contacts with the American government or with American activities. (Skolnikoff 1967, 148)

Walter McConaughy, then Assistant Secretary for Far Eastern Affairs at the State Department, expressed the containment motivation for the agreement quite clearly by introducing the agreement as follows:

The sudden callous resumption by the Soviet Union of nuclear explosions in the atmosphere reminds us that this new relationship [between the U.S. and Japan], this interdependency, has grown up in the era of the cold war. Some aspects of the relationship which has grown up between Japan and the United States are responsive to the threat posed by communism. We have, for instance, a special security relationship with Japan. Nevertheless, mutual security is only one part of the partnership between Japan and the United States. (McConaughy 1961, 635)

The situation had changed considerably by 1987, when representatives of Japan and the United States were negotiating a new science and technology agreement. This time, the U.S. was not so much interested in checking Communism as in curbing Japan's economic influence. There had developed considerable sympathy for protectionism applied to the exchange of ideas and not just material goods. As William Graham, Science Advisor to President Reagan, summarized the problem in Senate hearings on the agreement: "In the U.S., the current imbalance in our science and technology relationship with Japan gains heightened urgency and visibility in the context of the imbalance of our trade relationship and the widespread public sympathy for protectionist measures" (U.S. Senate 1987, 18).

Biotechnology figured prominently in discussions about U.S.-Japanese science and technology relations. For instance, one of the most frequently cited statistics in discussions about exchange imbalances was the fact that in 1986 there were 323 Japanese researchers working at the National Institutes of Health (NIH), while there were only three American NIH researchers working in Japan. Moreover, many of the Japanese researchers working at NIH were from Japanese industries. And, to add apparent insult to apparent injury, the Japanese researchers at NIH were actually supported by NIH. As Clyde Prestowitz, testified:

> And what is happening is that we are training a number of Japanese industrial researchers in the general principles and techniques of biotechnology, and they are returning to Japan and applying those specific products. . . .
> So to use an example of the kind of thing I would like to see us strive for, in return for having those 300 Japanese researchers at NIH, it would be nice for us to have the opportunity to take our industrial researchers, people from companies like Monsanto or Dupont, and send them to work in the laboratories of the companies in Japan who are sending their researchers to NIH.
> That would establish a more equitable balance in the flow of biotechnological research. (Ibid., 27–28; see also Prestowitz 1988, 326)

The U.S.-Japan Science and Technology Agreement, finally signed in 1988, took into account these concerns by providing more opportunities for U.S. researchers to work in Japan, and by providing special patent protections for countries in which ideas were developed in the course of intellectual exchanges.

Policing the exchange of scientific and technological information with Japan was not the only response to America's new national security problem. There were also legislative attempts to promote technological competitiveness by facilitating the transfer of new scientific and technological developments from government projects and laboratories to the private sector for commercialization. For instance, the Federal Technology Transfer Act of 1986 (PL 99–502) amended the Technology Innovation Act of 1980 (PL 96–480) to promote technology transfer *preferentially to American businesses.* The 1986 legislation also called for government labs to institute reward systems for personnel who are involved in technology transfers that actually result in commercialization. One particular series of legislative efforts to enhance technology transfer and technological competitiveness is especially important for understanding the history of the Human Genome Project. These efforts, spearheaded by New Mexico Senator Pete Domenici, will be discussed shortly.

The Human Genome Project was not conceived as an instrument of national security, although, as we shall see, it was soon thereafter promoted as such. Its conception, while not straightforwardly identifiable with any particular event, is often traced to the so-called "Alta Summit," a conference held in Alta, Utah in December 1984, sponsored by DOE, or more specifically by the Radiation Effects Research Foundation, the successor to the ABCC (Cook-Deegan 1989; 1994, 92–96). The purpose of the meeting was to convene a group of experts to discuss the possibility of more sensitive methods of detecting mutations among offspring of the survivors in Hiroshima and Nagasaki (see participant list, page 13 above). Recall that Neel and Schull's study had yielded "negative" results—in other words, there was no measurable increase in mutation rate. This is quite different from saying that there was in fact no increase in mutation rate. Neel and Schull believed that the radiation exposure must have caused some additional mutations, but that the design of their investigation was simply not sensitive enough to detect them.

The main problem with the study was its one-generation duration, coupled with the fact that mutations were screened at the gross phenotypic level. Under these circumstances, only dominant mutations could have been detected. And as was well known, dominant mutations are much less frequent than recessive mutations. Therefore, a small genetic effect could have been discovered only through the study of many more irradiated parents and offspring than were available. Only a very large genetic effect could have been detected. And fortunately, for the sake of the survivors and for the sake of diplomacy, no large effect was found.

Neel and Schull later employed a more sensitive test involving electrophoretic studies of proteins from blood samples of parents and offspring gathered during the early years of the ABCC. Electrophoretic techniques are capable of distinguishing the protein produced by a recessive version of a gene from the protein produced by a dominant version of that gene. By electrophoretically comparing an offspring's proteins with those of its parents, then, one can look for variants unique to the offspring, and hence attributable to recessive as well as dominant mutations. However, these findings were also negative, in spite of what Neel and Schull continued to be the fact of the matter, namely, that some small increase in mutation rate had occurred (Neel et al. 1988; 1989; 1990; 1994; Schull 1995).

The Alta Summit participants, including Neel, discussed the pos-siblity of yet more direct—that is to say, DNA-level—techniques of detecting mutations in humans. This would be of far greater use to DOE than just in evaluating the Hiroshima and Nagasaki populations. DOE's health research mandate covered the mutagenic effects of all energy sources. Various techniques were discussed, among them: di-rect sequencing of the DNA of parents and offspring, which would reveal base-pair level changes. This would be a suitably fine level of resolution, but unfortunately, as the participants agreed, it was not at the time feasible. Time and cost-efficient detection of mutations in humans at that level would require considerable advances in technol-ogy—in automation and informatics in particular. As one participant summarized the conference in a published report:

> The Workshop identified and analyzed six DNA methods for detec-tion of human heritable mutation, including several created at the meet-ing, and concluded that none of the methods combine sufficient feasibil-ity and efficiency to be recommended for general application. An in-creasing flow of innovative, exciting developments in DNA technology is expected to improve rapidly the prospects for practical methods of mutation measurement, and gives further emphasis to the virtue of not rushing to premature application. Significant resources will be required on an international scale for miniaturization, quantitation and automa-tion of DNA technology, and for the conception, development and eventual application of the mutation methods. Resolution of mutation at the DNA level should bring many new challenges to genetics, and may, for the first time, provide a sound basis for health and environmental pro-tection of human DNA (Mendelsohn 1986, 344; see also Delehanty et al. 1986).

Neel later addressed the potential contribution of the Human Genome Project, beyond what his ABCC studies could provide, to more accurate estimates of the effects of radiation and other muta-gens:

> We are at the end of a cycle of studies in Hiroshima and Nagasaki, and the question is: what next—if anything . . . ?
> My personal opinion is that if further studies are undertaken, they should not be taken but on a scale to advance substantially our under-standing of the genetic risks of radiation for human populations. . . . In this regard, I see such a major program on the nature and significance of radiation damage to DNA as profiting greatly from some of the develop-

ments of the Human Genome Project, providing an immediate application
for some of the data issuing from that Project. (Neel 1990, 337–49)

DOE representatives and supporters would argue in succeeding
years 1) that the agency needed the sequencing technology that would
make it possible to detect mutations at the DNA level, 2) that the
agency was, because of all the resources at its disposal in the form of
the national laboratories, in an excellent position to develop that technology, and 3) that the sequencing technology to be developed would
be valuable independently of its use in mutation detection, for broader
biotechnological uses, and would contribute to America's biotechnological competitiveness relative to Japan (Galas 1990; Wood 1990).

One of the major theses of Robert Cook-Deegan's fine history of
the Human Genome Project is that the need for a sensitive mutation
detection procedure was short-lived as the major rationale for DOE's
involvement in the HGP (Cook-Deegan 1994). The technology to be
developed became an end in itself—and a reasonable end—since the
national laboratories seemed not only capable of developing it, but
also dedicated to commercializing it in accordance with the new technology transfer and technological competitiveness legislation. In turn,
this provided a new mission for the national laboratories, and a timely
one at that. The waning of the Cold War made the future of the labs
very uncertain, since their main mission up to that point had been
weapons research. The Human Genome Project was a mission more
appropriate to the new national security concern: economic competitiveness. This is not to say that geneticists at the national labs abandoned the hope that the sequencing technology would be put to use in
detecting mutations (they also hope to use the technology to better understand the process of mutation, the function of mutation-repair
genes, and other general issues related to DOE's health research mandate). But these were not the sorts of reasons that played a major role
in convincing Congress to support DOE's involvement in the Human
Genome Project.

DOE had a strong supporter on Capitol Hill, Senator Pete
Domenici of New Mexico, whose concern for the future of the national labs was understandably motivated by his concern for the economic future of his own state (Los Alamos National Laboratory plays
a major role in the Human Genome Project). Cook-Deegan reports
how, in the mid-to-late eighties, Domenici was seeking a new mission

for the national laboratories—a mission that would enhance the labs' contribution to the local and national economy, and that was not so dependent on the Cold War. After all, "What happens if peace breaks out?" he asked a group of economists and science administrators whom he had invited to a meeting in 1987 to discuss the issue (ibid., 104).

The Human Genome Project, by now well along in the planning stages at Los Alamos, was an ideal new mission. Domenici incorporated it, along with research on superconductivity and semiconductors, into a bill (S. 1480) designed to further integrate private industry into the national lab system and thereby further promote technological competitiveness.[2] The measure failed in part because of growing NIH interest in the genome project, together with the feeling of various influential scientists, scientific organizations, and members of Congress that NIH was the appropriate home for the project (Cook-Deegan 1994, 135–60). But DOE representatives and their supporters forged ahead to secure at least a shared role. Domenici co-sponsored a second bill with Lawton Chiles, chairman of the NIH appropriations committee, and Edward Kennedy, chairman of the NIH authorization committee. This one, the "Biotechnology Competitiveness Act of 1987" (S. 1966) provided for joint DOE-NIH leadership in the project (U.S. Senate 1988). It passed the Senate but died in the House. But the technology-transfer and technological-competitiveness payoffs of the Human Genome Project had been sufficiently well established; funding for the DOE and NIH projects was ultimately made available through the usual DOE and NIH authorization and appropriations procedures.

The technological-competitiveness argument in favor of the Human Genome Project runs throughout the House and Senate authorization and appropriations hearings, and other Congressional hearings, not only with respect to DOE's but also NIH's role. Members of Congress have often posed the question of the project's significance in these terms, and representatives and supporters of the DOE's and NIH's involvement have been well prepared to answer the question in those same terms. For example, Senator Ernest Hollings:

---

2 For a nice, brief history of technology transfer legislation in connection with the Human Genome Project, see the remarks of Pete V. Domenici in U.S. Senate 1989, 12–14.

How can we guarantee that American firms get the first crack at the technology that the Human Genome Initiative will provide, so as to stay ahead of foreign competition in this critically important field? (U.S. Senate 1989, 5)

Representative John Dingell:

The project will be important to America's international and scientific competitiveness. A recent newspaper article asked, "will the first complete human genetic directory carry the copyright Made In Japan?" (U.S. House of Representatives 1988, 2)

Representative David Obey:

One of the DOE people was quoted as saying that the Japanese were five years ahead of us. Given the competitiveness issue which we have in this country, and the trade issue and things like that, it sound to me like this argument is about to be couched in terms of them versus us

What I am trying to figure out is what are the implications for this country if that assertion were to be true, that the Japanese were moving ahead of us in this field? What are the implications for us scientifically? What are the implications to us economically? (U.S. House of Representatives 1987, 34)

Robert Wood, Office of Health and Environmental Research, DOE:

I would like to underscore the fact that the Department is aggressively pursuing the involvement of American industry in the human genome program. Interactions that are developing between industry and the national laboratories, where the major art of the Department's effort is centered, are expected to facilitate the ultimate commercialization of innovative technologies. An example is the formal cooperative effort between industry and the Los Alamos National Laboratory Human Genome Center for shared funding and shared staffing of research related to computational sciences and instrumentation development. (U.S. House of Representatives 1989, 31)

Robert Moyzis, Center for Human Genome Studies, Los Alamos:

Because of its complexity and scale, biologists at Los Alamos have found great advantage in working side-by-side with their colleagues in the physical, engineering, and computational sciences. This interaction has produced novel solutions to the problems of complexity and scale, including robotics, automation, and computational algorithms that have freed humans from the "routine, boring" tasks the critics oppose. . . . Further, the technology and data produced by this project will have immediate applications in "small science" as well as the biotechnology industry. The "spin-offs" are real. A number of patents have been awarded or applied for based on Los Alamos Human Genome work. Some of these innovations are already moving into the commercial sector. As an example, the image processing capabilities developed by Los Alamos computational scientists to help construct our "interstate" map of human chromosomes, have immediate applications in many areas, ranging from clinical diagnostics to criminal forensics. (U.S. Senate 1990, 68–69)

Leroy Hood, California Institute of Technology:

As we all know, America is currently the world leader in biotechnology. This leadership is unequivocally being threatened by the Japanese. The human genome project, both through technology and the creation of a powerful infrastructure, is helping to insure this future world leadership. Let me make two points in this regard.

Number one, biotechnology offers America a real opportunity to redress our failure in industrial competitiveness and, accordingly, the negative trade balance. An example: in 1981, a small venture capital company called Applied Biosystems was initiated to actually commercialize several instruments for molecular biology that were developed in my laboratory. This company now employs 1,300 people. Today it is the world leader in molecular instrumentation for molecular biology. Its profits last year were $180 million; 60 percent of these profits came from abroad; primarily from Japan and from Europe.

Hence, when we are technologically competitive, we can generate a positive trade balance. The human genome project to develop a variety of new biological tools and technologies is going to spawn new industrial opportunities that will on the one hand create new industries and on the other hand will give the old industries new opportunities. And I would say that biocomputation will be one of the major industrial opportunities of the next 15 years.

Point number two is the biotechnology industry is not yet mature. It will not mature until the second or third decade of the next century. And what this means to me is, its potential is virtually unlimited, particularly if we keep in mind its applications to agriculture, to the environment, to animal health. And certainly it is an industry that is going to range far beyond the $100 billion range.

Hence, the human genome project will prime the American economic pump. It is a critical time to develop these new technologies. If we decline to do so, rest assured our competitors will fill the vacuum. (Ibid., 91–92)

The issue of U.S.-Japan competition in the area of biotechnology, and in connection with the Human Genome Project in particular, became heated in 1989 when James Watson, then head of the NIH genome project, wrote a testy letter to Kenichi Matsubara, a leader of molecular biology and biotechnology in Japan (Cook-Deegan 1994, 218–21). Watson sent copies of the letter to other scientists; one copy made its way to reporter Leslie Roberts at *Science*. In the letter, Watson complained about Japan's relatively small contribution to basic genome research, and Japan's complete failure to contribute to HUGO, the international Human Genome Organization, which had been formed in 1987 to coordinate the various national genome programs (e.g., to guard against unnecessary duplications). The implication was that Japan would let other countries pay for the research and then try to commercialize the results. But, Watson warned, "Japan should no longer expect to benefit from the generosity of other nations if it decides to remain outside the HUGO sphere" (Roberts 1989, 576). The conflict was portrayed as *"Watson* versus Japan" in the press, complete with a caricature of Watson glaring at a generic Japanese (ibid; also McGourty 1989, 679; see also Dreger essay, this volume).

Other scientists were quoted who disapproved of Watson's undiplomatic behavior. But in fact the Human Genome Project had been discussed and promoted from the very beginning in explicit connection with fears about growing trade deficits with Japan due to unfair Japanese commercialization of American ideas. Watson was hardly alone in this view of the matter.

CONCLUSION

What I have tried to do here is to highlight the post-Cold War significance of the Human Genome Project, in contrast with the Cold War significance of one of the genome project's intellectual ancestors, the Atomic Bomb Casualty Commission. In certain respects, this is a very limited study. For instance, I have really only considered the *anticipated* national security and diplomatic significance of each enterprise, not the actual success of each. It would be more ambitious to pursue the latter. An even more ambitious historical project would be to investigate how and to what extent the *conduct* of the Human Genome Project—ranging from the establishment of institutional infrastructure, to the formulation of objectives, to day-to-day practice—reflects or does not reflect its post-Cold War context, and similarly how the conduct of the genetics component of the ABCC reflected or failed to reflect its Cold War context.

Nonetheless, the post-Cold War significance of the Human Genome Project is striking in comparison with the Cold War rationale of its predessor. Consider, as one final indication of the difference, the use of the metaphor "Manhattan Project of biology" to describe the Human Genome Project (DeLisi 1988, 489; U. S. House of Representatives 1989, 1). This suggests, as has often been pointed out, that the Human Genome Project is an instance of "big science." But this metaphor, together with the metaphor "Apollo Project of biology" also clearly suggests (I would say that they are intended to suggest) that the Human Genome Project has an important role to play in national security.

The ABCC was also a very large-scale enterprise, with national security relevance as well. But it certainly would not have occurred to ABCC representatives to portray their project as the Manhattan Project of biology. The ABCC was intended, from a diplomatic point of view, to be an antidote to the Manhattan Project—a symbol of alliance with, not aggression against, Japan.

It is important to keep in mind the post-Cold War context of the Human Genome Project, not only in order to understand its history, but also its future. Two contributors to a 1986 DOE workshop on genome sequencing spoke of the international trade/economic issues as one of the main "forcing functions . . . at work to drive human genomic sequencing" (Anderson and Anderson 1986).

Most scholarly discussions of the Human Genome Project focus on its broad diagnostic and therapeutic significance—and reasonably so considering the gravity of these issues. But their gravity should not blind us to the role that economic issues have played in the Human Genome Project. It should be noted, along these lines, that economic competitiveness is not only a goal of the DOE labs but of the NIH as well. In 1992, the Bush Administration developed a new strategic plan for the NIH, according to which one of the Institute's main goals would be "to enhance the nation's economic well being" (Pear 1992).

In 1996, a year in which many government agencies saw their budgets cut significantly, the NIH was awarded a 5.8% increase, $175 million more than NIH officials had requested, "after Republicans concluded that biomedical research was an engine of economic growth" (Pear 1996, A-13). We are naive to the extent that we fail to factor these sorts of considerations into the future of the Human Genome Project and its applications.

6

Metaphors of Morality in the Human Genome Project

Alice Domurat Dreger

In 1989 a curious cartoon of Nobel Prize-winning biologist James Watson appeared in the British journal *Nature* as part of an article entitled "Speak Softly or Carry a Big Stick" (McGourty 1989). The article consisted of a short report of Watson's recent testimony before the House of Representatives Subcommittee on International Scientific Cooperation, a testimony fraught with Watson's apprehensions with regard to prompt sharing of scientific results. The drawing which accompanied the piece depicted Watson wrapping himself in an American flag and looking none too friendly (see Figure Four). Watson, one of the co-discoverers of the double-helical structure of DNA, was during this time (from 1988 to 1992) director of the National Institutes of Health Center for Genome Research, the NIH branch of the U.S. Human Genome Project. An enthusiastic and effective supporter of the HGP from the start, Watson's beliefs and style were not uncontroversial, at least among his British colleagues, as this cartoon shows. Also notable with respect to this *Nature* cartoon was that in it, the dark (usually red) stripes of the American flag were not sketched solid; instead they were drawn to be imitative of the banding of chromosomes, or of an electrophoresis analysis of DNA. The result was that, in this picture, not only was a "statesman of

155

science" wrapped in the flag of the U.S., genetics and the genome literally formed part of the fabric of America.

FIGURE FOUR

J. D. Watson Draped in American Flag. Reprinted with permission from *Nature*, 341: 679. Copyright 1989 Macmillan Magazines Limited.

What could all this have meant? In the late twentieth century, as scientific research grows more and more expensive and governments around the world face serious fiscal crises, scientists often find themselves having to explicitly justify their public funding to non-scientists. Professional researchers like Watson sometimes face the challenging task of convincing fellow nationals that their scientific projects are good not just for scientists, but good for the nation as a whole—good enough to deserve the precious funding needed. In practice this means demonstrating that the goals of the research at issue are in some important way parallel with—or at least not opposed to—national goals. Today, the more a large scientific project looks like it supports the values and goals of the nation (or at least the values and goals of the powers that be), the more likely it is to get funded.

Hence the caricature of a flag-wrapped Watson. In the mid-to-late 1980s, a small group of prominent U.S. scientists including James Watson decided to appeal directly to Congress for funds for human genome mapping rather than to continue obtaining funding piecemeal through peer-review agencies. The HGP did not by any means constitute the start of large-scale human genetic research in the U.S. As historians have often pointed out, and as Europeans put off by James Watson's claims of American priority and superiority in genetics like to note, productive research into human genetics had been going on for decades before the HGP proper took shape (see e.g., Fortun 1992; Judson 1992; Ferguson-Smith 1991). What changed when the U.S. Congress and subsequently governments of Japan and Europe became much more actively involved in genome mapping were the level of funding, the speed of technology development and gene-map production, and the style of organization. In short, gene-mapping funding and productivity increased significantly, and HGP researchers were organized in a top-down fashion unfamiliar to most biologists at that time. What also changed was that genome scientists felt obliged overtly and repeatedly to align their work on the genome with their respective national values, myths, and goals in a way they had not before—to wrap themselves in national flags.

This paper traces the ways in which early proponents of the U.S. HGP won moral and financial support from Congress largely by aligning their professional values and goals with dominant American values and goals. As shown below, Project advocates garnered backing chiefly by employing metaphors which portrayed the HGP as a natural and necessary part of the American way, as an extension—

indeed as an admirable manifestation—of the traditional American value system. Specifically I look at the use of two major tropes, first, the "genetic frontier" and second, "bad genes," and show how these were used to capture the public imagination and to make the HGP seem a necessary, even inevitable part of the American agenda—an integral part of the fabric of America.

Notably, as detailed below, this alignment of the scientist's values and goals with the national goals did not occur without friction and turmoil, for scientists as a group have their own peculiar tradition of values and goals, a tradition which does not necessarily easily mesh with those of any other ethnicity or nation, including the U.S. Indeed, I would argue—borrowing on the ideas of ethnomethodographers like Sharon Traweek, Bruno Latour, and Steve Woolgar—that in the late-twentieth century, scientists form something of a sub-culture, a sort of ethnic group. Although geographically dispersed throughout many nations of the world, rather like the Jewish diaspora, scientists share a culture in many ways distinctive from their surrounding national cultures. Like its ethnic counterparts, the scientific culture is comprised of a set of ideals, values, imperatives, a system of rewards and punishments, a hierarchy, a canonical history full of superhuman heros and great struggles, and even an origin myth. So while an individual scientist may clothe him or herself in the flag of one particular nation, s/he is often at some level culturally a "scientist." As shown below, when scientists are obligated to make themselves and their work follow the ethos of the nation (as Watson and other advocates of the HGP were), they may find their loyalty to the culture of science called into question by colleagues.

What we see, then, in the founding of the U.S. HGP is something of a dramatic morality play in which fundamental questions about what it meant to be an American and what it meant to be a scientist stood at the center of a struggle for national support of a Human Genome Project. We gain from this brief history some important insight into how the U.S. HGP came to be in spite of the serious challenges it faced at the start, and some more general insight into what happens to each when scientific culture clashes with national culture. I conclude with some tentative suggestions of how we might think about the binds and obligations of the national scientist, that is, what the morals of this morality play might be.

FRONTIER MYTHOLOGY

Since its inception, many images and metaphors not specifically "American" have been used to conjure up a sense of the Human Genome Project. To cite some of the most favored of perhaps dozens of possible examples, the Project has been described as: solving a jigsaw puzzle (Roberts and Aldhous 1992, 28); reading the book of man [sic] or the book of life (Palfreman 1989); constructing a blueprint for a human (Angier 1992); and carrying on a quest for the holy Grail of genetics (Gilbert 1992; remarks of Rep. Ralph M. Hall [Texas] in U. S. House of Representatives 1989, 1). Indeed, sometimes the images used to refer to contemporary human genetics research are piled so thick as to reach the level of the absurd; take for example the title of a 1991 book which contains at least four images: *The Human Blueprint: The Race to Unlock the Secrets of Our Genetic Script* (Shapiro 1991). Still, the most common images used to describe the genome and the HGP's relation to it have been and are (respectively) land and the mapping of that land. This was also the trope most favored by proponents of the Project who testified to Congress and the lay public about the worthiness of the HGP. It has also been the trope most quickly and enthusiastically taken up by Congressional members in favor of the Project.

The idea that scientists could "map" genes—i.e., that the genome was a sort of geography and geneticists its cartographers—was nothing knew to geneticists by the 1980s. The terminology of "gene mapping" had been around at least in its rough form for decades and the basic concept was understood by most geneticists to date back as far as the work of Thomas Hunt Morgan and his colleagues at the "Fly Room" of Columbia University in the 1910s. But the trope of "gene mapping" exploded in the late 1980s as a result of public discussions about funding of the HGP. Advocates of the genome project, for whom the metaphor of mapping made perfect sense anyway, took the image of maps and frontiers and provided non-scientists with textual and visual images of their quest, namely to survey and chart the "genetic frontier." The use of this trope (frontier mapping) greatly helped the cause of the HGP for several reasons, as demonstrated below. First, it made the project visualizable to non-scientists; lay people could with the help of map analogies imagine what the Project aimed to do. Second, it made the Project seem doable and manageable, almost simple, and it implied a definite, realizable goal whose progress could be measured and which would presumably not require endless funding. Third, it made the Project appear ethically unproblematic and value-neutral, since maps in themselves do not seem dangerous or imbued with particular values (on this point see Monmonier 1996). This was particularly important given the many questions of ethics raised about the Project, its structure, and its probable outcomes. Finally, and perhaps most crucially, it situated the Project rhetorically as part of the great American tradition of manifest destiny; it made the project "American" in style and gave Congress an imperative to beat other nations, especially Japan, to the promised and yet-unclaimed land.

Let us consider first this last issue, namely, the likening of the exploration of the "genetic frontier" to previous, famous (and invariably successful) journeys of discovery. It is notable that HGP literature commonly includes stylized images of old geographic maps. No doubt early HGP proponents, many of whom were geneticists, first found genome-as-unexplored-territory metaphors compelling because the language matched well their sense of the project. Namely, that what needed doing was an "epic" exploration and specifically a mapping of this unknown genetic "land." Certainly these metaphors were so regularly and consciously employed by Project advocates because the mapping trope also contained in it the sense of a great mission, one assumed to be readily appreciable by almost any American and certainly by members of Congress: the HGP was simply part of the great American tradition of bold exploration and settlement which stretched from the initial Anglo-European colonization of the Americas to the American "conquest of space" and beyond (e.g., Renato Dulbecco quoted in Kevles 1992, 22).[1]

Frequently, proponents chose specifically to equate the genome with the rich, mythical land of pre-U.S.-settlement—California—and

---

1 By contrast, when speaking among and to other scientists, genome scientists often characterize their work as the logical development in the "internal" progress of scientific research. Walter Gilbert, for example, has described the HGP as "a natural development of the current themes of biology as a whole" (Gilbert 1992, 83). Similarly, pro-HGP historian of science Horace Freeland Judson has opined: "Despite the sophistication and the sometimes stupefying intricacy of genetics today, the main line of its history is straightforward and casts a brilliant light upon the recent proliferation of methods and ambitious aims [like those of the HGP]" (Judson 1992, 38).

to compare the HGP to the exploratory journey of Lewis and Clark and the "opening" of the American west (James B. Wyngaarden in U.S. Congress, OTA Report 1988, 90; Robert M. Cook-Deegan in ibid., 20). For instance, in a 1990 guest editorial James Watson used this genome-as-California metaphor to excite the readers of *Omni* magazine:

> When the United States declared its independence from England in 1776, most maps of North America did not include California, now one of the most populous, diverse, and wealthy states in the nation. Only after the Civil War were California and the American West carefully and systematically mapped out. Expeditions and surveys laid the foundation for renewed exploration of the vast resources available to Americans in the great West.
>
> As the Nineties begin, we are poised to mount a similar effort of even greater importance—to construct the genetic map of humans; to chart the mysteries encoded in the 50,000 to 100,000 genes that make up our bodies. . . . (Watson 1990a, 6)

It speaks to the emotional strength of this historical metaphor that Watson opted not even to mention genetics until the second paragraph of his short article. By the end of the piece the enormous richness of genome seemed as much a given as that of California. Watson closed with the assurance that "the resulting mountain of knowledge will be mined for centuries, as we strive to conquer disease" (ibid.). How could any red-blooded American argue against the worthiness of this exciting new journey of discovery, settlement, and profit?

But some did. Thus, similarly to Watson and at about the same time, the biologist Robert Sinsheimer compared the HGP ("this grand, this unique opportunity") to the voyages of Columbus, and admonished skeptics of the HGP that, although "after 5 centuries the historical significance of his discovery is unquestioned," in Columbus' own time "his venture was sorely questioned and no doubt many thought the money could be better used" (Sinsheimer 1991, 2885). The rhetorical question was obvious: What if Columbus' support had been directed elsewhere—where would we be? Surely the HGP ought to be financed, for it was bound to stand, like Columbus' journey, as a "turning point in human history." Sinsheimer called on fellow Americans and biologists to rally around this latest "epic venture of discovery" (ibid.).

Such specific appeals to American frontier mythology were

common also in the HGP's iconography, a notable example of which was the cover drawing of the 1992 special issue of the journal *Los Alamos Science*, a public relations journal of the Department of Energy's branch of the HGP (Figure Five).

FIGURE FIVE

Los Alamos Science vol. 20, 1992. Reproduced courtesy of *Los Alamos Science*, Los Alamos National Laboratory.

In this colorful artist's rendition of the HGP, ordinary folks led by a white-coated scientist break through a group of mountains to behold the lush land of the genome—which in this case was depicted as endless rows of fertile chromosomes. This was the genome-as-California metaphor brought to life in full-color. The cover caption to this picture made verbally explicit the metaphor: "The human genome presents us with a *vast, largely unexplored frontier*, containing answers to many mysteries about how we evolved, how we are related to other living things, and how we differ from one another.

The Human Genome Project is an attempt to *open up this territory* to our understanding" (Los Alamos Natl. Lab. 1992, emphasis added). Such alignments of the HGP with American mythology invested the scientific project with American value and pride, and they simultaneously allowed Project scientists to argue for their freedom as paradigmatic brave American "pioneers" to journey out and "map" the "genetic frontier" (Cook-Deegan in U.S. Congress OTA Report 1988, 20).

Territory metaphors also went beyond specifically likening the genome to the mythological American frontier. For example, in its favorable survey of the HGP, the Office of Technology Assessment offered a user-friendly diagram of a "comparative mapping" which analogized population maps and genome maps (ibid.). In this schematic drawing the cell was likened to the earth, the chromosome to a country (the continental U.S., specifically), a chromosome fragment to a county, a gene to a town, and nucleotide base pairs to individual people (Figure Six).

A special 1993 update issue on the Genome Projects in *Science* employed similar, albeit more abstract geographic imagery for its front cover: a Vesalian-style muscle man gazed across a genetic landscape as a mouse peered back (Figure Seven). Likewise, inside the special HGP issue of *Los Alamos Science*, the chapter on gene mapping began with a stylized map-like drawing which was meant to depict the playing-field of the gene-mapping "game." In this picture (which looked like a cross between an Old World chart and a video golf game) a whimsical dragon swam in the ocean surrounding "island" chromosomes, and a compass indicated the way north. The legend revealed that the castles represented genes, the flags DNA markers, the white (blank) parts those genome "parts unknown," and so on.

This ongoing attraction to geography-genome metaphors must be partly due to the fact that these images make it possible literally to envision mapping. They make the genome appear very concrete, eminently manageable, seemingly quite self-evident, and the business of "mapping" therefore comprehensible, seemingly quite doable, and uncontroversial. This last advantage is particularly important when one considers the ethical quagmires in which the HGP has sometimes been said to be standing. Such clean, innocent images give the scientific research itself a feeling of clarity and relative innocence.

Land/mapping metaphors have also been employed to suggest the necessity of the HGP to the rapid success of all human genetic research and genetic medicine. In 1989, once the HGP had officially

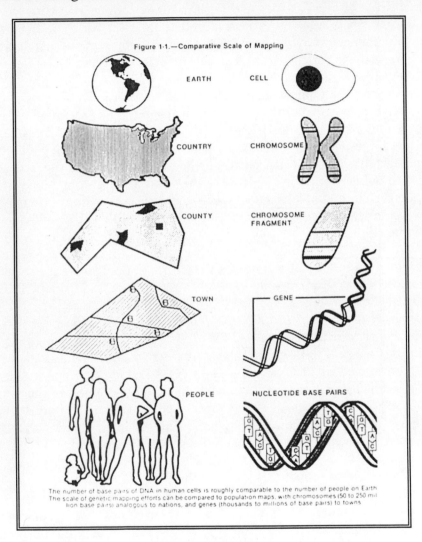

FIGURE SIX

Diagram from U.S. Congress, OTA Report 1988, 5. Reproduced with permission. Copyright 1988 The Johns Hopkins University Press.

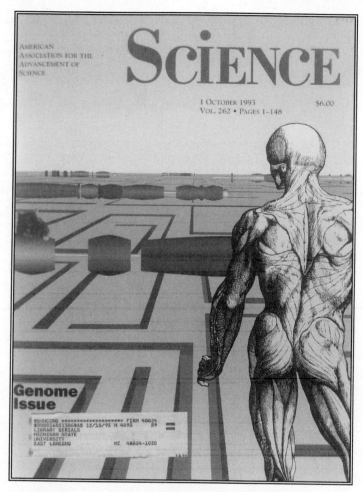

FIGURE SEVEN

Cover Drawing by Susan Nowoskwski *Science*, October 1, 1993.
Copyright 1993. Reprinted with permission from American
Association for the Advancement of Science.

begun, leaders of the Project found themselves faced with the
problem of "how to balance the public's desire for progress on genetic diseases with the [Program Advisory] committee's emphasis on
building a tool [i.e., the map] and not necessarily applying it"
(Roberts 1989, 167). Some HGP spokespersons found the solution to
the problem was to argue convincingly that the HGP would allow

faster and greater progress in "genetic disease" research. In other words, that the best, perhaps only, way to alleviate disease was to map the entire genome, and this argument was made largely through "HGP-as-road-map" metaphors. So for instance in a 1990 letter to the editor of *The New York Times*, James Watson and Norton Zinder (then both working on genome mapping at the National Institutes of Health) used a HGP/road-map metaphor to argue that

> the Human Genome Project cannot fail. Each step we take en route to our ultimate goals will make it easier for scientists to find genes. Like the system of interstate highways spanning our country, the map of the human genome will be completed stretch by stretch, beginning now. (Watson and Zinder 1990)

Watson and Zinder further explained that "our first major goal, to construct a fully connected system of maps of each of the human chromosomes, will guide researchers more quickly and cheaply through the labyrinth of human DNA to the genes they are so eager to find." The implication was that this would lead to rapid development of tests and treatments for those "thousands of human diseases [that] have their roots in malfunctioning genes, and many others [that] have strong genetic co-factors." If the reader was not yet convinced of the necessity of the HGP, Watson and Zinder added the warning that, "for diseases resulting from the interactions of more than one gene, discovering the underlying genetic factors will be impossible without a detailed map," i.e., impossible without the HGP (ibid.; see also U.S. Congress OTA 1988, 10).

Consequently, while proponents of the HGP usually did not claim that the project was a panacea *per se*, it was reckoned a trustworthy and badly needed "road map to health" (Wexler 1992, 240), a massive construction project to boost the "infrastructures of biology and medicine" in order to permit "ready access" to genes (Hood 1992, 138). Notably, unlike many other genetics research programs, the HGP did not set out to locate any particular genes, and certainly does not promise to develop any therapies for "genetic diseases." Instead, Watson and other genome scientists have offered the genome map as the concrete end-product, and interim maps as a way for Congress and the taxpayers to know they are getting something for their money. Indeed, the maps can and have been reproduced in the popular press periodically; progress can be *seen*, if not felt (see e.g., Angier 1992).

In short, the genetic frontier trope has allowed HGP advocates to convey to non-scientist Americans an understanding of, and deep appreciation for, the Project. Although many researchers, early in the history of the Project, objected to the Project—in part out of ethical concerns about the potential uses of the data or its applications, in part because it might take money away from other less glamorous projects, and in part because it seemed to some a waste of time, money, and talent to map and sequence the entire genome including the "junk"—detractors found it extremely difficult to have their objections heard in the public sphere. The lay imagination had been seized by the excitement of the journey to map a new land, and any ethical problems seemed the future problems of the settlers, not the explorers.

HUNTING DOWN BAD GENES

Appeals for support of the HGP were successful because alliance with American traditions and values took place on several fronts besides the frontier. Another important and prominent way in which genome scientists affiliated their work with American values was by promising that their work would aid the disadvantaged, particularly those individuals unjustly downtrodden by "genetic disease." Perhaps intentionally paraphrasing the inscription on the Statue of Liberty, in a 1989 editorial in *Science* Daniel Koshland claimed the HGP had to be done because it was bound to aid "the poor, the infirm, and the underprivileged"(Koshland 1989, 189). In reply to those critics who argued that the hundreds of millions in funding would better be spent on social welfare programs, HGP advocates promised the Project would directly address American social ills, assuming (as some HGP advocates suspected) social ills like alcoholism, violence, and homelessness were actually "genetic diseases." Koshland, for instance, insisted that "What these people [the critics] don't realize is that the homeless are impaired . . . . No group would benefit more from the application of human genetics" (Hubbard and Wald 1993, 187). In the same vein, in his testimony to Congress James D. Watson repeatedly argued that to *not* carry out a Human Genome Project would be "immoral" and "unethical"—essentially un-American (U.S. Senate

1989, 32; see also Angier 1990; Koshland 1989; Robert Moyzis in U.S. Senate 1990, 60). In 1988 Watson told a House subcommittee, "I believe that our scientific community could be judged irresponsible by our fellow citizens if we do not try to maximize as quickly as possible the intellectual resources devoted towards an all-out assault on all our important genetic diseases" (U.S. Congress OTA Report 1988, 40). This was of course an extremely compelling argument, especially in the context of traditional American values of helping the poor and the sick.

In this way, some HGP advocates construed financial support of the project as an issue of justice, another value basic to American mythology. Good people plagued with "bad genes" were said to be entitled to genetic justice. So, for instance, Nancy Wexler who trained as a clinical psychologist and led the search for the "Huntington's Disease gene" frequently likened "disease genes" to serial killers, and gene-hunters to courageous deputies of law enforcement:

> What we were proposing [namely a search for the "Huntington's disease gene"] was equivalent to looking for a killer somewhere in the United States with a map virtually devoid of landmarks—no states, cities, towns, rivers, or mountains, and certainly no street addresses or zip codes—with absolutely no points of demarcation by which to locate the murderer. Our critics said "wait until a more detailed genetic map is available, one with many more regularly spaced markers." This is, of course, a much better strategy if you have time to wait. But we are in a race against the Huntington's disease killer and have no time to spare. (Wexler 1992, 216)

The message was clear: innocent lives would be lost if maps were not produced by which killers could be found (and then hopefully stopped). Similarly, an inherent "goodness" was sometimes attributed to "disease-fighting" genes. For instance, in a *Time* magazine report on the genetics of retinoblastoma (a cancer of the eye), the "anti-cancer guardian genes" were depicted in an accompanying diagram as little knights with swords fighting off the dark, clawed hand that is cancer, the diagram following the way geneticists talked about the phenomenon.

The "bad/good gene" trope has, for Watson, Wexler, and many other HGP advocates, a normative as well as a descriptive component: the "bad gene" was not just physiologically "bad" (unhealthy); it also seemed to be morally evil, an entity that must be brought to justice. Not just science, but our moral fiber(s) were at stake. Accordingly, in a 1989 Congressional hearing Watson insisted that "we have a responsibility to use our science for human good" and a concomitant responsibility to fund science that would do human good (U.S. House 1989, 11; for similar remarks see George Cahill in U.S. Senate 1989, 75 and Doug Walgren in U.S. Congress OTA Report 1988, 78). It seemed downright un-American to refuse this call to justice.

Evelyn Fox Keller has suggested that the "bad gene" trope might be a historical successor to the ideas of "bad seed" and "bad blood" (Keller, personal communication). However, "bad genes" are more than just old wine in new bottles; an important change has occurred in the shift from the concept of bad blood or seed to the concept of bad genes. In the earlier metaphors, often the person alleged to be of "bad blood" or "bad seed" was denigrated as also "bad," that is, undesirable and unworthy. In the "bad gene" trope, by contrast, there is an intense and important break between the hereditary material—the "bad gene"—and the person possessing it. Indeed, the possessor is now seen as diametrically opposed to and unfairly possessed by the unfavorable gene. In other words, in the morality plays described by Wexler, Watson, and others, the person with the "bad gene" is in fact good and in need of direct rescue.

The "bad gene" trope seems, therefore, to be less in keeping with the "bad seed" or "bad blood" tropes and more in line with the popular tradition of "wars" against disease, including the War on Cancer. The "all-out assault" against these "bad genes" brings to my mind the epic battles of the "microbe hunters"—men like Pasteur, Koch, and other bacteriologists celebrated in the histories of Paul de Kruif (1926) and movies of de Kruif's generation. The victim of the "bad gene" seems, as in the stories of the microbe hunts, almost to be *invaded* by the evil, to be otherwise apart from it, at least in a moral sense. This good person/bad gene motif is also notably in keeping with the general trend in the U.S. toward a victim culture in which people are generally described as victims of—not originators of, not responsible for—their problems. It also escapes the old, ugly sort of eugenic implications, for we no longer focus our judgment on undesirable people, only on their unfortunate, undesirable genes.

THE VALUE OF "VALUE NEUTRALITY"

Despite the compelling appeals to American mythology and values detailed above, the HGP still faced in its early years problems in securing unmitigated Congressional support, for it was evident that scientific research had not always lead to what Watson called "human good," and that genetics research might well result in "unAmerican" outcomes. Many early observers feared the HGP might lead ultimately to the restriction of certain individual "liberties" like employment and insurance, perhaps to discrimination and restrictions based on genetic differences—even to the emergence of a "biological underclass," i.e., a pool of so-recognized "genetically-disadvantaged" people unable to keep up with the Jones' genes (U.S. House 1992, 21; *New York Times* 1991). Some elected and self-styled public advocates questioned whether or not instead restrictions ought be placed on genetic research in order to safeguard American freedoms.

At such junctures, in order to avoid restriction of research, HGP advocates turned away from alignment with American values and instead now asked non-scientist Americans to respect the culture of science. Genome scientists argued that restrictive regulation would go against the fundamental scientific ethos: to study freely without fear of what findings might mean, or how they might be used. Here the traditional imperatives of democratic America and those of science did not coincide neatly as they seemed to in the arena of frontier, Statue-of-Liberty, and justice metaphors, and to surmount this conflict of cultural imperatives, HGP advocates found it advantageous to employ a classic pure-science/applied-science distinction. In other words, in order to keep national concerns about the ethical, legal, and social implications (dubbed "ELSI") of the HGP from encroaching on scientists' desire for research autonomy, HGP proponents rhetorically decoupled genome research and its possible unfortunate applications from the start.

For example, when faced with the question of how scientists would prevent genetic research from being used in morally unsavory ways, with a hint of the discourse of gun-control opponents ("guns don't kill people; people kill people"), Nancy Wexler remarked that

> I've heard people say—including people in Congress and even some scientists—that the public can be hurt by genetic information. It's

true that in the past that information has been used against people. But genetic information itself is not going to hurt the public; what could hurt the public is existing social structures, policies, and prejudices against which information can ricochet [!] (Los Alamos Natl. Lab. 1992, 304)

This strategy of disassociating genetic research from its possibly harmful applications worked to deflect concerns and potential restrictions away from the HGP toward its technological spin-offs. Accordingly Robert Cook-Deegan, senior associate of the OTA and later head of the DOE's branch of the HGP, testified to Congress that public "policy concerns [should] really focus on the uses of genetic information, rather than on the process of discovery" because "these [ELSI] concerns do not bear directly on whether to support genome projects, but rather how to deal with the information and technologies that arise from them" (U.S. Congress OTA Report 1988,17, 22; see also U. S. House 1992, 28). Similarly, in their positive review of the scope and design of the HGP, a committee of the OTA bluntly stated: "Specific genetic information such as the location of a gene along a chromosome or the sequence of nucleotide bases composing a specific gene is value-neutral and as such is not ethically troublesome." (U.S. Congress OTA Report 1988, 79). Of course, the *argument* that, in policy-making, scientific research should be disassociated from its negative applications is not value-neutral, but value-specific to mainstream scientific culture. Genome scientists hoped that this might become a value of the American political culture, too.

This negotiation of cultural clashes was tricky, but genome scientists successfully navigated the choppy waters. While maintaining their loyalty to the freedom-of-research tenet of scientific tradition, genome scientists simultaneously pledged their allegiance to the American tradition of protecting individual rights by agreeing that research into ELSI should be funded, so long as it did not delay or detract from the business of mapping and sequencing. HGP proponents advocated a division of labor whereby ethical concerns would be handled by other "sub-cultures" of the nation. In fact, the portion of the HGP budget dedicated to ELSI research has turned out to be something of a jobs program for academics in the social sciences and humanities, and has tended to keep analyses focused on the potential applications of genetic research rather than on the nature, meaning, and propriety of the research itself. In the adopted scheme,

"As scientists and the public [note the alliances] anticipate the new knowledge, physicians, lawyers, social scientists, and philosophers are trying to anticipate the impact of the information on our institutions and our lives" (*Los Alamos Science* 1992 cover caption).

The rhetorical pure/applied-science distinction found its material analogue in the disciplinary division of labor incorporated into the design of the HGP. Scientists would worry about truth and let someone else worry about justice—though as far as scientists were concerned it would best be someone other than politicians. Genome scientists were clearly concerned that, in policy-making, politicians would fail to make the research/applications distinction, so critical to scientific autonomy. When one genome scientist was asked, "What role should Congress play in solving the ethical issues related to the genome mapping effort?", he responded "ONLY IN $ SUPPORT— LOOK AT THEIR PROBLEMS WITH ABORTION AND FETAL TISSUE—KEEP THEM OUT—PLEASE" (George Cahill in U.S. House 1989, 79, uppercase in original). Similarly in thorough-going frustration over potential delays caused by politicians with ELSI concerns, another American genome scientist remarked, "If the governments will leave us alone, we will be all right" (Norton Zinder as quoted in Lewin 1989). Of course, the idea was not really to be left entirely alone, but to be funded and left alone.

The more advocates could convince Congress that research should not be stalled because it might have harmful social effects, the better off the Project would be. In the end, in a somewhat bizarre turn, to achieve the funded autonomy they sought, genome scientists found themselves obligated to claim first that their work should be funded because it would surely have good applications for Americans, and second that any harmful impact on Americans should not be their concern. That is, they found themselves in the position of asserting simultaneously that their work was definitely applicable and definitely pure. Stepping back, one sees just how paradoxical the rhetoric of proponency became, though the contradictory nature of the claims was often masked because negotiation of goals and support occurred mostly through the mediation of abstract values and mythologies.

Rather ingeniously, some HGP advocates turned the basis of hesitation—that genome research might threaten American ideals—into an argument for proceeding promptly with the project, as they argued that the HGP would function as a solid test of the American spirit in the technoscience age. Arguing in favor of the HGP, Thomas Murray, Director of the Center for Biomedical Ethics at Case Western Reserve Medical School, summoned his fellow Americans to the challenge:

> We are a Nation built upon the premise stated so eloquently in the Declaration of Independence, that all men—let us say all persons—are created equal. . . .
> A challenge that will carry us into the next century is to learn how to keep the moral and political spirit alive in the face of the sciences of any quality. Those sciences, genetics among them, confront us with evidence of our differences. I think we can meet the challenge and how we meet it will tell us what kind of people we really are, how much we value our own dignity and that of our fellow humans. (U.S. Congress OTA Report 1988, 54–55)

Faith like Murray's, the faith that rights and freedoms would remain secure in the U.S. no matter what, pervaded pro-HGP testimonies. So, for instance, without further explanation, Walter Gilbert stated as a given that "the problems posed by the [forthcoming genetic] knowledge are not insurmountable and can be dealt with in a democratic society" (Gilbert 1992, 95). Using this logic, that everything will necessarily be alright in a democratic society, genome scientists argued that research ought not be delayed because of ELSI fears. Indeed, with subtle nods to the history of the cold war, the arms race, and military deterrence, ELSI concerns were reconfigured as all the more reason to *hurry* the American project—to beat other nations (especially Japan) to the end-product, since other nations could not be trusted fully to safeguard these ethical concerns (Macer 1992). To paraphrase CBS News's slogan, Americans seemed uniquely qualified to bring the world human genome research, as it could well involve important challenges to democracy and individual rights.

Largely through appeal to powerful American mythology and metaphors, HGP advocates succeeded in convincing non-scientists that the Project constituted, as then-Senator Albert Gore said in 1989, a necessary part of the "American agenda" (U. S. Senate 1989, 2; see also Watson 1990b, 44; and Watson in U.S. Senate 1989, 2). In the end the long "journey" into the "genetic frontier" seemed an important and almost inevitable part of the American tradition, and support of the HGP a vote for truth, justice, and the American way. Metaphorical allusions to the voyage of the HGP continued as Watson remarked, "As human beings it behooves us to support this highly promising research initiative.... It would be unthinkable not to move forward in this important area" (U.S. Senate 1989, 25–26). Senator John Kerry of Massachusetts concurred: "Clearly, this is a journey we must proceed on with. It will benefit our Nation, opening up whole new horizons.... So obviously, we must proceed" (ibid., 57).

In the end, questions about ELSI did not significantly delay the establishment and progress of the HGP, largely because non-scientists were persuaded that the HGP would result in tremendous good and very little uncontrollable harm (see, e.g., comments of Rep. Connie Morella in U.S. House 1989, 49). Indeed many key non-scientist players were finally convinced that to *not* fund the HGP would be (as Watson insisted repeatedly) "essentially immoral" and therefore un-American (quoted in Angier 1990). The HGP proposal to help the sick and genetically disadvantaged on a massive scale was extremely compelling, as it spoke to the core of American-style democracy—the protection of the individual, especially the downtrodden. Senator Pete Dominici of New Mexico (DOE HGP territory) voiced a common sentiment when he concluded to the Senate subcommittee on Science, Technology, and Space:

> I do not think we ought to let this research languish because it has problems of ethics and other issues. I think we ought to solve those, but clearly the pluses for humankind on this kind of program . . . in the field of wellness and health are probably the most significant in all of medical history, and we have to find ways to solve the other problems. (U.S. Senate 1989, 10)

Watson shared Dominici's conviction that, "as a nation, we just have to be fundamentally optimistic and if we can do something good with the procedure, we should get ahead and do it" (U.S. House 1989, 49). In the end American optimism did indeed win the day, and by 1990 the DOE branch of the HGP could confidently declare: "Obtaining a reference sequence of the human genome is now a national objective" (U.S. Department of Energy 1990, 4).

Watson and his colleagues were wrapped firmly in the American flag, and that flag had now come to support and incorporate the substance of the HGP.

THE ACCIDENTAL PRODUCTION OF TWO-HEADED SCIENTISTS

Genome scientists stuck to their guns and rallied around the values of the scientific culture whenever the issue of restricting research arose. Here their loyalty to the scientific ethos stood firm. But culture clashes between national/Congressional goals and scientific traditions arose on other fronts as well, and several of these clashes erupted into heated discussions among scientists about what the roles and ideals of a scientist should be today.

Two important ideals of the scientific culture have long been that, first, scientists should promptly and freely share their findings with other interested scientists, and, second, that scientists should not design or conduct their research out of a desire to personally and directly gain large financial rewards. But large-scale, nationally-funded gene mapping led many scientists around the world to rethink their loyalties to these tenets. In the U.S., the combination of the potentially very valuable applications of genetics research, national funding, international industrial competition, and capitalism led some Congressional representatives to encourage scientists to see that in their work they maximized profit for U.S. companies. It also led some U.S. scientists, including several prominent "statesmen of science," to reject openly these scientific ideals in favor of national competitiveness.

Although geography-genome analogies were originally used to elicit understanding and sympathy for genome research, it did not take long for the genome-as-fertile-land analogy to become a genome-is-fertile-land equivalence statement. The HGP had barely begun when the genome was declared "something to trade" (Watson in U.S. House 1989, 53), a prized commodity—*real* estate. Inevitably, soon a number of individuals and organizations sought the deed to the land they claimed to have found, a deed in the name of their nations or in their own names.

A governmental agency, the National Institutes of Health (NIH), was actually first to seek genome-sequence patents. Bernadine Healy, then director of the NIH, backed by the secretary of Health and Human Services, had the NIH first apply for patents on gene fragments in June of 1991. Healy and her patenting allies held the conviction that only if the information was patented and protected would private industry be willing to invest the time and money needed to develop important medical applications, because only then could exclusive licensing agreements be obtained (Roberts 1992, 912). Healy also insisted that "failure to patent the mysterious gene fragments would cause American taxpayers and biotech companies to lose out on the wealth expected to flow from a cascade of genetic discoveries" (Waldholz and Stout 1992, A1). Nationally-funded scientists, according to Healy, had a duty to forsake scientific ideals of "free information" in order to secure for their nations the profits at stake.

Within a short time of the start of large-scale gene mapping, the genome had truly become the new California. Because of the potentially enormous profits to be made from genome technologies, the search for valuable genes quickly lead to a "land grab" (Kolata 1992), one the Wall Street Journal dubbed the "biggest race for property since the great land rush of 1889" (Waldholz and Stout 1992). Small wonder when *Business Week* declared that the genetic map would lead to a "pot of gold" (Carey 1992). Many genome scientists apparently did not see the patenting craze coming, and some objected heartily when in February of 1992 the NIH, still under Healy's direction, applied for patents on 2,375 more gene fragments, a move that contributed to Watson's resignation as director of the NIH HGP. Watson and others critics feared this patenting blitz "would foster secrecy among scientists, destroy the essential—and fragile— international relations on which the Genome Project depends, and hamstring the biotech industry" (Roberts 1992, 912). Watson's HGP dream had been fundamentally scientific and personal: "to me it is crucial that we get the human genome now rather than twenty years from now, because I might be dead then and I don't want to miss out on learning how life works" (Watson 1992, 164–65). Watson was accordingly dismayed to discover that the alignment of the HGP with the agenda of American industrial capitalism meant a shift in the way the genome science would be done, with less free exchange of findings among colleagues, and perhaps even a slower production of the final genome map. Yet many pro-HGP scientists (including Watson) had initially encouraged the instantiation of the genome-as-territory metaphor—the move to thinking of the genome as valuable real estate—because they knew well that the possibility for development of valuable technologies would convince many Congressional representatives to support the HGP.

As Watson and other genome scientists quickly learned, the reality is that in the late-twentieth century, when money is tight and America's economic strength seems to be in decline, it is not necessarily enough for a project to be morally or scientifically worthy; too many projects are that. An eminently fundable project is one that also promises to make and keep America strong in image and in dollars. The potential for monetary profits from the HGP was a major reason for its establishment and survival in tough political and economic times. As Senator Domenici had told those gathered for a hearing on the HGP in 1989, "some scientists do not like to hear it, but . . . commercial value of science is getting very, very important and while we want most of them to be random [presumably meaning in their choice of research] and doing their thing, we might have to ask some what is the commercial value" (U. S. Senate 1989, 9). Watson and other genome scientists had been quick to assure Domenici and his colleagues that HGP scientists knew and respected their concern for capitalism: if the business of America was business, the business of American biology must also be business. Or, as Watson remarked at a Congressional hearing, "you know, it is the thing 'What is good for General Motors is good for the United States, or vice versa'" (ibid. 1989, 58; see also statements of Albert and Pearson in U. S. Congress OTA Report 1988, 4, 44–45; also Watson in U. S. Senate 1990, 23).

While a pure-science/applied-science distinction was well and good for forestalling ELSI concerns, it was promptly set aside when the question of international economic competition arose. Early HGP advocates found themselves spending much of their time showing how the HGP could allow American geneticists to stay in the forefront of their field while also "generat[ing] a positive trade balance" via the biotechnology industry (Hood in U. S. Senate 1990, 91; see also Cahill in U. S. House 1989, 45). It can only have helped the ongoing HGP that a 1990 survey conducted by the Industrial Research Institute showed that, among representatives of 257 research-intensive private companies polled, of the then-active U.S. "big science" projects the HGP ranked highest in popularity, beating out (in order) the Hypersonic Airplane, the Space Station, the Strategic Defense Initiative ("Star Wars"), and the Superconducting Super Collider (Rensberger 1990). In fact endorsements from industry leaders had supplemented early HGP scientific testimonies (e.g., statement of Richard Godown in U.S. Congress OTA Report 1988, 133), and Watson himself had told Congressional leaders that he wanted "to see the fruits of this program benefit the American people and its industry" (U.S. Senate 1990, 12). As a consequence, Watson should hardly have been surprised when Healy, his boss at the NIH, took very seriously the genome-as-territory metaphor he helped instantiate, and began seeking patents, those deeds to genome territories, to protect American economic interests.

This created a stir among scientists around the world as never before. Whose interests was the genome scientist to put first? The scientific culture's? Their own nation's? Their own? International debates over the right to own the genome quickly escalated into full-blown disputes as players in genome research soon discovered that "private maps like private armies induce fights" (Goodfellow 1992, 777). Given that the genome began as a lush uncharted territory and soon became the subject of Manifest Destiny and private settlement, surely it was only a matter of time before other nations would question the imperialism of the U.S. in the land of the genome. Consider then the following historical vignettes:

• Representatives of several nations argued over the wisdom of a "Balkanization" plan. They could not agree on whether it would result in peace or strife (Dickson 1989). Some thought a policy of "constructive balkanization" (Cantor 1992, 110) which would distribute territories would quiet disputes, but opponents of the scheme argued that it would lead to the big powers getting the best territories and the smaller powers being left to pick up the less valuable pieces.

• Debates over American isolationist politics raged on Capitol Hill, and a U.S. statesman announced that he was ready to "go to war" with Japan, if necessary (Roberts 1989b).

• In response, the French boasted that they knew how to keep "world peace" (Goodfellow 1992; see also Gorman 1993).

• But the Germans repeatedly threw a wrench into negotiation talks, refusing to budge on their demands (Coles 1988).

• The British called for open "diplomacy" ("Diplomacy Please" 1989), even while carrying on "secret" negotiations (Anderson and Alhous 1991).

• Meanwhile, Australians and South Africans argued among themselves over whether or not to get involved (Callen 1992; Jenkins 1992).

After a quick glance at the list of incidents one might think they describe the prelude to World War II, or the plot to Tom Clancy's latest international spy novel. But in fact these scenarios all depict real-life moments in the history of the international human genome mapping projects, as the genome ceased to be *like* a territory or *like* a commodity and became one, as scientists found themselves willingly or unwillingly abandoning the traditional values of the scientific culture.

Indeed, genome scientists of various nationalities soon found themselves at odds in disputes over who, if anyone, should have the right to pieces of the genome. For his part, James Watson unabashedly supported the idea of using strong-arm tactics to get the genome mapped and sequenced as quickly as possible. He joked to Congress about "sending the military in" to "[force] some cooperation from other nations" (U. S. House 1989, 57), and threatened to use informational protectionism against the Japanese, who, he argued, had a history of intellectual and economic free-loading. Watson refused to stand by to watch Americans do the basic research only to have the Japanese use American data to generate interesting analyses and profitable applications. Testifying before a Congressional subcommittee in 1989, he asked:

> . . .what happens if the other countries don't decide to really put up a significant share of the cost? Will we just pass out the data because we are good guys? Well, I am opposed to it. . . . I think it would be against America's national interest to essentially work out the human genome and then pass it free on to the rest of the world. (Ibid. 1989, 15)

Now, while *Nature* depicted Watson draped in the American flag, its American counterpart *Science* ran a cartoon of Watson in which a Japanese scientist held a bag of money out of Watson's reach as Watson hid the double-helix behind his back (Roberts 1989b).

Many Congressional representatives were quite sympathetic to Watson's fiercely-nationalistic protectionism, and the Japanese grew visibly disturbed. Curiously, in those early days of the HGP, the adamantly American Watson noted that some non-scientist Americans might feel more dedicated than he himself to the scientific ethos of freely sharing information:

> I know that my view is not held by all of my fellow Americans, who just think that the openness of scientific information is so important that we should not in any way hinder it and the United States has to lead by openness, but . . . I think it would be irresponsible for us . . . to subsidize the science in other countries. (U.S. House 1989, 15)

Watson thought his "sharp-edged nationalism" (Davis 1990) would win him support in Congress, and though it generally did, some members of Congress objected that scientists were duty bound to insist on scientific "free enterprise" (Rep. Ralph M. Hall in U. S. House 1989, 53). These protesters seemed more enamored with the mythology of scientific cooperation than even Watson, the prominent statesman of science, as they believed that only the free trade of results could lead to true progress (see, e.g., Ron Wyden in U.S. Congress OTA Report 1988, 30). So while Watson encouraged potential supporters to be American (to support the HGP), some of his audience expressed surprise and dismay that Watson was not more "scientist" in his orientation.

As in the case of ethics debates, in the realm of patenting disputes, many scientists struggled with the tensions of being ethnically scientific and ethnically American.[2] Should they operate in a manner in keeping with the scientific tradition and share their information freely without immediate concern for potential profits, or support the interests of their nation (or themselves) and retain valuable secrets? Whereas at one time the dictum in science might have been stated as "publish and profit," now in biotechnology, where exclusive knowledge can bring great wealth, the dictum might more accurately be construed as "publish *or* profit." As early as 1987 George Cahill, then a researcher at the Howard Hughes Institute, remarked on the

---

2 Genome scientists of other nations found themselves caught in similar internal "ethnic" conflicts as they had to choose whether to be loyal to traditional scientific ideals or the dominant ideals of their nation. For instance, West German scientists who proposed genome mapping found themselves criticized by parliamentarians who saw frightening analogies between genome research and the history of Nazi eugenics (Dickman 1988).

FIGURE EIGHT

Reproduced with permission of Dennis Renault of *The Sacramento Bee.*

dilemma of whether one ought to follow traditional scientific norms or the profit motive; Cahill confessed his belief that when a research project like genome mapping held such huge potential for profits, "we have to look at the bucks to ethics ratio" (quoted in Roberts 1987, 358).

Some scientists wished aloud that genome data could exist as "commons," open ground democratically and freely available to all, the plan so idealistically imagined by the Office of Technology Assessment in its early HGP review (U. S. Congress OTA 1988, 10).

Under such an ideal scientists could freely share data and be both good Americans and good scientists according to the mythological traditions of each culture. But the imperatives of American capitalism made such a happy double-identity well-nigh impossible. A cartoon from the *Sacramento Bee* depicted the dilemma of the scientist trying to be loyal to American-style capitalism and traditional science values (Figure Eight). In it one scientist displayed a fellow scientist to two visitors and informed them, "We've finally developed a scientist who can perform independent research for both the University and private industry without any conflict of interest." That new scientist had two heads, both of which looked rather dazed. How to be both a traditional scientist and American industrial capitalist? Only a mutation of what it meant to be a scientist could achieve this.

In an attempt to maintain their traditional scientific identities amidst territorial disputes over the genome, some scientists argued that, like Antarctica, the genome is a domain that could not or should not be owned. In 1989, George Cahill, then treasurer of the international Human Genome Organization (HUGO), begged Congress to agree that "The human genome belongs to the entire human race" (U. S. House 1989, 46). Originally this statement would have been a tautology—of course every human has a human genome!—but it became instead an opinion, indeed a plea, with the instantiation of genome-as-territory metaphor. By 1990 even Watson found himself insisting that "the nations of the world must see that the human genome belongs to the world's people, as opposed to its nations" (Watson 1990b, 48).

CONCLUSION

We have seen in this short history of the HGP how many scientists experience a culture of science which differs in some important ways from other cultures to which they belong. When obligated to seek public funding, scientists may be forced to confront what it means to be a member of those various cultures and to decide to which culture they will be loyal. We have also seen that the "selling" of the HGP to non-scientists in the U.S. occurred largely via appeal to common American mythologies and values, and that, to succeed in securing funding, proponents of the HGP had to find ways to show that genome scientists subscribed to those American values that would help their cause, and were exempt from those which would hamper it.

I hope this history will serve as something other than a how-to manual for marketing big scientific projects to Congress and the public, and something more than a source of comfort to scientists who find they are not alone in feeling torn between the values of dominant American culture and the values of science. Rather, I hope this history will lead us to consider and discuss what should be expected from scientists and science projects who seek public support. Some might argue that the proliferation in HGP testimonies of abstract political rhetoric—of values, mythologies, and so on—would stand as reason to keep non-scientists, and especially Congress, out of the business of deciding who gets what science funding, since such an encounter seems to invariably wind up politicizing science. But I think this would be impossible, and perhaps even wrong, given that it seems reasonable to allow those supporting scientific research to have some understanding of and some say in the direction of that research. We do have the ideal still in America of no taxation without representation.

Instead I would prefer to see this history as reason to encourage all participants in discussions of science funding to subscribe to one more American ideal—bravery—and to drop the heavy and emotional illusions to abstract values and mythologies in favor of honest and frank discussion about what is possible and probable in scientific research, with genuine attempts to understand each other's values, interests, abilities, and goals. I am sure we can all agree that it would be more heartening if, when scientists go to Congress, there was less converting of scientists into politicians, and more converting of politicians to an understanidng of what it means to be a working scientist.

Philosophers Lisa M. Heldke and Stephen H. Kellert have recently put forth the case for actively incorporating a notion of responsibility into our idea of objectivity. I do not have space to describe their entire concept here, but it is worth mentioning a few of the major points. Heldke and Kellert suggest we think of objectivity and responsibility not as stagnant states, in which a claim or action is or is not objective and/or responsible, but rather that we think of objectivity and responsibility as processes—that we think of being responsible as a necessary part of being objective. They explain that, under this way of thinking, "Inquiry is marked by objectivity to the extent that its participants acknowledge, fulfill and expand responsibility to the context of inquiry" (Heldke and Kellert, 361). Heldke and Kellert recognize what decades of history of science have shown, namely that,

> participants in inquiry are never completely 'impartial' or 'disinterested,' because all participants have some stake, some interest, in the outcome. . . . Objectivity is [instead therefore] fostered by acknowledging our concrete situations and interests, and by opening them up to ever-wider scrutiny." (Ibid., 374)

What is especially exciting about Heldke and Kellert's idea is that they do not limit this suggested way of being to scientists, but argue that for maximum benefit to accrue, all participants in an any endeavor act responsibly "by being explicit about their own values, interests, presuppositions and goals, and by actively seeking to expand the circle of those to whom they are responsible" (ibid.). Thus, in the case of the HGP for instance, not only should HGP proponents and participants be much more explicit about their values, interests, etc.; so too should non-scientists. Indeed, Heldke and Kellert would suggest that anyone potentially influenced by the HGP has a responsibility to learn (at least at the basic level) what the Project aims to do and what it reasonably can do.

This might sound like the old ethical leg-irons that genome scientists feared from the start, but I think instead this model would make life much easier for genome scientists. It would give them the opportunity to speak frankly about their goals, needs, and concerns—instead of having to cloak these in metaphors and myths—and it would require non-scientists to listen and to respond with equally frank discussion of their goals, needs, and concerns. Such a genuine "cultural" exchange, one that would occur because of a common value of good knowledge and responsible behavior, would surely be better for all concerned. I do not think it would result in a loss of science funding, but rather in a greater public respect and appreciation for the work of scientists. However, only if we commit ourselves to such a form of communication will we achieve it.

Part Two

The Genome Project and Eugenics
Introductory Comments

No single ethical issue surrounding the genome project has raised more concerns than the possibility it poses for a "high-technology" eugenics in the future. The generation of a massive amount of new information about the genetic foundations of numerous human traits would seem to offer many ways in which one could select either for or against certain traits deemed desirable by the wider society. The decision of the architects of the genome project to pursue the sequencing of the entire human genetic structure, rather than to focus their attention on identifying the genetic causes of known hereditary disorders such as Cystic Fibrosis, Tay-Sachs syndrome, or Huntington's disease, indirectly opens up the possibility of a widespread use of genetic knowledge for many non-medical uses.

The "eugenic ideal," proximately dating as the principle of a social movement from the work of Francis Galton (1822–1911) (Kevles 1995; 2000), has been a concept within western political discourse at least since Plato, who proposed in Book Five of the *Republic* (459d) the selective breeding of the guardians to ensure the most superior offspring (positive eugenics), coupled with the destruction of those deemed inferior (negative eugenics). As a powerful social force, the human eugenics movement assumed in the early decades of this century the character of a world-wide social force that for a time brought together

185

social progressives and conservatives, scientists and laypersons, in a movement that resulted in sterilization laws, and eventually the state-sponsored mandatory eugenics of the National Socialist regime in Germany (Kevles 1995; Ludmerer 1972; Weindling 1989).

Scholarly investigations in recent decades have decisively under-mined the older historical conclusion that this eugenics movement of the early decades of the century was simply "fringe" pseudo-science. The endorsement of eugenics was too widespread and the scientific authority of some its primary advocates—Charles Davenport, R. A. Fisher, H. J. Muller, Fritz Lenz—was too substantial to allow this conclusion. The remarkable feature of the Nazi eugenic measures en-acted initially in 1933 was not the novelty of these laws. It was only that they put political force behind proposals that had already been widespread in the industrial world, including the United States, and were initially received with endorsement in American journals (Popenoe 1933).

The papers in this section explore several facets of these questions and they juxtapose several different positions in a broad discussion of the ethical questions surrounding the notion of eugenical selection. The opening essay by Martin Pernick sets an American historical framework, and explores in detail the more popular social history of the American eugenics movement of the 1920s and 30s. This study displays the way in which eugenics was able to enter feature films and public media discussions. His paper also illuminates the complexity facing any definition of "normalcy" and "defectiveness" in the hereditary domain, a critical issue whenever proposals to select for or against an ideal genotype are considered.The strong social components in such definitions are highlighted by his study.

The issues surrounding the possibility of a new form of eugenics that plausibly will result from the HGP are then explored in papers by Arthur Caplan and Philip Kitcher, with commentaries by Timothy Murphy and Diane Paul. These discussions present new perspectives on the notion of both "positive" and "negative" eugenic selection. These discuss the possibility of implementing what Kitcher has termed "Utopian" eugenics with full awareness of the mistakes of the past and the social complexities of the present. The commentaries display some of the vigorous discussion that this topic generated at the conference.

7

Defining the Defective:
Eugenics, Esthetics, and Mass
Culture in Early Twentieth-Century
America

Martin S. Pernick

From 1915 to 1918 Chicago surgeon Harry Haiselden electrified the nation by allowing the deaths of at least six infants he diagnosed as "defectives." Seeking publicity for his efforts to eliminate those he considered "unfit," he displayed the dying infants to journalists, and wrote a book-length series about them for the Hearst newspapers. His campaign was front-page news for weeks at a time.[1] He also wrote and starred in a feature motion picture, *The Black Stork*, a fictionalized

---

1 This essay is based on material from my book (Pernick, 1996), used with permission of Oxford University Press. A different version of this essay has also appeared in Mitchell and Snyder 1997. For easily accessible and/or unique accounts see the following: Bollinger 1915; Meter 1917; *Chicago American* (hereafter *CA*) (24 July 1917): afternoon edition; *Medical Review of Reviews* 23 (1917), 697-98; Werder 1915; Grimshaw 1915; *Call* (13 December 1915); Hodzima 1917; *New York Sun* (13 November 1917); *New York Times* (16 November 1917), 4; *Chicago Herald* (20 November 1917); Stanke 1918. For cases prior to 1915, see *CA* (23 November 1915): 3. Extensive coverage may be found in newspapers nationwide, especially in the Midwest and East, for 12 November–30 December 1915; 5–7 February and 14–16 March 1916 (Bollinger, Roberts, Werder, and Grimshaw cases); 22–27 July and 12–20 November 1917 (Meter and Hodzima cases); 25–30 January 1918 (Stanke case); and 18–20 June 1919 (obituaries). Unless otherwise stated all cited newspaper articles begin on page one.

account of his cases.[2]

In the unprecedented debate prompted by Haiselden's actions, hundreds of Americans took a public stand. A majority of those quoted in the press *opposed* preserving the lives of "defectives." They included public health nurse Lillian Wald, family law pioneer Judge Ben Lindsey, civil rights lawyer Clarence Darrow,[3] historian Charles A. Beard, and even the blind and deaf reformer Helen Keller (Keller 1915a; 1915b).[4]

Yet despite these dramatic events, media coverage of the issue faded rapidly during the 1920s. By the time similar proposals surfaced again in the mid-1930s, Haiselden and his actions appear to have been almost totally forgotten.[5]

This story is important, not just for its novelty and drama, but because it vividly demonstrates the crucial though little-recognized role played by mass culture in constructing both the meanings and the memory of the early-twentieth century movement for hereditary improvement known as eugenics.[6]

---

2 I found and restored the only viewable print, of the 1927 version. It is available for research use at the University of Michigan Historical Health Film Collection. An unprojectable fragmentary paper print of the 1916 version is at the Library of Congress Motion Picture, Broadcasting, and Recorded Sound Division, Washington D.C. #LU-9978, Box 110. I am working with the library to arrange for reanimation of this version.

3 On Wald see *Independent* (3 January 1916):25; on Lindsey see *Chicago Herald* (18 November 1915); on Darrow, see *Washington Post* (18 November 1915); on Beard see *New York Times* (18 November 1915): 4.   Darrow later repudiated eugenics in Darrow 1926. I have identified 333 individuals who were quoted in the mass media on this issue. 167 (51%) opposed treating at least some types of impaired newborns, including fourteen advocates of active killing. Another thirty-two favored leaving the choice up to the doctor, without saying what they thought the doctor should do. Only 116 (35%) said doctors should try to save all infants. Of course, this was not a scientific public opinion sample, but it does reflect the image of public opinion that was presented by the press. See Pernick 1996, Table One.

4 Haiselden's critics initially cited her case to prove the social utility of preserving the lives of those with disabilities, e.g. *Washington Post* (18 November 1915): 6; *New York Sun* (18 November, 1915); but Keller strongly supported Haiselden. So did at least one other spokesperson for the disabled, see *Detroit News* (18 November 1915): 25.

5 I could find only three references to Haiselden in the *New York Times* and *Chicago Tribune* between 1920 and 1960.

6 The term "mass culture" includes any productions made for a mass audience, whether or not they were demonstrably "popular" in origin.

Eugenic leaders frequently attacked mass culture for what they considered its vulgar distortion of scientific ideas. Yale professor Irving Fisher complained, "Eugenics is one of the few cases in which a scientific term has come into popular use," but "it is subject to a great deal of misconception" (Fisher 1915, 64; Fisher and Fisk 1917, 294).[7]

Historians of eugenics focus on the movement's professional leadership, and many have implicitly adopted the leaders' views of mass culture. But I will argue that such an approach misses the vital role of mass culture as a battleground on which scientists, physicians, popularizers, journalists, censors, and audiences struggled to shape the meanings of "eugenics" and "heredity."[8]

The American eugenics movement supported a very diverse range of activities, including: advanced statistical analyses of human pedigrees, "better baby contests" modeled on rural livestock shows, compulsory sterilization of criminals and the retarded, and selective ethnic restrictions on immigration.[9] These efforts all were seen as "eugenic" because they all aimed at "improving human heredity." But the meanings of "eugenics" depended on the answers to at least four related but separate and distinct questions:

1. What does "improvement" mean?
2. What does "heredity" mean?
3. By what methods should heredity be improved?
4. Who has the authority to answer questions 1–3?

This paper examines the role of mass culture in defining eugenics, by providing one illustrative example of how that culture served as a battleground for competing answers to each of these key questions. First, mass culture's representations of beauty and ugliness illustrate

---

7 For many similar comments see Dewey 1914, 349; Guyer 1927, 426; Holmes 1915, 222–27; Kevles 1985, esp. 58; Robinson 1922, 91–93; Illinois State Charities Commission 1914, 65, as quoted in Curtis 1983, 91. For the occasional less hostile view of mass cultural meanings see NCRB 1914, 272; *Good Health Magazine* 50 (1915): 485–88, and 51 (1916): 594–95.

8 For a few important exceptions see Bogin 1990; Rydell 1984; Nelkin and Lindee 1995.

9 The literature on eugenics is vast. For an introduction to American eugenics, see Paul 1995; Ludmerer 1972; Haller 1963; Mehler 1988; Reilly 1991. To place America in comparative context, see Kevles 1985; Adams 1989; Allen 1975; Sapp 1983.

the construction of what counted as "improvement." The second example explores why in mass culture the meaning of "heredity" was not limited to traits caused by genes. The final section examines how eugenic methods came to include death for those judged defective, and the ironic role of the mass media in creating a professional monopoly over such decisions.[10]

In its heyday during the 1910s and 20s, eugenics was widely accepted as being an objective science, and when I use words like "defective" and "unfit," I am quoting what eugenicists believed to be purely technical terms. But while eugenics claimed to be purely objective, this paper will show that subjective values, such as esthetic standards of beauty and ugliness and moral attributions of responsibility, were central to eugenic constructions of hereditary disease and disability.

WHAT TRAITS ARE GOOD? EUGENICS AND ESTHETICS

Esthetic values played a critical though little-known role in eugenic constructions of fitness and defectiveness. Eugenics promised to make humanity not just strong and smart but beautiful as well.[11]

Efforts by leading scientists to explain the evolutionary role of beauty began in 1871 with Darwin's analysis of sexual selection in *The Descent of Man*. Darwin's cousin Francis Galton, who coined the term "eugenics," began his scientific career by compiling a "beauty map" of Britain, for which he calculated the ratio of attractive to plain and ugly women he encountered at various locations (Russett 1989, 78-92; Kevles 1985, 12; Wiggam 1924; Anon. 1915, 5; Birken 1989; Trombley 1988, 79; Wells 1905.)

But while these and other evolutionary scientists studied the esthetic component of eugenic "fitness," esthetic concerns appeared more frequently in the mass media.[12] One major eugenic popularizer,

---

10 A short paper can only briefly illustrate these points, but each is documented more fully in Pernick 1996.

11 I am grateful to Peter Laipson for his suggestions on this section.

12 Thus the esthetic dimension of eugenics has often been overlooked. In addition, eugenicists' insistence on the objectivity of their science also made the movement seem hostile to emotions and art. Such was the view of James Joyce's Steven Daedalus, who contrasted "eugenics" and "esthetic," and accused eugenics of trying to reduce beauty to its biological functions (Joyce, *Portrait of the Artist as a Young Man*, first published in 1916, quoted in Kevles 1985, 119). Historians of

Albert Wiggam, saw an attractive appearance as the *best* external indicator of overall hereditary fitness. He regarded health, intellect, morality, and beauty as "different phases of the same inner . . . forces." "Good-looking people are better morally, on the average, than ugly people." Thus he concluded, "If men and women should select mates solely for beauty, it would increase all the other good qualities of the race" (Wiggam 1924, 272, 279). In the most extreme version of this view, esthetic preferences were simply Nature's instinctive guide to finding the fittest mate, a view both Wiggam and Haiselden sometimes explicitly endorsed (ibid; Haiselden 1915a, [2 December]: 2).

But both eugenic leaders and popularizers were skeptical that truly healthy beauty could be recognized by the untrained eye. Scientists since Darwin found beauty problematic precisely because esthetic preferences in choosing a mate often did *not* seem to favor other adaptive traits (Cronin 1992; Gould 1993, 371-81).[13] Thus eugenicists did not simply endorse existing cultural preferences, but actively attempted to "improve" current standards. Irving Fisher explained that careful propaganda was needed, to "unconsciously favorably modif[y] the individual taste. . . in mate-choosing" (Fisher and Fisk, 1917, 322). Wiggam agreed. "If their ideals of human beauty are properly trained," young people will "unconsciously reject the ugly," and will "fill their homes with beautiful wives and handsome husbands"(Wiggam 1924, 275; see also Guyer 1927, 438).[14]

(contd.)
Germany pioneered the recognition of the esthetic dimension of eugenics. See especially Gilman, 1995; Mosse, 1978, 2; Mosse 1985; Weindling 1989; Saürländer 1994, 16–19; Tryster 1991, 13.

13 In addition, many professionals among the eugenic leaders felt that "the mind is more important than the body" (*New York Times* [26 November, 1915]: 8). Sociology professor Franklin H. Giddings explained more bluntly: "The idiotic child should mercifully be allowed to die. The child with a good brain, however crippled otherwise, should be saved" (*Independent* [3 January 1916]: 23). For others with similar views, see *New York Times* (18 November, 1915: 4; and 25 November, 1915); *Chicago Herald* (17 November, 1915); *Chicago Tribune* (18 November, 1915); *New York American* (19 November, 1915): 6; *New York Medical Journal* (26 December 1914): 1247, 1249; *The New Republic* (18 December, 1915): 174.

14 The Guyer passage is retained from the 1916 edition. For similar esthetic efforts in German eugenics see Weindling 1989, 410–13. The presumably unintended implication of communal living or polygamy in Wiggam's quote results from his diction, not my ellipsis.

To illustrate the content of this esthetic propaganda, I will focus on the only two surviving pro-eugenic full-length American motion pictures of the 1910s and 20s: a feature film called *The Black Stork* that dramatized Dr. Haiselden's crusade against saving impaired newborns, and the pioneering government-produced health education series *The Science of Life*, a twelve-reel survey of high-school biology, distributed by the U. S. Public Health Service from 1922 to 1937.[15] In

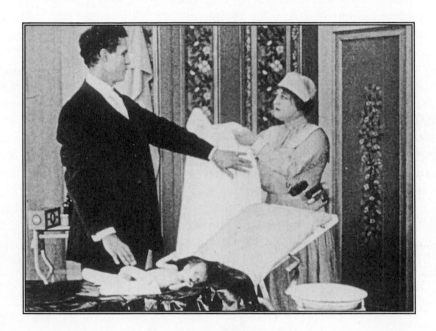

FIGURE NINE

The Black Stork. Doctor refuses to operate on disabled newborn. University of Michigan Historical Health Film Collection, used with permission of John E. Allen.

---

15 Individual school districts continued to use *The Science of Life* for decades after 1937. For more on the film see Pernick 1993. The last three reels focused on hygiene, human reproduction, VD, and eugenics. I have identified over 40 films shown in the U.S. between 1900 and 1930 that dealt with aspects of eugenics and/or human heredity. Most were short one- and two-reelers not yet subject-indexed in any reference guide, and so are largely unknown to scholars. Almost all these films are believed "lost." I have identified and analyzed them based on written records, in Pernick 1996, chap. Seven.

The Black Stork, Claude, who has an unnamed inherited disease, ignores graphic warnings from his doctor, played by Haiselden, and marries Anne. Their baby is born "defective" and needs immediate surgery to save its life, but the doctor refuses to operate (Figure Nine). After God provides a horrific vision of the child's future of misery and crime, Anne agrees to withhold treatment, and the baby's soul leaps into the arms of a waiting Jesus.

Both films equated beauty with fitness. "An attractive appearance goes hand in hand with health," explained *The Science of Life.*[16] And both attempted to influence audience concepts of beauty. But each presented internally-conflicting esthetic standards, an ambivalent mix of modernism and romanticism. *The Science of Life* emphasized stark mechanical images. It urged "THE WOMAN OF TOMORROW" to develop strength and beauty through vigorous exercises, demonstrated by a short-haired woman whose hard flat body was accentuated by stark black tights and knee-level photography (Figure Ten).

An attractive body also was explicitly equated with a sleek streamlined locomotive, whose beauty became manifest in powerful motion and efficient function. Photographed in a low-angle tilt shot that swept upward from wheel level, the engine's sharp clean lines and powerful mass appeared starkly silhouetted against the sky.

Motion pictures first made it possible to display the beauty of bodily action. The desire to depict the poetry and science of motion contributed to the development of cinema, while the use of film helped reshape modern beauty in terms of physiology not anatomy, as active function not just static form (Musser 1990; Dagognet 1992; Braun 1992).[17]

But *The Science of Life* also promoted older romantic concepts of beauty. Intercut with the starkly modern images were scenes in which

16 *Science of Life* reel 12, *General Hygiene,* National Archives Motion Picture Division, College Park MD (hereafter NA), reel number 90.26. The following discussion is also based on two versions of reel 11, *Personal Hygiene for Young Men* and *Personal Hygiene for Young Women,* NA reels 90.24 and 90.25. *New York Times,* (15 November 1923): 10; *American Journal of Public Health* 12 (December 1922): 1033, and 13 (September 1923): 737; *Journal of Social Hygiene* 14, (January 1928): 14; Motion Picture Division Scripts, New York State Archives, Albany, NY. Box 2565, Folders 12, 471 and 12,493, including a clipping from the *New York Herald* (15 April 1923); Records of the United States Public Health Service, National Archives, Record Group 90, File 1350. I thank Peter Laipson and Aloha South for locating the National Archives material.

17 I thank Rebecca Zurier for prompting these ideas.

FIGURE TEN

The Science of Life.   Fitness as modern hard-edged mass in efficient motion. Source: University of Michigan Historical Health Film Collection, from National Archives. Used by Permission.

health and fitness were represented by a long-haired round-cheeked young woman in calm repose, photographed as glowing with cleanliness and natural sunlight in gauzy soft focus (Figure Eleven).[18]

The Black Stork emphasized the more-naturalistic modernism of Thomas Eakins. Beauty was illustrated by athletic adolescents in outdoor settings: five naked boys diving into a swimming hole, a woman in a swimsuit doing handstands on the beach. The 1916 edition of *The Black Stork* explicitly *attacked* mechanical standards of beauty, using

---

18 This distinctive esthetic mix reached its peak in *Way to Strength and Beauty*, a 1925 German film shown in the U.S., which combined dramatically modern steep-angle cinematography with scenes of both classical and primitive Teutonic athletes. The film was produced by the UFA and is available from the Bundesarchiv in Cologne and the Library of Congress Motion Picture, Broadcasting, and Recorded Sound Division, Washington D.C. See also: *Fit: Episodes in the History of the Body* (Straight Ahead Films, 1993); Banner 1983; Banta 1987.

a speeding motor car to represent not modern esthetics but the false allure of the "fast" life. Yet, when the film was re-released in 1927, industrial modernism dominated. The sequence linking fast cars to loose living was deleted, while lengthy new scenes portrayed beauty as a massive new automobile, owned by a "professor of heredity."

FIGURE ELEVEN

The Science of Life. Fitness as soft-focus romantic beauty, cleanliness, and repose. Source: University of Michigan Historical Health Film Collection, from National Archives. Used by Permission.

This ambivalent esthetic vision exemplified what Thomas Mann called "a highly technological romanticism." Eugenics promised to create a romantic utopia by means of modern science, and its esthetic propaganda reflected this uneasy mix of goals.[19]

If eugenics equated fitness with beauty, it labelled ugliness a disability. "It was terribly ugly," Haiselden wrote of one baby. Such ugliness was not "light or superficial;" it was a true "handicap"

---

19 Thomas Mann (in Craig 1986, 21) had used the phrase to describe "the really characteristic and dangerous aspect of National Socialism."

(Haiselden 1915a, [29 November]: 2; [30 November]: 2).[20] Both films selectively highlighted the repulsive ugliness of the "unfit," as in several scenes comparing them to cattle (Figure Twelve).[21]

FIGURE TWELVE

The Black Stork. The "unfit" compared with cattle. Source: University of Michigan Historical Health Film Collection, used with permission of John E. Allen.

Again echoing Thomas Eakins, Haiselden described the case on which *The Black Stork* was based as "not a pretty one. It mars the pages of this book—as I intended it should mar them. It is better that the deformities of this tiny castaway should sear themselves into the minds of thinking men" (Haiselden 1915a, [2 December]: 1).[22]

---

20 For present debate over the role of esthetics in defining disease and disability see the conclusion below.

21 On photographic iconography, see Elks 1992. On the disabled in entertainment film see Norden 1994. On disability in recent horror films see Longmore 1985.

22 In many ways Haiselden's approach to art mimicked that of Thomas Eakins. On art and disfiguration of the canvas, see Fried 1987. For other comparisons see Pernick 1996, 60–71.

Eugenic popularizers promoted definitions of ugliness that reinforced their judgments on other human differences, including gender, class, race, and nationality. Wiggam lamented,

> We want ugly women in America and we are getting them in millions. . . . [T]hree or four shiploads have been landing at Ellis Island every week. . . . I have studied thousands of them. . . . They are broadhipped, short, stout-legged with big feet; . . . flat-chested . . . and with faces expressionless and devoid of beauty. (Wiggam 1924, 262, 273–74)

These "draft horses" were rapidly replacing "the beautiful women of the old American stocks, the Daughters of the Revolution" (ibid.).

FIGURE THIRTEEN

The Black Stork. Physical disability linked with race. Source: University of Michigan Historical Health Film Collection, used with permission of John E. Allen.

Both films linked esthetics, disability, and race. The only identifiable blacks in each were photographed as repulsive defectives, as in this example from *The Black Stork* (Figure Thirteen), and Haiselden repeatedly linked "blackness" with "ugliness" (Haiselden 1915a, [24 November and 30 November]: 2).

Like race, economics also shaped the esthetic distinctions between the fit and the defective. *The Science of Life* promised that both "health and success" awaited the visually-attractive, and in both films, couples who illustrate "wise mating" wear tastefully conservative but up-to-date business suits and dresses, on their prosperously-stout bodies. Portraying "others" as ugly was central to labelling them defective, while diagnosing "others" as diseased reinforced the perception of them as repulsive.

Eugenicists insisted that their diagnoses were based entirely on objective science. Even Helen Keller, the famed reformer who had become blind and deaf in childhood, believed objective science could determine which mentally-impaired infants should be eliminated. "A jury of physicians considering the case of an idiot would be exact and scientific. Their findings would be free from the prejudice and inaccuracy of untrained observation" (Keller 1915a, 173–74).

Many eugenicists admitted that the distinction between "fit" and "defective" relied on esthetic values, yet they denied that such classifications were therefore subjective or unscientific. They did not claim to be "value-free," but rather that their esthetic values had been validated by objective scientific methods (Kevles 1985, 12).[23]

Eugenic esthetic judgements were based on broad cultural values that were not unique to eugenicists. But eugenics did not simply reflect cultural values indiscriminately. The movement's technocratic utopianism attracted mostly middle-class, native born segments of the population who brought with them a specific set of values shared with others of their background, and these selected preferences in turn shaped eugenic images of beauty and ugliness.

Thus, eugenic popularizers promised to make people more attractive while they intervened in mass culture to selectively enhance the attractiveness of what they considered beautiful. They offered to eliminate ugliness while depicting as ugly everything they wished to eliminate. Their media efforts reveal the internal tensions and circu-

---

23 For Karl Pearson's insistence that his diagnosis of Jewish inferiority was based on "the cold light of statistical inquiry" not "prejudice," see his paper in the 1925 inaugural issue of *Annals of Eugenics*, quoted in Gould 1983, 296–98. See also Wiggam 1924, 272–79.

lar logic that characterized their constructions of the "fit" and the "defective."

These contradictions were heightened by the ease with which viewers found alternative meanings in such propaganda.[24] For example, many reviewers of *The Black Stork* reported that the film's exhibition of deformed bodies evoked pity, or even fascination, rather than disgust. Disregarding the titles that labelled Claude and Anne's son an "abyss of abnormality" filled with "criminal desires," critics consistently praised the actor for making this character appealing, even noble.[25]

Others found that scenes intended to make disabled people look repulsive instead made the film itself "repellent" and "revolting." These critics often praised the film's educational and social value, but they found it esthetically unacceptable: "grim," "depressing and unpleasant,"[26] "repulsive."[27] Louella Parsons complained that it was "neither a pretty nor a pleasant picture," because "it shows poor, misshapen bodies of miserable little children" (Parsons 1917).[28] Critics concluded that anyone who wanted to see such films must be sick, suffering from a "morbid" perversion of the esthetic senses. These reviewers shared the eugenic desire to pathologize ugliness, but feared that displaying ugly diseases would only create diseased audiences.[29]

---

24 While the responses of ordinary viewers are hard to document, movie reviewers and film censors demonstrate that such "unauthorized" interpretations were common.

25 *Chicago Tribune* (2 April 1917): 18; *New York Dramatic Mirror* (17 February, 1917): 32; *Wid's Film Daily* (5 April 1917): 220; *Motion Picture News*, (24 February, 1917): 1256.

26 *Exhibitors' Trade Review* (24 February, 1917): 836; *Motography* (24 February 1917): 424. For similar mixed reviews see *New York Dramatic Mirror*, (17 February 1917): 32; *Motion Picture News* (24 February 1917): 1256.

27 *Wid's* (5 April, 1917): 220–21.

28 Rival critic Kitty Kelly called it the "most repellent picture" she had ever seen (*Chicago Examiner* [4 April 1917]: 8). The *Chicago Tribune* admitted the "ideas may be all right," but found the film "as pleasant to look at as a running sore." Pursuing such clinical metaphors to the limit, *Photoplay* called the screenplay "so slimy that it reminds us of nothing save the residue of a capital operation." *Chicago Tribune* (2 April 1917): 18; *Photoplay* (12 June 1917): 155.

29 The National Board of Review of Motion Pictures' advisors repeatedly used such language to describe the film's audiences as sick. See: Andrew Edson of New York City's Education Department, 17 November; U.G. Manning, 18 November; Jonathan Dean, 18 November; Ernest Batchelder, 22 November; Maude Levy, 20 November; W. L. Percy, 21 November; and Robbins Gilman, 23 November; all

These unintended esthetic responses were one important reason films about eugenics were often banned.[30] Film censors went far beyond policing sexual morality, to include what I term "esthetic censorship," much of which was aimed at eliminating unpleasant medical topics from entertainment films.[31] From the perspective of such esthetic censors, both pro- and anti-eugenics films were unacceptably ugly. The powerful New York state film board banned both *The Black Stork* and *Tomorrow's Children*, a 1934 anti-eugenic melodrama, because eugenics was too "disgusting" a topic. They rejected *The Black Stork* in April 1923, not only for its "inhuman" position on euthanasia, but its "most unpleasant," "very distressing," "most revolting" depictions of disease. About *Tomorrow's Children* they concluded, "The sterilization of human beings is not a decent subject for public entertainment." The attempt to make the disabled look ugly made eugenics seem repulsive as well.[32]

(contd.)
1916 and all in National Board of Review of Motion Pictures Records, Controversial Film Correspondence, Rare Books and Manuscripts Division, The New York Public Library, Box 103.

30 *Chicago News* (6 April 1917): 21.

31 The year it first rejected *The Black Stork*, the influential early Pennsylvania state film censor board adopted a list of esthetic offenses that included any films about eugenics and a range of other medical topics (Pennsylvania 1918,15-17). The first Production Code of the Motion Picture Producers and Directors of America (1930), which synthesized this and similar state lists of forbidden topics, labeled "surgical operations" a "repellent subject," and included a catch-all restriction on all other "disgusting, unpleasant, though not necessarily evil, subjects," that was used to eliminate most other graphic or unpleasant depictions of medical issues (Jowett 1976, chaps. FIve, Seven, Ten). The code of 1930 is reprinted in ibid., 468–72. On pre-code films and the rise of censorship see also Couvares 1992; Vaughn 1990; De Grazia and Newman 1982. For esthetic censorship of Tod Browning's *Freaks*, see *New York Times* (9 July 1932): 7 (Bogdan 1988; Fiedler 1978).

32 For censorship history of *The Black Stork*, see National Board of Review of Motion Pictures Records, Controversial Film Correspondence, Rare Books and Manuscripts Division, The New York Public Library, Box 103; Motion Picture Division Scripts, New York State Archives, Albany NY, Box 2565 Folder 383 and 12, 421. Quotations are from letter of disapproval, Commissioner to H. J. Brooks, 4 April 1923, Motion Picture Division Scripts, New York State Archives, Albany NY, Box 2565, Folder 383. In a private straw poll of community leaders from across the country conducted by a film industry voluntary rating agency, the National Board of Review of Motion Pictures, 9 of the 52 respondents explicitly cited esthetic objections as a major reason for not approving *The Black Stork*, National Board of Review of Motion Pictures Records, Controversial Film Correspondence, Rare Books and Manuscripts Division, The New York Public Library, Box 103. On

WHAT IS HEREDITY?: EUGENICS AND MORAL RESPONSIBILITY

Between the 1880s and 1910s, scientific concepts of heredity changed dramatically. The nineteenth century "Lamarckian" view that environmentally-caused changes in individuals could be passed on to their offspring was gradually supplanted by August Weismann's doctrine that individual heredity was permanent and unaffected by environmental forces.[33]

Yet in mass culture, the terms "heredity" and "eugenics" continued to be applied to traits that most scientists now attributed to environmental causes, from infections like tuberculosis and syphilis, to bad prenatal care and malnutrition. For example, the only movie made by the "Eugenic Film Company," the 1917 film *Birth*, never mentioned genetics but simply provided detailed pregnancy and child care advice.[34] A 1917 feature *Parentage* was advertised as strongly "eugenic." It contrasted the families raised by good and by bad parents,

---

(contd.)
Tomorrow's Children see "Memo on Behalf of the Motion Picture Division to the Commissioner of Education," 4; and "Court of Appeals Brief for the Respondent," 6, both in Motion Picture Division Scripts, New York State Archives, Albany NY, Box 333, Folder 28, 361. The censors initially also declared the film "immoral" for showing audiences "methods that . . . prevent conception" but this argument was soon dropped, see letter of Irving Esmond, 24 August 1934, in Box 296, Folder 27, 387. For the role of VD and sex education films in the growth of film censorship, see De Grazia and Newman 1982; Kuhn 1988.

33 For the importance of this change see Degler 1991, part One. The new definition meant individual heredity was unchangeable. For the first time, science viewed heredity and environment as distinct and exclusive categories, and selective reproduction now became the only mechanism for changing the future genetic composition of the population.

34 A scene script of *Birth* is in the Library of Congress Motion Picture, Broadcasting, and Recorded Sound Division, Washington D.C., copyright records #MU-835; quotes are from *Wid's Film Daily* (19 April 1917): 244–45. See also *Moving Picture World* (28 April 1917): 609; *Motion Picture News* (28 April 1917): 2687; *New York American* (8 April 1917): 7M, and (15 April 1917): 4M; *Motography* (28 April 1917): 915; *Detroit News* (29 April 1917): 6; and idem. (6 May 1917): 5. See also *New York Evening Journal* (9 April 1917): 8, and ads (11–14 April 1917): movie page; American Film Institute 1989, 69–70. See also the Children's Bureau film *Well Born, Child Health Magazine* (December 1923): 571–72 and (September 1924): 407; *Educational Screen*, (February 1924): 80; *American Journal of Public Health* 14 (1924): 276; *Bulletin of the National Tuberculosis Association* (January 1924): 4. Other Children's Bureau films of the 1920s are available at the National Archives, but no surviving copies of this one are known.

explicitly conflating the effects of genetics and environment.[35]

Historians of eugenics usually attribute such examples to mass culture's scientific illiteracy (Haller 1963, 141–43; Kevles 1985, 100). But in doing so they miss a complex interplay among scientists, popularizers, and the public in defining "heredity." First, many scientists argued that Weismann's theory left room for environmental contributions to genetic disease; such views were neither unscientific nor limited to mass culture.[36]

Second, and most significantly, in mass culture and occasionally in the scientific literature, the term "hereditary" was not limited to genetics, but meant that you "got it from your parents," regardless of whether "it" was transmitted by genes, germs, precepts or probate. Thus *The Science of Life* defined a man's heredity as "what he receives from his ancestors."[37]

Such definitions were not based on faulty science, but on a different set of concerns. On this view, what defined "heredity" was the parents' moral responsibility for causing the trait, not the technical mechanism through which parental causation was transmitted. By this definition of "heredity," "eugenics" meant not just having good genes but being a good parent.[38]

---

35 Quotes from *Wid's* (14 June 1917): 369–71. For promotion see *Motion Picture News,* June through August 1917, passim.

36 What were called environmental "germ poisons" were widely believed to cause inheritable mutations. Conversely, genetic factors might determine who was most susceptible to environmental damage. And some reputable scientists remained "Lamarckians" into the 1930s. On germ poisons see: *Popular Science Monthly* 88 (1916): 84–85; Brandt 1987, 14–15; Saleeby 1923, 309; Sadley 1923, 346–47. Anti-Lamarckian prohibitionist John Harvey Kellogg gave extensive publicity to alcohol as a germ poison: *Good Health Magazine* 51 (1916): 75–76; 52 (1917), 502–503; 54 (1919), 164, 219–24, 273–78, 717. For a Lamarckian view of this research see Warthin 1930, 57–58; Sournia 1990, chp. 7; Crowe 1985; Warner and Rosett 1975. On environmental susceptibility see: Dubos 1953, 28–43, 125–28; Proctor 1988, 215–17. For persistent Lamarckian beliefs in the 1930s see Warthin 1930. Thus eminent scientists like psychologist G. Stanley Hall considered germ-fighting part of eugenics. His plan for a "Department of Eugenics" specifically included infectious diseases and milk inspection among its responsibilities (Ross 1972, 362–63, 413). I thank Alice Smuts for this reference. Similar examples include: Fisher and Fisk 1917, 293–94; *Good Health Magazine* 54 (1919): 658; *Social Hygiene Bulletin* 2 (January 1916): 3; and 3 (November 1916): 4.

37 This meaning is the subject of the old joke that "insanity is hereditary—you get it from your children." See also *Oxford English Dictionary* "Heredity."

38 Many motion pictures explicitly defined "eugenics" as meaning "fit to marry." *The Black Stork* itself was retitled *Are You Fit to Marry?* when it was

Mass culture hardly ignored Weismann's science. At least a dozen films of the 1910s dichotomized heredity and environment, usually by portraying the life of a child whose foster parents differed radically from the biologic parents. Most such films concluded that key human differences were caused by environment, not heredity (at least for girls), but almost all presumed Weismann's radical disjuncture between nurture and nature.[39] Thus mass culture clearly reflected scientific concepts of heredity.

But in turn, leading scientists sometimes used "heredity" to mean "parental responsibility." The eminent British statistical geneticist R. A. Fisher argued that syphilis was a eugenic concern because it ran in families. "There may be something very much like inheritance [of syphilis], in the practical sense. Whether there is inheritance in the biological sense is not the only matter. We are anxious to make a more perfect mankind and we are interested in the practical side" (Fisher 1923, 318, 464–65).

Even eugenic leaders who limited "heredity" to biological inheritance defined their movement as seeking good parenting, not simply good genes. New Jersey physician Theodore Robie declared at the Third International Eugenics Congress in 1932 that "it would . . . be

---

(contd.)
rereleased in 1918–19 and in 1927. On the link between causality and morality see Haskell 1977, esp. chap. Eleven; and Tesh 1988.

39 Environment wins out in: *Are They Born or Made?* (Warner 1917); *A Daughter's Strange Inheritance* (Broadway Star-Vitagraph 1917); *A Victim of Heredity* (Kalem 1913); *The Power of Mind* (Mutual 1916); *A Disciple of Nietzsche* (Thanhouser 1915); and *The Red Circle* (Pathe 1915). Heredity wins in: *Heredity* (Biograph 1912); *Heredity* (Broadway Star 1915); *Inherited Sin* (Universal 1915); *The Power of Heredity* (Rex 1913); and *The Second Generation* (Pathe 1914). One of the first commercial melodramas to be billed as an explicitly "eugenic" film was D. W. Griffith's *The Escape* (Reliance-Majestic 1914). Following a prologue by Dr. Daniel Carson Goodman that called for breeding humans as carefully as livestock, it traces in gruesome detail the awful consequences of human "mis-mating." The fictional "Joyce" family compresses into two generations all the defects and deviance found among two centuries of Jukes and Kallikaks. The film's conclusion may have been based on Lamarckian concepts of heredity (critics disagreed on this point. See *Moving Picture World* [13 June 1914]: 1515 versus *New York Dramatic Mirror* [10 June 1914]: 42). The physician-hero cures a lunatic strangler surgically, and redeems the lunatic's prostitute sister by marrying her. But whatever definition of heredity it was using, what made the film "eugenic" was its dramatization of the effects of bad parenting (*American Film Institute* 1989, 244; *Variety* [5 June 1914]: 19; *New York Times* [2 June 1914]: 11). See Everson 1978, 76; Connelly 1986, 74.

204 Martin S. Pernick

conducive to racial improvement to sterilize even those feeble minded who do not necessarily fall in the hereditary group," since *"mental defectives tend to maintain inferior homes in inferior environments, and they quite generally rear their children in an inferior manner"* (Robie 1934, 202).[40] Whether or not all traits caused by parents were labelled "hereditary," any trait caused by parents was part of "eugenics." Identifying the parents as the cause made the parents morally as well as medically responsible.[41]

ELIMINATING THE UNFIT: EUTHANASIA AS EUGENIC METHOD

A few prominent scientists like German zoologist Ernst Haeckel favored death for the unfit as early as 1868 (Haeckel [1868] 1876, I, 170–71; 1905, 21, 114–20; van der Sluis 1979, Gasman 1971, 91).[42] But prior to Dr. Haiselden's cases such ideas rarely won public en-

---

40 For similar views of the American Eugenics Society in 1935 see Huntington 1935, 41–42, quoted in Mehler 1987, 16.

41 These concerns retain their influence today, even among those leading the return to supposedly-biological explanations. Thus Richard Herrnstein and Charles Murray assert in *The Bell Curve*, "If women with low scores are reproducing more rapidly than women with high scores, the distribution of scores will, other things equal, decline, *no matter whether the women with the low scores came by them through nature or nurture.*" [emphasis added]. Quoted in Browne 1994, 3.

42 On Haeckel's follower Ploetz, see Trombley 1988, 71. Other eugenic professionals who advocated such views prior to 1910 included Hungarian welfare expert Sigmund Engel, British physicians Robert Rentoul and Charles E. Goddard, Chicago dentist Eugene Talbot, Yale Law Professor Simeon Baldwin, physicians William D. McKim and Edward Wallace Lee, Chicago surgeon G. Frank Lydston, psychologist G. Stanley Hall, and psychiatrist Walter Kempster. The first effort to legislate such proposals was introduced in Michigan in 1903, and similar bills were debated in Iowa and Ohio in 1906 (Engel 1912). For discussion of Goddard, see Russell 1975, 59. For Rentoul, see Trombley 1988,19; Baldwin 1899, quoted in Cohen 1985, 87; and Talbot, [1898]), 3–4. For other U.S. proposals to kill the mentally-retarded as early as 1883, see Hollander 1989; McKim 1900, 188–92; Hall, 1910; Ross 1972, 318–19; Reilly 1991, 37–38; for Lydston, see Kuepper 1981, 65; for Lee see *New York Medical Journal* (26 December 1914): 1251. On the Michigan plan, see Curtis 1983, 69–70; *Chicago Examiner* (22 May 1903); *Detroit News* (22 May,1903). On Rodgers see *Michigan State Gazeteer* (Detroit: R. L. Polk, 1903); *Detroit News* (22 May 1903): 3. For Kempster, *New York Times* (26 January 1906); for Dr. R. H. Gregory, see van der Sluis (1979, 135). Infanticide for reasons including elimination of sickly infants had been practiced in the ancient world and in many non-Western cultures.

dorsement from eugenic leaders. Most advocated selective breeding, not the death of those already born with defects. Charles Davenport, perhaps the foremost American eugenic researcher of the period, insisted in 1911 that eugenics did "not imply the destruction of the unfit either before or after birth" (Davenport 1911, 4; in Kuepper 1981, 62). Irving Fisher echoed Karl Pearson's "fundamental doctrine . . . that everyone, being born, has the right to live," but not the right "to reproduce his kind."[43]

Yet when Haiselden moved the issue from theory to practice, these same leaders proclaimed him a eugenic pioneer. Fisher now wrote to "emphatically approve" Haiselden's action. "I hope the time may come when it will be a commonplace that. . .defective babies be allowed to die." Davenport likewise now urged doctors not to "unduly restrict the operation of what is one of Nature's greatest racial blessings—death." If medicine prevented the death of defectives "it may conceivably destroy the race."[44]

Haiselden's attention-grabbing actions were a calculated effort to radicalize the eugenic leadership, a strategy anarchists of the time popularized as "propaganda of the deed." "Eugenics had a million theories . . . . But it lacked drive," he explained. "[T]he times were crying for some one central deed—some decisive action that would draw together all these theories . . . into one definite crusade" (Haiselden 1915b, 16).[45] Haiselden was only one doctor, but by gaining extensive media coverage of his dramatic acts, he was able to reshape the leadership's definition of eugenic methods.

Ironically, while mass culture provided a key battleground on which competing groups struggled to define eugenics, many powerful figures in the film and journalism industries opposed public involvement in making these life-and-death decisions. *The New York Times* demanded that the power to selectively withhold treatment from impaired infants be "kept strictly within professional circles," free from "unenlightened sentimentality" (*NYT* [13 November 1917]: 12). One common suggestion was to create special medical committees, what

43 Fisher in National Conference on Race Betterment 1914, 472, 475. See also ibid., 477, 500–501; 1915, 89–90, addenda slip for p. 61.
44 Fisher and Davenport, quoted in *Independent* (3 January 1916): 23. The same article also contained an endorsement from Raymond Pearl.
45 See Woodcock 1962, 328, 336, 462. Parallels with the approach of Dr. Jack Kevorkian today are developed in Pernick 1996, 170–71.

Helen Keller called "physicians' juries for defective babies" (Keller 1915a, 173–74). Mass culture thus played an important role in promoting the expansion of professional power.[46]

This media deference to professional expertise in turn contributed to the rapid decline in coverage of Haiselden's crusade and its erasure from public memory. Even people who demanded the death of the unfit opposed publicizing the issue. "I think all monstrosities should be permitted to die," wrote university president Frank H. H. Roberts, "but I do condemn the physician for making such a public ado about the matter" (*Independent*, 3 January 1916: 26). In an editorial entitled "He Forgets Silence is Golden," the *New York Times* endorsed Haiselden's right to let infants die, but denounced his use of media publicity. "If he is wise, as most doctors are, he settles the question for himself . . . and the incident does not become a subject of public discussion" (*NYT* 16 July 1917: 10).[47]

---

46 See *Chicago Herald* (20 November 1915): 2, and (21 November 1915): 3; *New York Call* (29 November, 1915): 3. A Los Angeles proposal was reported as early as November 20, see *Chicago American* (24 November 1915): 3. For Haiselden's use of consultants to confirm his non-treatment decisions, see *Chicago American* (22 December 1915): magazine page; *Chicago Examiner* (24 July 1917); *New York Times* (16 November 1917): 4; *Medical Review of Reviews* 23 (1917): 607; *Chicago American* (16 November 1917): 3. For modern parallels see Mahowald 1988. Occupational health pioneer Dr. Alice Hamilton noted the irony that mass culture demanded expanding the power of the profession. "Curiously enough it is not the medical profession which is seeking an extension of its rights; it is the laity which is trying to force upon physicians a power over life and death which they themselves shrink from" (Hamilton 1915, 266). But popular support for giving doctors this particular power itself depended on a broader progressive-era faith in the methods of science, a faith which was actively promoted by medical and eugenic leaders.

47 Although the *Times* shifted its position on non-treatment, the editors consistently maintained that in practice, the "wise" physician should make such decisions silently (*New York Times* [18 November 1915]: 8; [22 November 1915]: 14; [29 November 1915]: pt. II, 10.) Columbia University sociology chairman Franklin H. Giddings applauded the death of "molasses-minded" mental defectives, but felt it was a "question that should be considered soberly, thoughtfully and by rigorous intellectual processes. To put it up to the general public in all the emotional and imaginative setting of a photo-play is, in my judgment, an utterly wrong thing to do." A series of legal investigations upheld Haiselden's refusal to treat impaired newborns, but he was expelled from the Chicago Medical Society for publicizing his actions. *New York Times* (18 November 1915): 4; and Giddings to W. D. McGuire, 20 November 1916, National Board of Review of Motion Pictures Records, Controversial Film Correspondence, Rare Books and Manuscripts Division, The New York Public Library, Box 103.

This growing support for professional power and secrecy, combined with the growth of esthetic objections to eugenic subjects, drastically curtailed media coverage of Haiselden's activities. By 1918, Haiselden's last reported euthanasia case received only a single column-inch buried deep inside the *Chicago Tribune*, a paper that had supported him editorially and given front-page coverage to all his previous cases. Mass media preoccupation with novelty and impatience with complex issues clearly played a role in this change, as did the altered agenda of war-time and post-war politics. But in part, the disappearance of public debate appears to reflect a deliberate decision that the topic itself was unfit to discuss in public.[48]

As soon as the media attention flagged, eugenic leaders resumed their prior assertions that they opposed euthanasia, as if Haiselden had never existed. Irving Fisher started to distance himself as early as 1917, reiterating his earlier view that "eugenics does not require the old Spartan practise of infanticide," while simply ignoring his recent accolades for Haiselden (Fisher and Fisk 1917, 294).[49]

Thus, by 1930, Haiselden and his cases were not simply forgotten but intentionally erased from the history of eugenics. Both the immediate success and the eventual erasure of his efforts were the sometimes ironic product of the struggle to shape how mass culture portrayed the meanings of eugenics.

CONCLUSIONS

Mass culture was a battleground on which elite and other concepts of eugenics competed and interacted. Esthetic and moral values played a key role in eugenic constructions of the hereditarily "fit" and the "defective," and these values were products of complex struggles to impose meaning on mass culture. Eugenic popularizers tried to use mass culture to promote their complex and circular esthetic vision, but their efforts often had unintended consequences. The scientists' focus on the genetic mechanisms of heredity coexisted in uneasy ten-

---

48 *Chicago Tribune* (28 January 1918): 12. See Pernick 1996, chaps.Six, Nine.

49 Even Dr. William J. Robinson, one of Haiselden's most vigorous supporters in 1915–16, wrote in 1917 that "no eugenic considerations will induce us to adopt Spartan-like methods and to neglect or kill off the weak and puny . . . . Every child that is born . . . is entitled to the very best of care" (Robinson 1917, 138). See also 73–76. For similar disavowals see Wiggam 1924, 283; Paul 1917, 142.

208 Martin S. Pernick

sion with a broader, explicitly moral language in which hereditary causation meant parental responsibility. And while mass culture revealed wide support for empowering doctors to let the unfit die, it provoked even stronger opposition to talking about it in public.

Finally, by making these points I do *not* mean to imply that eugenics was either *uniquely* value-laden or peculiarly influenced by mass culture. Nor am I claiming that mass culture corrupted "pure" genetic science by infecting it with "extraneous" esthetic and moral concerns. Rather, I believe the history of eugenics is valuable because it makes so dramatically visible the cultural values that are inevitably part of defining any human difference as a disease or a disability, and identifying any specific factors as "the" cause (Steinfels 1973; Rosenberg 1992; Engelhardt 1975; Gilman 1985).[50]

The role of esthetic judgments in the definition of "disease" and "disability" is still a highly controversial issue today: should laws protecting the disabled against discrimination apply to those who are simply judged unattractive; should health insurance cover "cosmetic" surgery? (T.R.B. 1987, 4). Do esthetic values create disability, in the same way that high stairs and other physical barriers do? Could changing such values create a more accessible culture? I believe the history of eugenics shows the futility of trying to draw a sharp line between "objective" physical diseases and "subjective" values. Any time a culture defines disease or its causation, it is making a partly-subjective value-based judgment. Greater awareness of the inevitability of the value-based component of these debates might or might not help reach more satisfactory decisions. But pretending that such decisions can ever be made without values only de-legitimates and prevents the necessary critical analysis of the implicit values at stake.

---

50 Pioneering work on these issues was done by Veatch (1976) and Kuhn (1962). Early eugenicists did not claim to have "value-free" definitions of disease and health, but did claim that their values had been proven objectively true by value-free scientific methods.

8

What's Morally Wrong with Eugenics?

Arthur L. Caplan

The topic of eugenics cannot be discussed for long without encountering the Holocaust. This is as it should be. When contemporary geneticists, genetics counselors and clinical geneticists wonder, as they sometimes do, why it is that genetics receives special attention from those concerned with ethics, the answer is simple—history.

The events which led to the sterilization, torture and murder of millions of Jews, Gypsies, Slavs, and children of mixed racial heritage in the years just before and during the era of the Third Reich in Germany were rooted firmly in the science of genetics (Muller-Hill 1988). Rooted not in fringe, lunatic science but, in the mainstream of reputable genetics in what was indisputably the most advanced scientific and technological society of its day, the pursuit of genetic purity led directly to Dachau, Treblinka, Ravensbruck and Auschwitz.

As early as 1931 influential geneticists such as Fritz Lenz, Fritz were referring to National Socialism as "applied biology" in their textbooks (Caplan 1992). As difficult as it is for many contemporary scientists to accept (ibid.; Kater 1992), mainstream science provided a good deal of enthusiastic scientific support for the virulent racism that fueled the killing machine of the Third Reich.

When the Nazis came to power they were obsessed with securing the racial purity of the German people. The medical and biomedical communities in Germany not only endorsed this concern with "negative eugenics," they had fostered it. Race hygiene swept through German biology, public health, medicine and anthropology in the

1920s and 1930s, long before the Nazis came to power (Weiss 1987; Muller-Hill 1988; Proctor 1988; Kater 1992). Many in the medical profession urged the Nazi leadership to undertake social policies that might lead to enhancing or increasing the genetic fitness of the German people (Kater 1992).

Eugenics consumed the German medical, biological and social scientific communities in the decade before World War II. Many physicians and scientists were frantic about threats they saw to the genetic health of the nation posed by the presence of inferior populations such as Jews, Gypsies, Slavs, and to a lesser extent because the threat was more distant, African peoples (Adams 1990). The steps they took to protect against the public health disaster of a 'polluted' racial stock were so awful, so immoral and so heinous that they have, rightly, shaped all subsequent discussion of the ethics of both human genetics and eugenics.

NEGATIVE VS. POSITIVE EUGENICS

Steps to eliminate unfit or undesirable genes by prohibitions on sexual relations, restrictions on marriage, sterilization or killing, are all forms of negative population eugenics (Kevles 1985). Nazi judges and scientists ordered children killed or sterilized who had parents of different racial backgrounds or were thought to have genetic predispositions toward mental illness, alcoholism, retardation or other disabilities. This was done to remove the threat such children posed to the genetic stock of the nation and to avoid having to pay the costs associated with institutionalization and hospitalization (Caplan 1992). Laws were enacted prohibiting marriages between those whom Nazi race hygiene theory held were likely to produce degenerate offspring.

Conversely, on a smaller scale, the Nazis tried to encourage those who satisfied Nazi racial ideals to have more children. The most extreme form of encouraging eugenic mating was the *Lebensborn* program: National Socialist program, which gave money, medals, housing, and other rewards to persuade "ideal" mothers and fathers to have large numbers of children in order to create a super-race of Aryan children (Proctor 1988). The provision of rewards, incentives and benefits to encourage the increased representation of certain genes in the gene pool of future generations constitutes positive population eugenics (Kevles 1985).

Nazi race hygiene theories were false. There is no evidence to support the biological views of the inherent inferiority of races or the biological superiority of specific ethnic groups which underlay the eugenics efforts of the Third Reich. There is not even any firm basis for differentiating groups into races on the basis of genetics (see various selections in Part Two of Harding 1993). The negative eugenics programs race hygiene spawned were not only patently unethical, since they were completely involuntary and coercive, they were also based upon assumptions about genes and race that are not true. The Nazi drive to design future generations based on what can now be understood as invalid science skewed by racism led to concentration camps, forced sterilization, infanticide and genocide.

THE LEGACY OF GERMAN EUGENICS

The rapid evolution of clinical genetics in the post-World War II era has been accompanied by a strong moral commitment to the autonomy of the patient. Those who work in clinical genetics are resolute in their belief that the purpose of their work is not to tell people what reproductive choices to make but to supply them with information which will empower them to make more informed decisions (Bartels, LeRoy, and Caplan 1993; Bartels, Leroy, McCarthy, and Caplan 1997). Whether or not value neutrality really does characterize the practice of counseling (ibid.), the centrality of autonomy in the normative ethos of professionals practicing clinical genetics is a direct response to the coercive horrors of Nazi Germany and the abuses carried out in the name of eugenics and public health in the United States in the first half of the twentieth century (Reilly 1991; Kevles 1985; Haller 1993; Caplan 1993b; Pernick this volume).

In recent years the desire to maintain a distance between the events in Germany fifty years ago and today's effort to map the human genome and apply the knowledge gained to human beings has added a new normative twist. Most of those prominently involved now in actually mapping the human genome and in attempting early forms of gene therapy are adamant in stating that they have no interest in modifying the human germline. By forswearing any interest in germline modification, those involved in the mapping and sequencing of the human genome can more easily deflect the kinds of moral concerns that would otherwise be directed toward the project as a result of the tragic history of Germany's involvement with eugenics (Anderson 1989; 1992; Garver 1991; Munson and Davis 1992; Duster 1990; Lewontin 1992; Danks 1994).

However, whether or not particular scientists or clinicians are serious or merely being prudent in publicly forswearing any interest in germline eugenics, the fact is that there is tremendous interest in American society and in other nations in using genetic information for eugenic purposes (Klass 1989; Munson and Davis 1992; Herrnstein and Murray 1994; Bobrow 1995; Kitcher 1996; Kitcher this volume).

Raising the issue of the application of eugenic knowledge to human reproduction is sometimes dismissed as histrionic moral grandstanding. Why worry about this issue when it is not now possible and is unlikely to be possible for many years to come? But the legitimacy of worrying about the impossible seems a bit easier to defend when squared against the pace of recent developments in genetics and genetic engineering.

Recent advances in the understanding of spermatogenesis as well as in the fields of animal cloning and assisted reproductive technology point toward methods that would permit the systematic alteration of genetic information in reproductive cells (Brinster and Zimmerman 1994). Eugenic goals could also be advanced through the use of embryo biopsy and the selective elimination of embryos or the selection of sperm or embryos known to be endowed with certain traits (Caplan 1995; 1998a; 1998b).

Some within the disabilities community have noted, that the use of genetic information which results in the prevention of the birth of children with certain traits or behavioral dispositions can be construed as a form of eugenics. Large-scale screening programs to prevent the transmission of congenital diseases do exist. The state of California has for many years encouraged women to be tested during their pregnancies for fetal neural tube defects. The success of the program is evaluated not in terms of information given to mothers but rather in terms of the number of children with handicaps who were not born. When number of births prevented is the measure of a public health intervention it is hard to say that anything other than negative eugenics fuels support for such programs (Duster 1990; Murphy and Lappe 1994).

Ethical debates about eugenics must acknowledge the horrors perpetrated in the name of eugenics in this century. But, despite the evil that has been done in the name of eugenics, the debate cannot end there. The moral permissibility of eugenic goals must be addressed on its own terms. For while arguments based upon history are instructive and important, those who see no analogies between our times and earlier times are unlikely to find warnings about the past sufficiently forceful to shape future behavior or public policy (Caplan 1992; 1994). And while the fear of the imposition of eugenic programs by a totalitarian regime must be taken seriously, it is not the only path eugenics might follow.

INDIVIDUAL VS. POPULATION EUGENICS

Improvement of the genetic makeup of a population can be sought through negative or positive eugenics. What is less widely noted is that either strategy can be pursued at the level of individuals and their direct, lineal offspring or for large groups or populations. Efforts aimed at improving or enhancing the properties of large-scale populations such as by providing incentives for large numbers of individuals with particular traits or abilities to marry and have many children or encouraging public health testing for neural tube defects constitute versions of population eugenics. The goal of such activities is to shift the makeup of the gene pool of future generations in particular directions.

Positive and negative eugenics can also be carried out by individual couples who are not interested in nor motivated by the overall effect of their actions may have on the societal gene pool. Activities intended to permit individuals to endow their children and their subsequent offspring with desired traits are instances of individual eugenics. A decision to try to implant a bit of DNA associated with a desirable trait into an egg so as to have the trait present in one's child is an instance of individual eugenics. So would a decision to clone oneself for simple reasons of vanity and self-perpetuation. So too may be a decision to abort a pregnancy when a fetus is found to have cystic fibrosis or spina bifida.

Attempting to choose the genetic makeup of one's offspring with an eye toward creating a tall child is to engage in individual eugenics.

214 Arthur L. Caplan

A government program with the goal of creating large numbers of tall people is an example of population eugenics.

Individual and population eugenics are conceptually distinct, but procreative decisions can be motivated by both concerns at the same time. Those who want to pursue population eugenics may be able to do so either by encouraging those interested in individual genetic enhancement to pursue their individual goals, or by efforts to control or change the overall reproductive behaviors of large numbers of people. If a dictator wants to create a future population with a higher IQ, programs might be created to discourage people with low IQs from marrying (Caplan 1995). The Chinese government passed a law to this effect in 1995 (Bobrow 1995). The same goal can also be reached by discouraging people perceived as having poor genetic endowments from reproducing by changing welfare, housing and educational policies (Herrnstein and Murray 1994). Encouraging individual couples to use genetic testing or embryo biopsy and then embryo selection in order to help insure that their children have high IQs is still another road to the same destination (Caplan 1998a). If enough families pursue individual eugenic goals, a population eugenic goal may result.

In China, policies which restrict individual couples to having a single child combined with social attitudes that prevail among many Chinese favoring boys over girls have led many couples to take steps that will insure that their child is a boy. While the Chinese government has not instituted any policies that explicitly favor the creation of male rather than female children, the combination of choices made by individual couples appears to have been powerful enough to produce a shift in the overall gender composition of the Chinese population toward more males.

Population eugenics need not be coercive, but, historically, it almost always has been. A great deal of social pressure was applied in the German *Lebensborn* programs of the 1940s. More recent efforts to shift the genetic norms of populations exemplified by the attempt to encourage those with the 'right' racial makeup to reproduce, as is evident in the ethnically selective pronatalist policies espoused by governments in many parts of the world, are less obviously coercive but still involve a great deal of cultural and societal pressure. The stated policies of some religious bodies, such as certain Orthodox Jewish sects or some priests of the Greek Orthodox church, that they will not bless marriages where no genetic testing for diseases has been done, constitute examples of possible coercion for population eugenic goals by non-governmental powers.

Instances of individual eugenics are harder to find. But, decisions by families to try and have boys rather than girls may represent examples of individual eugenic thinking (Caplan 1995). And there are many instances in which persons try to avoid having a child with what they perceive as a burdensome defect or disease by using genetic testing.

INDIVIDUAL EUGENICS, THE WAVE OF THE FUTURE?

It will not be long before science makes it easier to put eugenic aspirations for populations or individuals into practice. The manipulation of gamete production to identify and eliminate unwanted traits and the ability to accurately insert genetic information directly into sperm or embryo are the subjects of investigation around the world. Embryo biopsy wherein cells are removed from an embryo, cultured and their DNA content analyzed for various propensities are all ready being touted by a number of infertility clinics. When these techniques are identified and refined, the chance to pick the biological endowments of our offspring, to give them the "best" possible start in life, will have enormous appeal to many.

The hope of using science and medicine to create children who get the best possible start in their lives is very different from the forced use of medical and scientific knowledge to solve society's perceived ills by creating biologically superior populations or simply killing those deemed inferior. There are those who would agree with the population eugenic goals espoused by the founder of the so-called "Nobel Prize" sperm bank, Robert Graham, the Director of the Repository for Germinal Choice in Escondido, California, that we owe it to future generations to try and maximize the genetic endowment of at least some of its yet-to-be-born members by carefully selecting which genes we pass along from us to them (Caplan 1995). But, the real impact of new techniques, such as the transplantation of sperm stem cells, embryo biopsy and genetic testing of sperm and eggs, is likely to be seen in the conduct of individual parents seeking to fulfill their aspirations and dreams for their children.

Genetic enhancement in the future is much more likely to be the product of the norm that good parents make sure that their children have the best chance possible to succeed in life, than it is the imposition of a governmental mandate that all must procreate with the goal of enhancing the presence in the gene pool of persons with a particular biological phenotype. In Western societies with their strong normative commitments to autonomy and privacy, individual eugenics will have a rosier future than the harsh and intrusive steps required to implement and sustain public policies aimed at population eugenic goals.

The day when we need to decide whether it is wrong to choose the genetic makeup of our children is not very far off. Some argue that we lack the wisdom to choose well (Lewontin 1992). But, that hardly stops parents today from seeking to better the lot of their children through environmentally mediated efforts at enhancement. In a society that places so much emphasis on maximizing opportunities and achieving the most efficient use of resources, it is hard to believe that pressures will not quickly arise on prospective parents to use genetic information and techniques for manipulating genes to better the lot of their children or of future generations of children.

For some, the historical abuses committed in this century in the name of eugenics are sufficient grounds for prohibiting or banning any efforts at any form of eugenics, positive or negative; individual or group. However, negative population eugenics is not individual positive eugenics. If most people agree that parents have a right, if not a duty, to try and maximize the well-being and happiness of their offspring, then it is not likely that the record of historical abuses carried out in the name of negative population eugenics will hinder efforts to incorporate genetic information into procreative decisions about our children and their immediate descendants. As it stands today, most parents, particularly those in the middle and upper classes, would probably be more troubled by failing to use genetic information to try and improve the lot of their offspring than they would by doing so.

If that is so, then what values should influence our ideas about human normality, perfection and impairment? Can parents really be trusted to choose the characteristics of their children? Is the only reason that the eugenical aspirations of people such as Robert Graham are dismissed as kooky is that those who espouse them are holding out false hope because there is no guarantee artificial insemination using the sperm of men selected for desired traits will produce these traits in their offspring? Making false promises to people is certainly wrong, but surely there are more fundamental issues involved in the ethics of intentionally designing "better" babies than false advertising.

Should We Pick The Traits Of Our Children?

Suppose it becomes easier to achieve conception reliably outside the womb, making the analysis of the genetic makeup of embryos a simple task. Would there be any reason not to allow prospective mothers and fathers to select the biological endowments of their children? This question must be answered in light of the degree to which it is now possible for parents to obtain information about the genetic makeup of their unborn children in order to minimize the chance of having a child with a lethal or disabling medical problem (Kolker and Burke 1994).

What should medicine do in the face of the strong desire most parents feel to bring their child into this world healthy? Is there any criteria or definition that would allow the sorting of human traits into desirable and undesirable categories? Even if such a classification could be done, is it part of medicine's professional responsibility to allow parents to pick and choose among the desirable and undesirable characteristics they wish their child to have?

Health, Disease, Disability And The Aims Of Medicine

When most people think about health, they think about it as referring to the absence of illness or disease. If you are not sick, then you are healthy. But this way of thinking about health, about what is normal and abnormal, is confusing. It is possible to be free of disease and still not be seen by others as healthy. You can be out-of-shape, nervous, on-edge, lacking in self-confidence, have no stamina, feel completely awkward and unsure of yourself, and still, on the view that the absence of disease constitutes health, be considered healthy. But that seems to stretch the meaning of health to include a bit too much. Health is not simply the absence of disease and dysfunction (Caplan 1993a; 1998b).

Health means something more. Health refers to a state in which a person is flourishing, in which bodies and minds are working, not at adequate levels but, at optimal levels. Health makes essential reference to a concept of optimizing, not merely reaching, some level of minimal functioning.

If it is true that health refers to optimizing the functions of our bodies and minds then it is easy to see why health is such an elusive goal. Health is an ideal, not an average or a minimal threshold. When parents say they want a healthy child, they may well mean that they only want their children not to be sick, meaning to be disease free or able to function without serious impairments or disorders. But, they may really mean they want a healthy child, meaning they want their children to enjoy the best possible physical and mental functioning. They want their yet-to-be-born children to function at optimal levels. This is precisely the sort of wish, the pursuit of perfection, that leads some men and women to want to use the services of a facility such as the Repository for Germinal Choice.

The desire for health, in the full sense of the concept, is behind parental decisions to place their children in the elite nursery schools, expensive private high schools, tennis camps or special music or art classes. These institutions are valued not because they prevent disease or dysfunction. They are valued because many parents see them as the means to developing the best possible skills and abilities in their children. They are valued as a part of health, not as a means of avoiding disease.

Presume that those who do seek out eugenic services want the healthiest possible offspring. Is there anything morally wrong with pursuing perfection as the goal of reproduction?

Is There A Persuasive Argument Against Individual Eugenics?

Arguments that are commonly made against the morality of trying to design perfect children fall roughly into three categories; the unavoidable presence of force or compulsion, the imposition of a standard of perfection, or worries about inequity arising from eugenic choice. The first worry is not one that seems appropriate to individual eugenic choice. The latter two may not be especially telling against individual eugenic wishes either.

Coercion

Certainly it is morally objectionable for governments or institutions to compel or coerce the reproductive behavior of persons (Reilly 1991). The right to reproduce without interference from third parties is one of the fundamental freedoms recognized by international law and moral theories from a host of ethical traditions. It is also morally wrong to allow the state to impose its vision of the future by force. However, the goal of obtaining perfection or pursuing health with respect to individual eugenics is not made objectionable by these arguments. What is morally wrong is coercion, compulsion or the use of force with respect to reproductive decisions.

The Repository for Germinal Choice and genetics counseling programs at academic medical centers and private clinics make a special point of avoiding any hint of coercion or compulsion in their activities. It would seem those who correctly find the reproductive policies of the Nazi regime in Germany during the Second World War, the government of South Africa prior to the creation of democracy in 1993 or the current population policies of China ethically abhorrent are repulsed more by the means than the goals involved in efforts to design future generations.

The Subjectivity of Perfection

Some who find the pursuit of perfection morally objectionable worry about more than coercion. They note that it is simply not clear which traits or attributes are properly perceived as perfect or optimal. The decision about what trait or behavior is good or healthy depends upon the environment and circumstances that a child will face. To pick traits, features or attributes in the abstract is to simply reify prejudice as optimality (Rapp 1988; Harding 1993).

Views about what is perfect or desirable in a human being are more often than not matters of taste, culture and bias. But they are not always simply the product of subjective feelings.

There are certain traits; physical stamina, strength, speed, mathematical ability, dexterity, and acuity of vision to name only a few, which are related to health in ways that command universal assent in almost any cultural or social setting imaginable. It would be hard to argue that a parent who wanted a child with better memory or greater physical dexterity was simply indulging his or her biases or prejudices. As long as there is no coercion or force used to compel persons to make choices about their children that are in conformity with particular visions of what is good or bad, healthy or unhealthy, there would seem to be enough consensus about the relationship between certain physical and mental attributes and health to permit parents to choose certain traits, features and capacities for their unborn children in the name of their health. And if no coercion of compulsion were involved, it could even be argued that parents should be free to pick the eye or hair color of their children or other equally innocuous traits as long as their selection imposed no risk for the child and did not compromise the child's chance of maximizing his or her opportunities.

A parent might concede that their vision of perfection is to some degree subjective but still insist upon the right to pursue their own values. Since we accept this point of view with respect to child rearing, allowing parents to teach their children religious values, hobbies, and customs as they see fit, with almost no restrictions short of imperiling the life of the child, it would be difficult to reject it as overly subjective when matters turn to the selection of a genetic endowment for one's child.

A different set of objections commonly raised against the morality of trying to achieve perfection hinge on concerns about slippery slopes of various types. Some worry that allowing parents to pick the traits of their children will lead inevitably to government forcing its vision of perfection upon anyone who wants to have children. But, this argument has problems as well. For one, it flies in the face of a number of facts about the pursuit of perfection in other areas of health care.

For many years cosmetic surgeons, psychoanalysts, and sports medicine specialists have been plying their trades without any slope having developed in American society to the effect that those with big noses or poor posture must visit a specialists and have these traits altered. Some choose to avail themselves of these specialists in the pursuit of perfection. Many do not. If there is a slippery-slope from permitting individual choice of one's child's traits to limiting the choices available to parents, it is a slope that does not start with individual choice. And if there is a problem of a slope, then it must be shown why it is morally permissible to seek perfection after one is born, but why such efforts would also be wrong if engaged in prenatally.

It is certainly and sadly true that twentieth-century history brims with instances of genocide, mass murder and ethnic cleansing. These are, nevertheless, problems of politics, government and ideology. There is nothing inherent in the decision to indulge one's preferences about the traits of one's child that is morally wrong as long as those preferences do nothing to hurt or impair the child. If there are slippery slope problems that confound the morality of eugenics, they lie in the flaws of politics as well as misunderstandings about the nature of population genetics and diversity, not in the desire to have a "better" baby.

Equality

Another objection to allowing eugenic desires to influence parenting is that it will lead to fundamental social inequalities (Kitcher 1996; Kitcher this volume). Allowing parental choice about the genetic makeup of their children may lead to the creation of a genetic "overclass" which has unfair advantages over those who parents did not or could not afford to endow them with the right biological dispositions and traits. Or it may lead to too much homogenization in society where diversity and difference disappear in a rush to produce only perfect people, leaving anyone with the slightest disability or deficiency at a distinct disadvantage.

Equity and fairness are certainly important concepts in societies that are committed to the equality of opportunity for all citizens. However, a belief that everyone deserves a fair chance may mean that society must do what it can to insure that the means to implementing eugenic choices are available to all who desire them. It may also mean that a strong obligation exists to try and compensate for any differences in biological endowment with special programs and educational opportunities. It is hard to argue, in a world that tolerates so much inequity in the circumstances under which children are brought into being, that there is something more offensive or more morally problematic about biological advantages as opposed to social and economic advantages.

It is also difficult to argue in a world that tolerates large numbers of privileged persons the right to pursue the best education for their children in situations and contexts that may well produce homogeneity in the end-results, that the pursuit of perfection or enhancement at the cost of homogeneity is allowed in schools, music lessons or summer camps when the intervention is environmental but not when it is biological. The fact that kids with privileged social backgrounds go on to similar sorts of educational and life experiences does not seem sufficient reason to prohibit the parenting practices of the upper class. It should also be pointed out that similarity is a matter of degree and that what looks homogeneous to some may appear to be very different to others. Small differences can make a big difference even in a world in which some people pursue goals that will lead toward some degree of biological homogeneity.

There does not appear to be a persuasive, in principle ethical reason to condemn individual eugenic goals. At least none of those canvassed in this paper provide a basis for ruling any and all eugenic goals out of bounds as self-evidently immoral. While force and coercion, compulsion and threat have no place in procreative choice, it is not so clear that it is any less ethical to allow a parent to pick the eye color of their child or to try and create a fetus with a propensity for mathematics then it is to permit them to teach their children the values of a particular religion, try to inculcate a love of sports by taking them to games and exhibitions or to require them to play the piano in order that they acquire a skill.

If there is an argument to be made against eugenics, it would seem to be most persuasive against group or population eugenics. Efforts to shift the composition of the gene pool would seem to require or be more prone to slip toward the imposition of a vision by government or other powerful institutions. In so far as coercion and force are absent and individual choice is allowed to hold sway, then, presuming fairness in the access to the means of enhancing our offspring, it is hard to see what exactly is wrong with trying to create more perfect babies or better adults.

Commentary on "What's Morally Wrong with Eugenics?"

Timothy F. Murphy

Arthur Caplan has rightly pointed out that for historical reasons coercion is the fulcrum of any analysis of the evils of eugenics. We do not like and should not accept coercion in, for example, reproductive decisions. What complicates objections to eugenics is that in principle we already accept many of the goals in whose name eugenics goes forward: the eradication of genetic disabilities, the promotion of resistance to disease, the cultivation of native intellect, and in general protecting human beings from the vicissitudes of fate. It is sometimes hard to see why coercion toward those important goals would be necessarily a bad thing, especially if we are tempted to improve the poor, the despised, and the politically weak. If we otherwise worry about eugenics, it is because of the evils that have transpired when the values of specific persons—Francis Galton, Charles Davenport, Fritz Lenz or even a "reform" eugenicist like Hermann J. Muller—have been mistaken for the values and goals of human beings generally (Kevles 1985; Weiss 1990). But we think—rightly or wrongly—that we know this latter problem well enough to avoid it, so coercion remains the central problem for analysis in individual efforts to control the traits of children through interventions having eugenic effects.

In this regard it remains worthwhile to problematize eugenic goals not only by analyzing their effects, but also by attending closely to their motives. The first issue I want to mention is one that is already familiar to cultural critics who note that it is hard to separate out individual choice from the larger social currents in which we all swim. What may be wrong with eugenics is that these social currents coerce in ways we scarcely notice, our choices becoming thereby mere instantiations of prevailing cultural imperatives about what bodies, traits of gender, and capacities of intellects are desirable. One way to see this conflation of social value and personal choice is to consider whether a child has a duty to exhibit any particular trait. To make things interesting, I would like to raise the question of whether parents should be recognized as having the right to control the sexual orientation of their children. The value and desirability of homoeroticism is a matter much debated, and it will be illustrative for our purposes here.

One might think that the genetic disposition of at least some people toward being gay or lesbian is reasonable in light of the striking concordance of pedigree studies, DNA linkage studies, and even the finger skin ridge studies that have emerged recently, all of which typically conclude with a statement to the effect that their findings are "consistent with a biological contribution to sexual orientation" (LeVay 1991; Bailey and Pillard 1991; Bailey, Neale, and Agyei 1993; Hamer et al. 1993; Pool 1993; Hall and Kimura 1994; see also Schaffner, Manier this volume). There are many limitations to these studies and certainly not all homoerotic behavior or identities are robotic consequences of biological causality, but these studies provide enough evidence for the thought experiment I want to conduct here: suppose parents are interested in controlling the sexual orientation of their children and think they can do so in light of recent research.[1] The media think this is a live issue: questions of prenatal diagnosis, selective abortion, and genetic therapy routinely accompany reports

---

1 I am not convinced that the research reports by LeVay (1991), Bailey and colleagues (1991, 1993), Hamer and colleagues (1993), and Hall and Kimura (1994) in fact offer any sort of mechanism by which to identify, let alone manipulate, the likely sexual orientation of a child, either prenatally or postnatally. If some sort of prenatal marker for sexual orientation is discovered, it would likely prove to be probabilistic and not offer definitive identification of a child's eventual sexual orientation. It also does not follow that such a marker would by itself permit direct control over the sexual orientation of children, for a marker is not a treatment. Short of a willingness to abort children with unwanted markers, a prenatal diagnostic test would not by itself guarantee control over sexual orientation.

about biological studies of sexual orientation. Would ethical analysis show eugenic efforts to control sexual orientation in children to be permissible or impermissible?

One way of getting at the question of whether parents should have the right to control the sexual orientation of their children is to ask whether children have a duty to have a particular sexual orientation. It is not clear that they do. One of the first conditions of an observable duty is that it must be within an agent's power to achieve. As it is unclear at present how developmental processes—including all relevant biological and psychological events—lead to one particular sexual orientation over another, it is unreasonable to assert that children could have a duty to observe an unknowable standard. These same limitations would also impede a parent's surrogate duty on the child's behalf since parents are no better situated to select a course of psychosexual development certain to end in a particular sexual orientation. Even if one wanted to argue that there was some sort of attenuated duty on the part of a child in regard to sexual orientation—because a child ought to be striving toward a moral ideal or some such reason— it does not follow that his or her duty always requires a sexual orientation consonant with parental expectations. Parental desire for particular traits in their children does not for that reason alone create a duty in children, however attractive those traits might be to parents. Relative to the variable circumstances of a child's life, there might conceivably be a duty to be, for example, homosexual—if a gay sexual orientation proved of greater benefit to the child or served some social obligation incumbent on a child.[2]

For many of the reasons just mentioned, it does not seem to me that a child could claim the right to a particular sexual orientation, a right that imposes a duty on parents to observe. Even if sexual orientation of children were under parental control, it does not follow that

---

2 The sort of benefit to the child I have in mind involves the discordance between expected sexual orientation and actual orientation which sharpens a child's intellectual powers in a way that would not have occurred had there been complete concordance between parental expectation and the child's actual sexual orientation. I take it that discordance of this kind is not entirely without benefit to a child. As a benefit to society I have in mind the putative benefits that obtain in society with a certain percentage of gay people. I do not wish to endorse sociobiology's explanation of the evolutionary emergence of homoeroticism necessarily, but it is the kind of benefit I see as sustaining the idea of the social utility (and possible derivative obligations) of gay people.

a child has a right to a particular sexual orientation. If homoeroticism is, as I will stipulate, no inherent deprivation but compatible with important human achievements and happiness,[3] it does not follow that failure to achieve a straight sexual orientation would count as parental dereliction of duty or an infringement of a child's right. The same would hold true of any putative right of a child to be gay since neither is heteroeroticism any sort of inherent deprivation.

The question of entitlement to control the sexual orientation of children might be looked at another way: might parents have a right to a child of a particular sexual orientation? Unfortunately this line of argumentation will not succeed because it is unclear against whom such a right might be asserted. Not the child, for a child has no duty to have any particular trait, as I have said. Not a physician nor any other identifiable party for no one knows how to control sexual orientation with the kind of certainty required to ground a moral duty. Perhaps a parental "right" could mean only a political right of non-interference by any individual or social group: parents should have a right assertable against society to control the sexual orientation of their children. One might bolster this argument by saying that parents have a duty of beneficence toward their children and that, ordinarily, this duty requires making them straight. One sexual orientation therapist put the matter this way: "A young child's natural instinct might be to just eat salty or sugary food. But every parent knows that is bad for them. They'll have a healthier life if they have a balanced diet. And emotionally they'll have an easier life if they are heterosexual" (Rafferty 1995). It does not follow, though, that parental beneficence need take the form of avoiding gay and lesbian children. Parental beneficence could justifiably take this form: raising children in an atmosphere not hostile to gay people and taking up political advocacy or educational reform to diminish the social harms that attach to homoeroticism. Protecting children from social harm need not take the form of erasing the "offending" trait when other equally beneficent measures might protect children from the disvalues that attach to homoerotic interests.[4]

---

3 I take Suppe's (1984) account to be convincing on this point.

4 I do not assume that parents have unlimited responsibilities of beneficence toward their children. It seems to me that a parent would satisfy any reasonable obligation of beneficence toward his or her children in regard to sexual orientation by raising the child in an atmosphere not hostile to gay people and/or by trying to make

Parental motives for wanting straight children can also prove morally suspect. A parent might express the desire for straight children by saying he or she wants to protect the child from homophobic society or as a way of preparing the child's success in relationships and society. In these instances, it is not the child's sexual orientation per se that is at stake, but its instrumentality in access to social goods such as prestige, jobs, sexual and romantic partners, and grandchildren. Protestations of beneficence notwithstanding, it is heterosexist to believe that as a class gay people must necessarily suffer an inferior and abject personal and social fate. In fact, many if not most gay people achieve fulfilling relationships; meaningful work and important jobs; desired religious vocations and ecclesiastical office; social acceptance if not prominence; good relations with their parents; and more and more single and coupled gay people are having children and their side effect: doting grandparents.[5] I suspect, therefore, that most motives for which parents wished to avoid gay children would not pass moral muster. There is no one social fate of homoeroticism, and a general presumption that parents can be beneficent only by trying to secure the heterosexuality of their children will not be generally convincing.

I do nevertheless believe that parents would be within their political rights to use any genetic or other controls that afforded them the possibility of control over their children's sexual orientation. This is not to say that a parent has the right to a straight child. It is merely to say that as a matter of social policy it would be unwise to impose barriers to the use by parents of prenatal diagnostics and interventions. In the future, if Crocker Pharmaceutical Corporation could guarantee straight children with a "magic pill" (Crocker 1979), or The Gunter Dörner Prenatal Clinic perfected a technique for assuring straight children by means of a hormonal intervention at just the right moment of fetal brain development,[6] parents would be entirely within their rights to use these services so long as they were demonstrably safe and efficacious. By the same token, I think that parents would be

---

(contd.)
the world a better and safer place for gay people.

5 In fact, even before gay people undertook adoption, artificial insemination, and even the more elaborate mechanisms of in vitro fertilization and embryo transfer, many of them did in fact have children (Bell and Weinberg 1978, 164–65, 391).

6 Dörner (1976) speculated that homoeroticism in human males could be thus eliminated.

equally within their rights to intervene to secure gay children, and it seems to me that—among gay and lesbian separatists especially—there would be a market for exactly such interventions.

There is every reason to believe that eugenic choices will come whether we want them or not, as biomedicine extends its dominion not only by concerted and grandiose undertakings like the Human Genome Project, but also by its ordinary work-a-day science. What is exactly wrong with eugenics is not that we undertake to make a better world, but the assumption that the primary construction material for that world is eugenically improved children and that any motive for such children is morally sound. It is unclear that children have a duty to any particular trait or that parents have a right to children of a certain kind. Nevertheless genetic science cannot but empower more and more choices over children. Will parental choices impede moral progress as much as a state imposed eugenics program might—by the gratuitous assumption that we will get to a better world if only we have better children, that is, children without the traits we do not like? If the state were to impose a requirement that all women use a genetic intervention to insure that their children were straight, we should all object to this intrusive interference of government. If all women freely used such an intervention because they believed that homoeroticism is bad for their children, and no gay children were born, how much better off would we be than if the state had imposed its rule? We would have respected the principle of non-interference but eliminated a class of persons who do not in fact experience life as the disvalue used to justify the intervention. In fact, in numbers that would be called a landslide in any political election, gay people say they would not want a magic pill given to them at birth to insure that they were straight (Bell and Weinberg 1978, 339). Far from opening an ever-receding horizon of choices, some eugenic practices would seem to force choices, to shrink the range of acceptable options, and unfailingly serve prevailing social orthodoxies.

Arthur Caplan has pointed out the ways in which eugenics is unobjectionable to the extent it does not violate rules of non-interference and does not have consequences detrimental to the population as a whole. It may be, though, that some of what is wrong with eugenics escapes analysis on these criteria. It seems to me that we should attend therefore to the motives of eugenic practices as much as their consequences before we render any final conclusions about what eugenics is worth to children and the world in whose name we improve them.

9

Utopian Eugenics
and Social Inequality

Philip Kitcher

SECTION ONE

It is helpful to divide the philosophical issues surrounding the Human Genome Project into three main groups.[1] First, there are questions about the scientific significance of the project. Second, the project raises immediate practical problems, notably in decisions about genetic screening and in connection with the potential release of genetic information about individuals. Third, there are long-term concerns about the desirability of applying our new ability to identify the genotypes of the unborn, concerns often posed in the accusation that we are on the verge of a new eugenics. In this paper, I shall primarily be concerned with the third cluster of issues, for it is here, I believe, that the hardest *philosophical* problems lie. But, in order to frame my discussion, I want to begin by reviewing perspectives on the first two groups, perspectives that I have defended in some detail elsewhere.

Many critics of the HGP have offered what might be called the "boggle argument."[2] This consists in announcing that the end-product of the project will be a huge list of As, Cs, Gs, and Ts, and asking,

---

1 This essay draws on my previous efforts to organize the issues around the HGP. See, in particular, Kitcher 1996 and 1995.

2 Alex Rosenberg offers an entertaining version of this argument in his contribution to *PSA 1994* Volume 2 (Rosenberg 1995).

rhetorically, what possible point there could be in any such list. Although the "boggle argument" is perfectly appropriate as a corrective to some of the more grandiose statements made in defense of the HGP, it is quite irrelevant to the actual practice of the project. First, at present, virtually all of the research on *human* genomes is directed at constructing better maps (both genetic and physical), and this research is making ever more efficient the strategy of positional cloning. Second, the principal current sequencing efforts are focused on other organisms (the bacteria recently sequenced by Craig Venter, *E. coli, S. cerevisiae*, and *C. elegans*), and these efforts are revealing many new genes (including many that will be homologous to genes in our own species) and properties of the organization of genes. Third, while the problem of hunting for genes in pages of sequence data has no simple, elegant solution, and while the problem of understanding the conformation of proteins is unsolved in general, the construction of databases from results about genes in a variety of organisms and about the forms of well-known proteins enables *ad hoc* solutions that are likely to be able to recover an extremely high percentage of the genes and proteins from future bodies of human sequence data. Fourth, if sequencing becomes sufficiently easy during the next decade, sequencing the entire three billion base pairs of the human genome may be the most efficient way to identify (almost) all our genes; if large-scale sequencing remains problematic, then the smaller efforts on non-human organisms will be combined with "quick-and-dirty" techniques for finding human genes, to provide almost as much information about our own species. What is right about the "boggle argument" is that there is nothing sacred and wonderful about achieving the full list of As, Cs, Gs, and Ts that might stand for "the" human genome. But the conclusion ought to be simply that the emphasis on the *Human* Genome Project is faulty advertising for an exciting venture, the "Genomes Project," that will reveal masses of interesting and useful things about many organisms including human beings.[3]

When we reflect on the kind of information that will be achieved in the near future, we are bound to confront the second group of questions. The development of detailed maps has enabled gene hunters to clone and sequence genes that are implicated in a number of human

---

3 These points are developed and defended in detail in chap. Four of Kitcher 1996.

diseases. These achievements can typically be translated, relatively quickly, into genetic tests, tests that might be useful in a variety of contexts: diagnosing a disease, identifying the particular form of a disease already diagnosed, identifying the risks of future disease or disability, exposing the genotype of a fetus, and discovering if a person carries a recessive allele. It is worth reminding ourselves that the first two contexts, although typically uncelebrated in the literature, are likely to provide real benefits without significant problems.[4] The principal difficulties with the power to engage in genetic testing concern the last three contexts, especially the possibility of identifying risks of future disability. As many authors have pointed out, there is little good to be done in telling people that they are at high risk for a disease when there is nothing that can be done to alleviate that risk (or nothing to be done that would not have been recommended whether or not the people had tested positive).[5] Testing is only useful when there is something special that those who test positive should do, something that would be inappropriate for the rest of the population.[6] Of course, medical lore does know of cases in which this necessary condition obtains: PKU is the obvious, much-cited, example (see Diane Paul's commentary following).

From the medical point of view, PKU is a perfect case because there is a diet, necessary for normal functioning for people who have the abnormal genotype, and harmful to people who have the normal genotype. Yet when we expand our horizons, there are grounds for worry. Because the diet is unpleasant and expensive, many children who need it do not receive it for as long as they should. As Diane Paul has lucidly argued, we really don't know if PKU testing has done more good than harm (Paul 1994). We do know that, in the early stages, children who should not have received the PKU diet were given it (with severe disruptions of development), that some people have not stayed on the diet long enough to avoid the deleterious build-up of phenylalanine, that mothers who had come off the diet as young

---

4 This is a point that Eric Lander has emphasized in public lectures.
5 One might reply that knowledge increases the autonomy of the subject. However, this is too stern an ideal—for many people could easily be crushed by news that they will inevitably suffer from a late-onset disease. This is most clearly shown in the studies of people's reactions to testing for Huntington's disease. See Andrews et. al. 1994, and Nancy Wexler's essay in Kevles and Hood 1992, 211–43.
6 A point stressed by Hubbard and Wald 1993.

adults gave birth to gravely damaged babies—but we don't know how to balance these individual tragedies against the successes that have come from early intervention. At risk of committing myself to an untenable dualism, I suggest that those who appeal to PKU testing as a shining example are right *so long as we consider the problem as a purely biomedical one.* The trouble comes from the nature of the social context in which the applications of biomedical knowledge are made. Because we do not provide economic support for families in which a child is diagnosed with PKU (as well as intensive counseling), we blunder away the possibilities of making a real difference to the quality of human lives.

The case of PKU testing has important implications for the future of genetic testing. Biomedical researchers are quite justified in thinking that some of their discoveries will have the potential to enable people to reduce their chances of future disease or disability. Actualizing that potential, however, may require significant changes in the ways in which medicine is practised. So long as patients (and, very often, the doctors who advise them) are buffeted by the forces of the market, so long as many segments of the population only have very limited access to advanced medical technology, so long as there is no attention to providing support (both economic and personal) for people who require expensive and uninviting procedures, what can be done in principle is unlikely to be achieved in practice.[7]

Just as the promise of new molecular knowledge must be evaluated by considering the social surround, so too the principal concerns about the flow of genetic information could be addressed by modifying features of contemporary medical practice. The idea that genetic information should be private appeals to us not because of the intrinsic value of keeping our genotypes to ourselves, but because, as things now stand, we could suffer serious harm if others came to know cer-

---

7 This was dramatized at a recent meeting in Bethesda, MD, organized jointly by the NIH/DOE working group on the Ethical, Legal, and Social Implications of the HGP and the National Action Plan on Breast Cancer. At that meeting, several women who had had positive diagnoses for familial early-onset breast and ovarian cancer recounted their struggles with insurance companies and employers. Insurers refused to cover prophylactic mastectomies that the women, and their doctors, saw as measures required to avoid grave risks to life. Some women lost their jobs. All have had great difficulty obtaining subsequent health coverage. Indeed, in order to protect anonymity, they spoke at the meeting under assumed names, and video recording of their presentations was not allowed.

tain things about the alleles we carry. Envisaging the era of genetic testing simply reinforces straightforward arguments for ensuring that all members of our society, irrespective of genotype or economic condition, have access to affordable medical care. By the same token, genetic testing in the workplace should be restricted to those situations in which it is possible to argue that applicants with particular genotypes are less qualified for the jobs they seek. *Principled* solutions to problems about the flow of genetic information are not hard to come by. The real difficulties are not philosophical but practical. How can we ensure that the principles are embodied in practice? Or, if it seems impossible to achieve principled solutions, how can we broker some form of acceptable compromise?

During the years since the HGP began, there have been numerous, often repetitive, discussions about genetic screening, about the use of genetic information in insurance, and about genetic testing in the workplace.[8] I believe that philosophical analysis is important to these discussions because of its delineation of ideal solutions, and that, as philosophical problems go, the construction of such solutions is relatively straightforward. Debates go on only because political pressures seem to make the ideal solutions impossible. Once this is recognized, we face a choice. One can either expose with maximal clarity the rationale for the ideal solutions, hoping thereby to engender a change of view that will remove political constraints, or one can resign oneself to working within the current system and attempt to find a compromise that will not ride too harshly over the rights and aspirations of vulnerable people. Political scientists and lawyers naturally gravitate to the second option; philosophers, I hope, see their vocation in the first.

With these preliminaries, I turn to the main discussion of this paper. The gulf between an ideal solution and the practical realities is most apparent when we consider the use of genetic knowledge in prenatal decision-making. The deepest, most disturbing, debate about the HGP emerges from consideration of the possibility that we are committed to a revival of eugenics. My aim is to show that defenders of the project are committed to a benign enterprise, *Utopian Eugenics*, whose ultimate rationale stems from concern for the quality of future human lives. However, once that rationale is exposed, consid-

---

8 For some of the best discussions, see Holtzman 1989; Nelkin and Tancredi 1989.

eration of existing social inequalities and of the political pressures that support those inequalities yields two important lines of argument. According to the first, there is a deep inconsistency in promoting the HGP and the utopian eugenics it fosters, without attending to the social causes of human inequality. According to the second, we should recognize the fragility of utopian eugenics within any society that fails to eradicate widespread social inequalities. I shall suggest that the most important philosophical problems about the HGP arise from these arguments, and that the sources of skepticism about the project are really grounded in sensitivity to broad questions in social philosophy.

SECTION TWO

To tag an enterprise with a name that carries the burden of a terrible history sometimes short-circuits important discussions. So it is with that part of the application of human molecular genetics that focuses on pre-natal choices. Many authors have worried that the possibility of discovering the genetic characteristics of a fetus will spawn a new eugenics, and the stigma associated with 'eugenics' is powerful enough to end the conversation right there. There is no doubt that many past ventures in eugenics are truly appalling (see essays by Pernick and Caplan this volume). However, if we are to condemn a future enterprise of choosing people, based on advanced knowledge of their genotypes, then we should try to identify the properties that have made previous forms of eugenics morally monstrous and see if these are shared by the envisaged practice of pre-natal testing.

From the start, we need to be quite clear about what is in the offing. Philosophers who have considered the role of molecular genetics in designing future generations have often indulged in science fiction, and ignored the more prosaic facts. Speculations about "perfect babies," "wonderwomen," and "supermen" are rife, as if there were realistic prospects of "engineering" people to our specifications.[9] Quite apart from the commonplace, but never-sufficiently-reemphasized, point that genotypes and environments work together to shape phenotypes, the technical difficulties of gene replacement therapy make *positive* programs of designing our descendants impossible.

9 See, for example, Harris 1993 which discusses the ethical issues in a sensitive fashion but seems at a far remove from the scientific realities.

There are two major problems: delivering DNA to the right cells and regulating its expression within the cell. Even the most successful current procedures for DNA delivery only reach a fraction of the target cells (fortunately, in some cases, as with the most lucky of the Severe Combined Immune Deficiency (SCID) children, this is enough to restore roughly normal functioning), and the techniques used so far, which address only the crudest types of gene regulation (the alleles inserted are permanently expressed), show clear effects of interference with other cellular processes (often the percentage of cells expressing the inserted protein drops rapidly). Add to this the significant problem of delivering DNA to the brain, and the fact that nobody knows how to eliminate the mutant alleles that are present. The result is a picture, familiar to everyone who practises molecular medicine and to anyone who takes the time to look at the details of the practice: doctors inject patients with molecules they need, producing, when things go well, a cobbled-together approximation to normal functioning, usually rigged with all kinds of devices to ward off unwanted side-effects. That picture is a long way from the naive dream of precisely replacing, in all pertinent cells, the alleles one doesn't want with the alleles one does. The reality depicted is full of promise for ameliorating severe disease and disability by finding inefficient and cumbersome methods of managing tolerable levels of the right proteins (Kitcher 1996, chap. Five). It is no way to design a baby. *Positive* selection isn't the issue. What is possible on a small scale now, and will be possible on a large scale within a decade, will be *negative* selection. Provided that parents have the genotypic potential to yield the combination of alleles they want in a child, they can use pre-natal testing to filter out unwanted progeny. From the very beginning of the HGP, it has been abundantly clear—although not the sort of thing champions broadcast in the halls of Congress—that one principal application of new knowledge in human molecular genetics would be a greatly increased number of abortions. Unless abortion is opposed on principle, most of these abortions would probably be viewed as benign and humane. As we learn more about the genetic bases of various diseases, not only the rare conditions that doom their bearers to permanent hospitalization but also more common cases that limit human functioning, prospective parents will have the ability to learn in advance whether the fetus one of them carries bears a relevant genotype, and, on that basis, they will be able to choose to continue the pregnancy or to terminate it. There is no doubt that this will avert some human tragedies: doctors, counsellors, and relatives of people who have had children with genetic diseases, can recount numerous instances in which lives of promise were destroyed because a pregnancy proceeded to term. Nonetheless, as we envisage the many kinds of genetic tests that will become possible in the next decades, it is easy to fear that a benign policy of forestalling disease may become a program for enforcing social prejudice, a new eugenics.

Although there are some objections to *any* use of pre-natal testing—even for diseases like Tay-Sachs or for the most disruptive trisomies and translocations—I shall take it for granted, for present purposes, that there are some instances in which identifying affected fetuses and aborting them would be morally justifiable. Doubts arise when we consider terminating pregnancies because the fetus has the "wrong" eye color or hair texture, when an allele "for" obesity or same-sex preference is present, when there are too few of the "alleles for intelligence." Now these doubts intertwine a number of issues which it is important to separate. But their great force comes from a resonance with the eugenic past, for these are the kinds of abuses that have figured largely in earlier attempts to apply human genetics.

A eugenic practice is an attempt, by some group of people, to shape the genetic composition of their descendants according to some ideal. Despite the large body of excellent literature on the history of eugenics, very little has been done to develop this characterization.[10] My own approach (which accords with that taken in a pioneering article by Diane Paul 1992) recognizes four dimensions of any eugenic practice. The first dimension identifies a subpopulation whose reproductive activity is to produce the desired results. The second dimension specifies the degree to which members of this subpopulation make their own reproductive decisions (or, conversely, the degree to which they are coerced by the ideals of others). The third dimension picks out the characteristics according to which the choices are made. The fourth appraises the quality of the genetic information that is used in making the reproductive decisions.

Nazi eugenics (and, to a lesser extent, the American eugenic ventures of the 1920s and 1930s—see Pernick this volume) occupies a

---

10 Particularly outstanding is Kevles 1985.

particularly loathsome region of the resultant four-dimensional space. Based on prejudices about desirable and undesirable human types, the Nazis selected particular subpopulations in which to promote breeding and particular subpopulations in which to reduce (or eliminate) reproduction. People in the latter groups experienced extreme forms of coercion, under the surgeon's knife or at the end of the storm-trooper's gun. The characteristics for which the future population was to be chosen (the glorification of the "pure Aryan" type) reflected a morally distorted and factually inaccurate account of human worth. Finally, the underlying genetic claims were frequently wildly wrong.

If the terrible evils of the past are in one corner of eugenic space, the envisaged future is at the opposite extreme. When human geneticists ponder the application of pre-natal testing, they imagine that everyone in the population will have the same opportunities for shaping the descendant gene pool—there will be no prejudicial selection along the first dimension. They suppose that each couple will make their own free decisions—there will be no coercion along the second dimension. They envisage that those decisions will be made reflectively, that they will be educated and freed from superstition and prejudice. Finally, they assume that the underlying genetic information will be accurate. What they endorse is a vision of eugenics as different from Nazi eugenics as could be. I shall call it "Utopian Eugenics."

But, even if we grant that there is considerable difference between utopian eugenics and the horrors of the past, we should still ask why we should occupy any position in eugenic space. There is a straightforward answer. Once we know how to identify the genotypes of future people, eugenics is the only option. It is quite illusory to think that we have a non-eugenic alternative, and the illusion can be exposed by considering exactly what it would be. Suppose, for example, that we said that nobody was to draw on any information from human molecular genetics in making reproductive decisions. This would be to institute a eugenic practice that has two highly problematic characteristics: first, its standard for the desirability of the descendant gene pool is that any such gene pool is preferable if it is brought about from ignorant decisions about the properties of offspring, if nature is left free to take its course; second, its position on the second dimension is highly coercive. But the fundamental point is that this is a *eugenic* practice. In a situation in which we have the option of bringing about future populations with various genetic characteristics, it commends one of those options, the one in which we act as we would have acted before the advent of detailed genetic knowledge. Once we lose our genetic innocence, we have alternatives, and, because we have to elect one of the alternatives, we have to practise eugenics.

It should now be evident that labelling the HGP as opening the door to eugenics settles nothing. Everything depends on whether the form of eugenics that will result is better than the alternatives. Champions of the HGP claim two things: first, that utopian eugenics is the form that will be actualized; second, that this is superior to its rivals, and, specifically, superior to the coercive policy of pretending that we are genetically innocent.

As we shall discover, defending these claims is far more complicated than might initially appear. There is little doubt about the value of three of the dimensions of utopian genetics. It is eminently reasonable to claim that a policy of non-discrimination along the first dimension is morally justified, that leaving people free to make their own reproductive decisions is a good thing, and that the genetic information we can expect to apply will be accurate, so long as we avoid the pits into which behavioral genetics has so often fallen in the past. The major worries surround the specification of the third dimension, the delineation of the moral ideal that should inform free and educated reproductive decisions.

Once again, I shall abbreviate a story I have told in far more detail elsewhere.[11] A practice of pre-natal testing should be oriented, I believe, toward a concern for the quality of future lives. In contemplating the quality of future lives, three kinds of factors must be considered. First, we should assess the chances that the person who would be brought into being would be able freely to develop a conception of what matters in his/her life, a sense of what is worth achieving and striving for. Granted that we can imagine the future person forming a life conception of this kind, we can identify the desires that are most centrally associated with it, and ask about the extent to which we could expect these desires to be satisfied. Third, and finally, we can evaluate the hedonic quality of the life, the balance between pleasure and pain that it would bring.

Now the most obvious worry about children with mutant alleles is that they will suffer, and, in some instances, pain is untreatable.

---

11 Kitcher 1996, chap. Thirteen. My discussion there draws on ideas of Dworkin 1993; Griffin 1985.

However, many of the principal uses of pre-natal testing are not like this: palliatives are available, seizures can be brought under control. The real horror is the massive disruption of development, the neurodegeneration, the permanent confinement to a hospital bed, the need for daily grooming, assisted feeding, the absence of any cognitive life. What is missing in such tragic cases is the possibility that the person will ever form a sense of what matters to her/him, so that the envisaged life is viewed as having intolerably low quality when assessed according to the first factor. For other genetic conditions—such as the various kinds of muscular dystrophy—we recognize that the future person's choices of what matters in life will either be severely constrained or else that the resultant central desires will almost certainly be frustrated.

If decisions about the quality of future lives are genuinely to be free, then it is important that social prejudices, or lack of social support, should not coerce prospective parents into decisions to terminate pregnancies that might have issued in lives of significant potential. Societies that cramp the lives of people with particular conditions (for example, those who do not meet prevailing standards of bodily shape or those sexually attracted to members of their own sex) effectively undercut the freedom of choice of prospective parents who find that the fetus carries an allelic combination associated (I shall assume correctly) with the despised condition. In an even more obvious way, a society may impose its standards by denying people with particular conditions the opportunities to develop in ways that bring them much closer to normal functioning. Because of imaginative programs, many Down syndrome children have achieved levels of development considered impossible thirty years ago. A society committed to providing such programs promotes the freedom of reproductive decision. Hence, I shall take it as a cornerstone of utopian eugenics that prospective parents can make their reproductive decisions in assurance that children born with debilitating conditions will receive the best support known, and that their children's lives will not be limited by social prejudices.

Utopian eugenics is obviously a program for a highly idealized world. However, since we are committed to some form of eugenics, it is worth being as clear as we can about the ideal and its demands. I think that utopian eugenics embodies a worthy ideal, and that the allure of this ideal is responsible for the view that human molecular genetics can be liberating.[12] The most important criticisms of the HGP do not stem from the faulty conflation of all forms of eugenics or from a failure to appreciate the worthiness of the utopian ideal. They result from advancing two theses: first, that the ideal commits us to much more than an enterprise in applied human genetics; second, that our failure to carry through with that commitment signals an acceptance of social inequalities that will reduce utopian eugenics, in practice, to a much darker and more morally problematic, program. In the rest of this essay, I shall be concerned with the merits of these two theses.

SECTION THREE

We are prepared, rightly I believe, to take steps to ensure that an unlucky legacy, passed on through the genes, does not doom a child to pain or to a pathetic life of restricted opportunities. But there are obvious questions: why should we rush to treat the unfortunate genetic inheritance of the few, while ignoring the unlucky social inheritance of the many? Shouldn't we commit ourselves to learning how to change the environments that break young lives as surely as defective proteins? While we should acknowledge our ignorance of the consequences of social interventions, it is worth remembering that the results of many forms of molecular therapy, are uncertain—and that, when the condition is sufficiently grave, we are prepared to take risks (as in gene replacement therapy for SCID). How bad must the plight of discarded children be to justify an analogous *social* experiment?

Critical rhetoric sometimes overstates the case, deriding the advertisement of future relief of human suffering. The anguish produced by genetic afflictions of all types, from Tay-Sachs and Lesch-Nyhan to the forms of cancer, is real and tragic. The potential of molecular medicine is equally real. It is right to celebrate the promise of human molecular genetics to enhance the quality of the lives that are led.

---

12 Negative discussions of the HGP often overlook its potential to avoid real tragedies. At a symposium in Washington D.C. in the Spring of 1994, a man from the audience responded to the worries of three speakers (Patricia King, Eric Lander, and me) by telling us of his daughter, who had given birth to two children afflicted with neurofibromatosis. After describing, in a very restrained way, how her life, and that of her husband, had been blighted, he asked why discussions of the HGP so quickly gravitate to the potentials for danger, and never dwell on the real promise. That reminder seems to be highly salutary.

Nonetheless, the questions I have posed are serious. If our moral venture rests—as I have suggested—on a concern for the quality of human lives, does consistency require us to undertake a more systematic assault on the pressures that shrink people's hopes and opportunities? Or are we properly conscious of our limitations, thankful for biological expertise and regretting the ineptness of social policy-making?

Our lives are the products of many lotteries, and only one of them shuffles and distributes pieces of DNA. Behind the often acrimonious controversy about the value of molecular genetics is a deep disagreement about the implications of this fact, a disagreement dividing people who may appropriately be called "pragmatists" from others whom I shall henceforth dub "idealists."[13] Idealists think that, when the underlying rationale for applying molecular genetics in pre-natal testing is exposed, we should become aware of a commitment to the quality of nascent lives that ought to be reflected more broadly in social action. Pragmatists maintain that we should not hold the local good we can do (by developing and applying our biological knowledge) hostage to quixotic ventures (doomed to uselessness by our ineptness at social engineering). The dispute between these groups intertwines two large classes of questions. What is the extent of our obligation to aid people whose initial circumstances greatly reduce the quality of the lives they can expect to lead? What are the practical possibilities for meeting these obligations, specifically for combatting the environmental causes of pinched and painful lives?

I shall primarily be concerned with the first kind of question, the moral issue; I lack the expertise in social science that would be required to treat the factual questions in the detail they demand. My aim will be to scrutinize an analogy that moves idealists, to clarify an important debate, not to resolve it. The starting point for the analogy juxtaposes the fact that our lives are, in large measure, the products of lotteries, both genetic and environmental, which deal fortunes unequally, with an attractive social ideal, the ideal that people should have (in some sense that is to be specified) equal opportunities to live happy and rewarding lives. Mismatch between ideal and reality

---

13 Paradigm idealists are Richard Lewontin 1993 and Ruth Hubbard (Hubbard and Wald 1993). Because the debate has not been explicitly presented in the form I adopt here, it is hard to identify pragmatists—but I suspect that many defenders of the HGP would embrace pragmatism.

prompts citizens of affluent democracies to believe that justice demands some attempt to remedy the unequal accidents of birth and childhood. Some type of help is required. Help typically calls for money, demanding, in consequence, some form of redistribution of resources.

But which form should we aim for, and how far should we go? Because money can easily be counted and differences in financial resources readily compared, it is natural to frame discussions of social ideals in terms of economic assets.[14] Natural, but, I think, mistaken. As the deep rationale for utopian eugenics reveals, what is of primary importance to us is something far more nebulous, something with several dimensions, and something that is not easily compared across individuals—think of rating the qualities of the lives of your friends, or of famous people. In trying to clarify our intuitive conception of the demands on us, we do better to focus on what is fundamental, the quality of lives, acknowledging that comparisons will often have to be imprecise. Of course, it would be naive to neglect entirely the connection between economic status and the quality of life. Although decreasing differences in assets is not an end in itself, it may well be an indispensable means to realizing worthy goals.

If we could assign precise numbers that measure exactly the quality of lives, then it would be possible to identify the *expected quality of a person's life* by proceeding as gamblers do when they figure the odds. Sophisticated bettors calculate what they can expect to receive from various betting arrangements by multiplying the returns from an outcome by the probability of the outcome and adding terms corresponding to each eventuality. In analogous fashion, we could consider the person's current state, associate each of the possible lives that might ensue with the probability of its occurrence, multiply the probability by the measure of the quality, and sum across all the possibilities. Life is far too complex for any such calculations. We cannot quantify the various dimensions, we cannot weigh one dimension against another with any precision, and we are ignorant of the exact values of the pertinent probabilities. However, that does not interfere

---

14 Or perhaps with "primary goods," as with John Rawls' rightly influential Rawls 1971. My own approach is much closer to that of Amartya Sen, and his emphasis on equalizing capabilities. See in particular Sen 1992. Comparisons between genetic and environmental differences provide the basis for many interesting defenses of Sen's points, but I shall not try to articulate them here.

with our ability to make some comparisons of expected quality of life. Suppose that two children are born into roughly equivalent social environments, one bearing two copies of the common mutation for cystic fibrosis (delta 508), the other carrying two copies of the normal allele. There is little hesitation in declaring that the expected quality of life for the first child is lower than that of the second; the possible futures for the first child divide into two classes, those in which the phenotypic expression of the CF alleles is mild (or in which harmful effects are mitigated through treatment) and the life is roughly equivalent in quality to that of the second child, and those in which the phenotypic expression of the CF alleles restricts activities and curtails possibilities; if the child experiences a future of the first type, the quality of her life will not be higher than that of the second child; if she experiences a future of the second type, the quality of her life will be lower; because there is a significant (although unknown) chance that her future will fall into the second class, the overall expectation is lower.

Nor do we have trouble in making some more complex comparisons. Imagine that the child bearing the two CF alleles is born into an affluent family, with two parents who are devoted to her and determined she shall thrive. By contrast, the child with the normal alleles is the sixth son of a single mother, who is unemployed and struggling to resist the pressures of a dangerous urban environment. Now there are clearly some possible futures for the girl in which her activities are so severely reduced, despite all her parents' efforts, that the quality of her life would fall below that we would expect for the boy. Nevertheless, with high probability, her range of opportunities and her ability to satisfy her central wishes will greatly exceed what is overwhelmingly likely to be available to the boy, so that there is a high probability of a large positive difference, outweighing the lesser chances of negative outcomes. Overall, her expected quality of life is superior.

It is now possible to state the idealist argument more carefully. We have obligations to improve the lives of the less fortunate, obligations that are honored in our commitments to attack genetic disease. By redistributing resources, taxing the well-to-do and using the revenues to provide compassionate care, as well as to invest in molecular research (both by seeking new ways of treating genetic disease and disability, and by providing support so that the resulting treatments are available to those who need them), affluent societies conform to a principle of social justice. Idealists conclude that the principle requires more of us, that we are morally obligated to intervene in other ways to raise the expected quality of life of people who have been victims of social roulette, that assets currently used to enhance the quality of lives already at a high level might more justly be employed elsewhere. To evaluate their case we need to have a clearer view of the principle on which they rely, and of its implications.

One way to think about social justice is to conceive society, highly abstractly, as a collection of individuals, each endowed with a particular amount of resources and each with a particular expected quality of life. Expected quality of life is fixed by many factors entirely beyond the person's control, the combination of genes that have been transmitted, the characteristics of the parents and the environment into which a child is born, and, of course, the resources that will be available during development. Redistributing resources would have an impact on the expected quality of the lives of all citizens. What ideals should guide us among the numerous options?

Three abstract principles have loomed large in democratic political theory, each of which seems to offer pure expression of an attractive ideal. One opposes schemes of redistribution in the name of individual autonomy: however beneficial the consequences may be, we cannot compel someone to give up assets which are his or hers to control. The second would seek to redistribute resources so as to make the total quality of life as high as possible: we are to imagine summing the measures of expected quality for the lives of all citizens, given all possible distributions of resources among them, and choosing that distribution that gives rise to the largest grand sum. Last is an explicitly egalitarian principle commending the distribution that comes closest to securing equal expected quality of life for all.

It is not hard to see that each of these principles is flawed. Although respect for individual liberty is a worthy ideal, that ideal cannot properly be expressed in a "hands off" attitude towards redistribution of assets. Once we realize that the freedom that counts—the freedom to choose that conception of what is significant in one's life—is a matter of degree, depending on the extent to which we have the opportunity to be guided by alternative possibilities, then redistribution might decrease the autonomy of the privileged by only a slight amount, while greatly enhancing the autonomy of the underprivileged.[15] The directive not to demand assets from the well-to-do, far from manifesting a concern for the autonomy of all, would be more accurately advertised as a maxim to respect the liberty of the winners in the lotteries that fix initial circumstances.

Maximizing total expected quality of life is heir to familiar troubles.[16] The grand sum is no respecter of the individual contributions. Everything comes out in the wash, and the highest total may easily be achieved within a society in which the majority have a very high expected quality of life, gained at the expense of a small number of people whose lot is doomed to be miserable. Perhaps we would increase the total expected quality of life within affluent democracies by withdrawing funds currently used to care for the most devastating genetic diseases, and spending them in quite different ways, possibly by building wonderful public sports facilities. It would surely be morally obscene to do so. Moreover, even if the arithmetic does favor our current allocations—maybe because families would be upset if the standard of care were dramatically reduced—that is surely not the correct moral basis for our determination to do what we can for those afflicted with genetic disease. The quality of their lives must be considered individually, not simply as some term in a colossal sum, and, despite the fact that expensive efforts do not increase expected quality by very much, both justice and compassion demand that we do what we can.

The third pure ideal is vulnerable because it insists on equality irrespective of total well-being. Anti-egalitarians frequently voice the "dog in the manger" objection, claiming that appeals for equality are based on the envious suggestion that nobody shall have anything that somebody lacks. Despite the difficulties of specifying the pure form of egalitarianism—just how do we compare how well unequal distributions approximate perfect equality?—it would appear that there is always one way of securing *exactly* equal expected quality of life for all, to wit by reducing the expected quality of each person's life to some very low value. Since this may well be the *only* way to achieve exact equality in expected quality of life, and since it is plainly unac-

---

15 This argument is a highly compressed version of a line of reasoning that I develop at much greater length in the last three chapters of Kitcher 1996.

16 There are any number of good formulations of the point. See, for example, Bernard Williams' contribution to Smart and Williams 1973.

ceptable, it is apparent that the ideal of equality must be compromised by attention to overall well-being.

Although we have to tolerate some inequalities, our respect for others might suggest that resources should be distributed to raise the expected quality of life of each person above a certain minimum level.[17] Depending on how the minimum is fixed, this requirement will either be toothless or unsatisfiable. Those afflicted with the most devastating genetic diseases will have an extremely low expected quality of life no matter what efforts we make, no matter what resources we assign to their care. If the minimum is set above the level they can attain, even when supplied with unlimited resources, then the requirement to redistribute so as to provide for everyone an expected quality of life above the minimum cannot be honored. If, however, the minimum is set lower, then the expected quality of life we demand for all will be so low that, except in the very gravest cases, redistribution will prove unnecessary. Even untreated, unsupported, children with cystic fibrosis have higher expected quality of life than those with Canavan's disease or Lesch-Nyhan syndrome, whatever resources we invest in them.

In the clearest cases of redistribution, greater equality and increased total expected quality of life go hand in hand. Guaranteeing public funds to provide the special diet for children with PKU would lower the expected quality of life of the affluent by a tiny amount, but would raise the expected quality of life of the PKU children enormously, producing both a more equal distribution and a larger total. However, it is sometimes just for the fortunate to make relatively large sacrifices to bring far smaller benefits to victims of circumstance. Perhaps the expected quality of the lives of people with the most serious genetic diseases would only be marginally reduced by providing a much cheaper form of care, and perhaps the savings could be used to offer much larger benefits to those who already live well. Even so, it would be morally indefensible to favor that distribution. Small gains for the unlucky count far more than larger improvements for the fortunate.

Reflection on these points suggests another way of formulating a social ideal. Our goal should be to redistribute resources so as to bring people's expected quality of life above a minimal level; when

---

17 The discussion of this paragraph is influenced by Derek Parfit's illuminating exploration of related problems in part Four of Parfit 1984.

this cannot be achieved, it is just to transfer resources so that the expected quality of life of some people is decreased, provided that the result both raises the expected quality of life of people below the minimal level and does not depress beneath the minimum the quality of life of those who give. This formulation would sanction our support for the genetically disadvantaged, even though that support costs more, in expected quality of life, than it yields. (A pictorial presentation of the effects of the redistribution is given in Figure Fourteen.)[18]

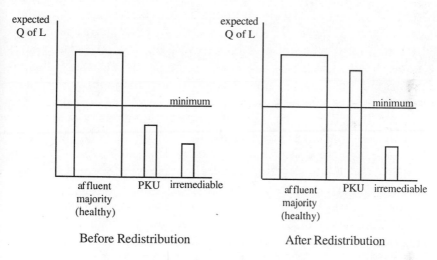

Before Redistribution   After Redistribution

FIGURE FOURTEEN

Possible effects of redistribution of resources on expected quality of life.

However, the new formulation, like its predecessors, is inadequate. It is at least possible, even quite probable, that every increase we could make in funding for biomedical research would raise, if only infinitesimally, the expected quality of life of people who suffer the severest genetic diseases. Would it therefore be just to withdraw resources from other citizens, up to the point at which their expected quality of life approached the minimum level, in order to make these minute increases? Intuitively, it would not: the sacrifices made are so

---

18 This figure, and those that follow, adapt to the present context the style of representation used by Parfit in part Four of Parfit 1984. See also Temkin 1993.

disproportionate to the gains that the redistribution appears quixotic rather than compassionate.[19]

It is a familiar point in general social philosophy that our ideals of justice need somehow to be combined, and my appeal to instances of genetic disease is only to highlight the fact that this point is plainly applicable to utopian eugenics. Respect for liberty must be tempered with concern for equality, the yearning for equality must attend to considerations of average well-being.

Attempts to make further headway in clarifying the idealist's case can usefully begin with the problem of deciding the level of support for those with genes that debar them from lives of any acceptable quality. We think it right to divert assets from people whose lives flourish to improve the lot of the genetically unfortunate, even though the positive effects are small in comparison with the benefits the affluent forego; yet it would not be right to proceed indefinitely, radically reducing the quality of life of the healthy to obtain vanishingly small expected gains for the genetically disadvantaged. How far, then, should we go? When is enough enough?

Here is an obvious thought. The expected quality of life of a person afflicted with a devastating genetic disease will inevitably fall beneath a particular value, the associated *quality of life ceiling*, however great the resources we make available for care and for research into the condition. More exactly, for a group of people who share a common genetic condition, I take the quality of life ceiling to be the maximum expected level of well-being they would attain, given any distribution of the resources available within the society.[20] Using this notion, we can explain the limits on a policy of giving aid to the genetically least fortunate. Moderate sacrifices on the part of those whose lives go well, which, while depressing the expected quality of their lives still retain it at a high level, can enable us to bring the genetically unfortunate close to their ceiling—we can make their lives

---

19 It might naively be thought that this is at odds with Rawls' "difference principle." However, trying to apply the difference principle to the present context would be quite misguided, for several reasons—not least because the principle is intended to govern "basic social institutions" and thus to function at a much more abstract level. It is an interesting question whether the genetic example of the text subverts an intuition on which Rawls' treatment of justice draws.

20 After I had reached this way of approaching the issue, I discovered, through Sen's discussion (Sen 1992, 89–93), that a formally similar idea has been developed by welfare economists. See, in particular, Atkinson 1970.

go almost as well as it is possible for them to go. Large sacrifices appear quixotic because they generate a huge disparity between the expected quality of life of the fortunate and their corresponding quality of life ceiling. If those sacrifices are made, then some lives will be very much worse than they might have been. (See Figure Fifteen for a graphical representation of these points.)

The farther people are below their quality of life ceiling, the more they deserve our support; the lower their expected quality of life, the more they deserve our support. Each of these principles, I believe, represents an important moral impulse, corresponding to our urge to remedy lost potential (whatever the absolute level) and to our commitment to help those who are less fortunate (whatever their potential). Sometimes the principles are in harmony, and we would rightly divert resources from the affluent to provide special diets for children

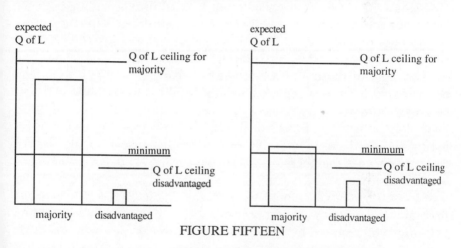

FIGURE FIFTEEN

When large sacrifices can do little for the disadvantaged

with PKU and compassionate care for those afflicted with Tay-Sachs. Sometimes they conflict, as when we envisage spending all societal resources in a massive effort to improve the lot of those who presently have the lowest expected quality of life.

A more complex illustration shows how the principles work together, and how they bear on utopian eugenics. Imagine that society consists of three groups of people: many are healthy, and all of them possess enough assets to raise their expected quality of life well above the acceptable minimum; a few suffer from a genetic condition, which, left untreated, will yield an expected quality of life well below the minimum, but which is susceptible to an expensive form of treatment that will raise their expected quality of life almost to the level of the healthy majority (think of a disease like PKU); most members of this second group lack the resources to purchase the treatment themselves; finally, there is a smaller group consisting of people who suffer from an even more severe genetic disease; even if unlimited resources were expended on them, their expected quality of life would always remain below the level of untreated members of the second group; they can be brought very close to their quality of life ceiling, if they are given extensive (and expensive) care; further investment of resources would increase their expected quality of life by minute amounts. The affluent members of the healthy majority are sufficiently numerous so that some of their assets could be diverted, with only small impact on their expected quality of life, providing both for the treatment of all members of the second group and for the extensive care of people in the third group. Redistributing resources in this fashion appears preferable to either of two alternatives: concentrating on the plight of those who are worst off, using all resources that can be spared without reducing the expected quality of life of the majority below the minimum until the expected quality of life of the third group has been raised to that of the second (which is impossible); ignoring the third group entirely, and using some of the assets of the majority to treat members of the second group. It is wrong to focus on the most needy and neglect people whose actual prospects are far below their potential, people who might obtain significant benefits from aid. Equally, it is wrong to think solely in terms of lost potential that might be restored by redistributing resources, using the fact that some members of society may have a wretched existence, whatever is done, as an excuse for forgetting about them. (The options and their relations are presented graphically in Figure Sixteen.)

The example just considered represents, in schematic form, our predicament with respect to (some) victims of genetic disease. To a first approximation, the population of an affluent democracy divides into three groups: those with genetic conditions for which we can do nothing except to provide humane care, those with genetic conditions that we can reasonably hope to alleviate, and the large majority. Because the minorities are so tiny in comparison with the majority, because the majority has great resources, and because we already know (or have well-grounded hopes of knowing) how to intervene effectively, both in providing care and in ameliorating the conditions, it is possible to act in accordance with the principles without demanding

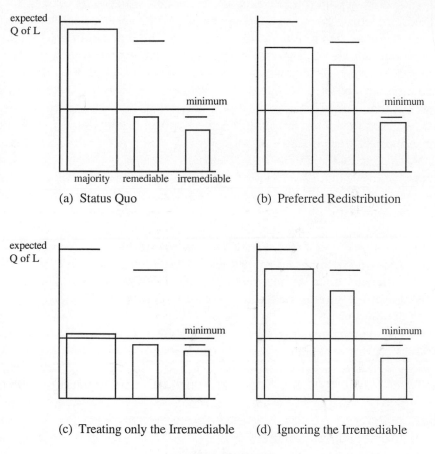

(a) Status Quo

(b) Preferred Redistribution

(c) Treating only the Irremediable

(d) Ignoring the Irremediable

FIGURE SIXTEEN

Resource allocation and quality of life under different redistribution models

large sacrifices. Indeed, many affluent democracies pursue appropriate policies, providing care and investing in biomedical research. Idealists think there is a direct lesson for social policy. With the basis of their analogy in clearer focus, we can now ask if they are right.

The most straightforward way of developing an argument by analogy is to envisage, as before, a society in which there are three groups. The first, once again, will be a large majority of healthy people, many of whom are affluent. The second will consist of children who, given the home environment they currently inhabit, have an expected quality of life below the acceptable minimum; each of the children in this group *could* achieve a life of far higher quality; if any of them were removed from the current brutalizing environment and provided with affection and stability, the expected quality of life would increase, perhaps, in some cases, to that typically attained in the first group. In the final group are children who, because of the environments into which they have been cast, have a quality of life ceiling that lies below the expected level of the second group, even if no resources are allocated to members of that group; with extensive care, these people can be brought close to their quality of life ceiling, and providing this care would not seriously affect the quality of life of the first group. To fix ideas, we can think of the third group as comprising those children whose cognitive and emotional functions have been irreparably damaged by environmental brutality. The second includes the children whose current environments severely limit their cognitive, emotional and social development, but who could still be rescued if their environments were radically changed. There remains the first group, the relatively healthy and affluent remainder of the population.

In the genetic case considered previously, it appears right to redistribute resources from the affluent majority to provide care for those whose genes set a very low quality of life ceiling and treatment for children whose quality of life can be restored. Do we have similar obligations when social circumstances cause analogous relationships among the expected quality of life of three groups? Pragmatists will surely reply that the examples are not analogous. There is no form of social intervention (the analogue of the PKU diet) which we know how to carry out, whose cost is small, and which would restore the expected quality of life of members of the second group to the level enjoyed by members of the majority. Their objections intertwine a number of considerations. Most prominent is a pessimistic assessment of the effects of social intervention: many people, reflecting on the history of attempts to care for the disadvantaged, conclude that the problems are intractable, that well-meaning efforts to engineer solutions often fail to help those who are supposed to benefit, and, worse, have unforeseen consequences that generate further social disasters;

the road to social catastrophe may be paved with the best of intentions. Additionally, pragmatists may stress the inefficiency of typical social interventions, claiming that, even where policies do bring benefits, they require expenditures much larger than the gains achieved. This thought alone would not prove telling—for, after all, we are willing to provide expensive care for victims of the most devastating genetic diseases, despite the fact that their quality of life is only raised by a small amount. However, where inefficiency on a small scale may be tolerable, inefficiency on a far larger scale might prove ruinous. The size of the second group in the social case is far greater than the size of the second group in the genetic case: there are far more children who suffer because of desolate or threatening environments than children with the genes for PKU, or even children who are victims of genetic disease. Pragmatists thus view the idealist proposals for social policies based on redistribution of assets as quixotic ventures, liable to produce only uncertain benefits for members of the second group, in danger of generating damaging side-effects, and requiring, because of the scale of the problem and the inefficiency of social interventions, sacrifices by the majority that would lower their expected quality of life by an amount too large to tolerate. (See Figure Seventeen (a) for

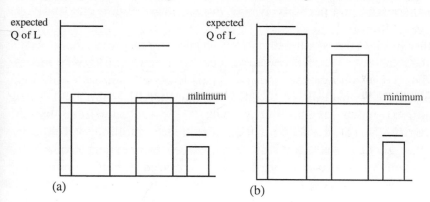

(a)   (b)

FIGURE SEVENTEEN

In (a) (the pragmatist's vision) large sacrifices are needed to raise the intermediate group above the minimum. In (b), the social analogue of the case of genetic diseases, small sacrifices produce dramatic gains for the intermediate group and bring the disadvantaged close to their ceiling.

a representation of the outcome, as the pragmatist sees it; Figure Seventeen (b) depicts the outcome as it would be were the cases analogous.)

Idealists have a number of ways of trying to counter pragmatist pessimism. There is likely to be disagreement about the consequences of social interventions. Disclaiming the need for any sophisticated social science, idealists may contend that we know the sorts of things children need: better schools, safer streets and playgrounds, stable homes, and realistic prospects for jobs. Infusing money into job-training programs, teacher salaries, school renovation, group homes for neglected and abused children, and developing an effective police force, trusted by local residents, would make an enormous difference to the expected quality of lives. In response to the charge that there may be unanticipated harmful consequences, idealists may suggest that grave problems require us to take risks—just as we might consider risky interventions when confronted with a devastating disease. How many lives could be improved, and by how much, by the kinds of measures idealists recommend? Most of us have opinions—and perhaps some experts have answers. But prior to the wrangling over facts is a moral question. Because the effects of social programs are uncertain, and because, in the United States and, increasingly, in other affluent democracies, the scale of the problem is large, following idealist recommendations is unlikely to prove cheap. What is the extent of the sacrifice the affluent are morally obliged to make?

Depending on a host of factual details, reallocating resources might raise the expected quality of life of the second group by varying amounts, while demanding greater or lesser sacrifices by the majority. (See Figure Eighteen a-c for three possibilities.) Reflection on the genetic cases teaches us that small losses in expected quality of life for the majority are legitimately required when we can raise the truly unfortunate close to their quality of life ceiling, and provide for the slightly less unfortunate lives of quality comparable to those of the majority. A whole spectrum of moral positions is consistent with this judgment. At one pole are those who place greatest emphasis on avoiding losses in expected quality of life: for them a redistribution that would cause the quality of lives of a large number of people to drop significantly below their quality of life ceiling is unjustifiable. At the opposite extreme are those who are most moved by the ideal of raising the expected quality of life of people whose prospects are currently bleak: even a large total loss in expected quality of life (computed by summing the changes in expected quality of life for each member of the society) should be tolerated in the interests of raising the expected quality of life of people currently below

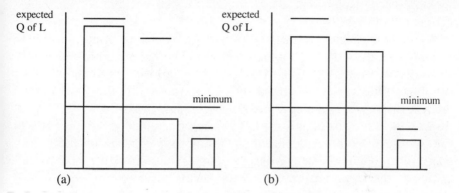

(a)   (b)

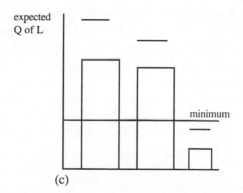

(c)

FIGURE EIGHTEEN

(a) shows very little redistribution of income; (b) displays redistribution with small sacrifices by the majority and large gains by the second group; (c) shows considerable sacrifices for more modest gains. Egalitarian commitments can be caused by judgments about which kinds of depicted states are preferable.

the acceptable minimum. Between the extremes are all the possibilities for balancing the two principles that emerged from the earlier discussion: the principle to avoid loss of potential quality of life (irrespective of actual level) and the principle to raise the quality of lives that currently have lowest expected quality (irrespective of potential). There is a *moral spectrum* along which different people, including pragmatists and idealists of various stripes, place themselves differently.

Molecular genetics does not confront us with the difficulty of positioning ourselves on this moral spectrum. The people who will be affected by the redistribution of assets—the children who receive care or whose lives are transformed by interventions like the PKU diet— are few enough in number to demand only small sacrifices. When we turn to evaluate the idealist analogy, however, disagreements about the pertinent social facts and about the proper position on the moral spectrum are interwoven. Many idealists, including the most vocal critics of the Genome Project, would surely offer an explanation for the apparent recalcitrance of social problems. Our failure, they believe, can be traced to the limitations of a moral perspective that embraces the redistribution of assets only timidly, withdrawing when our efforts encounter obstacles. Perhaps the remedy for our ignorance is to try a variety of approaches and learn which ones are beneficial. That can only be done, of course, if we are willing to commit substantial resources, prepared to position ourselves further toward the egalitarian end of the moral spectrum.

I have been exploring the most straightforward way of developing the idealist analogy—but the most straightforward way is not necessarily the right way, and the comparison so far considered may make it easier for pragmatists to dismiss idealists as starry-eyed optimists. In presenting the genetic example, I imagined the second group to consist of children whose quality of life could be significantly improved by applying our molecular knowledge, for example children carrying the genes for PKU. But PKU looms so largely in discussions of the fruits of human molecular genetics because it is a special case, something molecular medicine hopes to emulate repeatedly in coming decades. The promise of applications of molecular genetics is real, but that it is likely to issue not in sweeping solutions to problems of major families of disease—total victory in the war against cancer, say—but to a motley of techniques, useful to varying degrees in a broad range of cases. The idealist analogy can be developed differently by combining the concern for the quality of the lives of the least fortunate with a counterpart for this therapeutic pragmatism.

If we were to reconceive the second group of people in the genetic example, thinking of them as sufferers from genetic disease who may benefit, to varying extents, from future molecular medicine, then the original pragmatist response would have to be slightly modified. No longer would it be possible to claim that there is a known way of raising the quality of life of members of this group, effective across the entire class. Public commitment to funding programs of molecular research would be grounded in the prospect of an assortment of more or less useful therapies. Pragmatists would have to contend that nothing similar can be expected in the social case, that we cannot anticipate any such range of successful local interventions.

Statistics on rates of violent crime, unemployment, drug abuse, and education levels in some regions of urban America are sufficiently horrifying to inspire idealists to call for sweeping plans of social action—or, perhaps, sufficiently numbing to cause pragmatists to resign themselves to the impossibility of tackling problems of such vast scope. Yet it is worth taking a look behind the statistics. I shall draw, very briefly, on two recent discussions of the consequences of social inequality in the lives of American children.

In a moving book on the lives of some young boys in the Chicago housing projects, a book aptly titled *There Are No Children Here*, Alex Kotlowitz exposes quite specific needs that could be addressed, very directly, by a commitment of public funds (Kotlowitz 1991). The public defender who represents one boy does not have time to listen to his story because of the overload of cases; the energetic and inspiring teacher who encourages the boy's younger brother has inadequate funds for books and supplies; parts of the projects are always dark because the city has given up trying to devise a system of lighting that would be proof against vandalism; and, perhaps most importantly, the resourceful administrator who successfully sweeps a few buildings for drugs and guns (earning the respect and gratitude of the inhabitants) is unable to extend his operation to twenty other complexes—"money from the Department of Housing and Urban Development has not been forthcoming." It is hard to maintain that these difficulties are any more insuperable than those biologists face in trying to repair the damage done by defective, or missing, enzymes

—and hard not to believe that solutions to these mundane problems would bring real improvements in the quality of human lives.

Jonathan Kozol's *Savage Inequalities* is an equally scathing indictment of the conditions in which citizens and their elected representatives allow some American children to grow up. Kozol's book begins with the children of East St. Louis, a town in Illinois that has served as the home to chemical plants and to companies specializing in the burning of hazardous wastes. He quotes a St. Louis health official, who compares the conditions in which East St. Louis children live to the Third World (Kozol 1992, 10). The incidence of lead poisoning is high, but "[t]he budget of the city's department of lead-poison control, however, has been slashed, and one person now does the work once done by six" (ibid., 11). Despite the fact that the schools stand on grounds that have been heavily polluted by chemical spills (from the plants that no longer employ the parents of the schoolchildren), and despite the recurrent overflows and flooding by raw sewage, the state government has refused money to help clean up the mess. (The state government blames the local administration for its inefficiency. Finger-pointing, of course, fails to remedy the plight of children who become sick from the poisonous fumes or who repeatedly find their schools closed.)

Nobody can read these books without recognizing clear possibilities for concrete solutions to some of the problems that reduce the expected quality of life for poor children in the United States, solutions that money can buy just as easily as a special diet for PKU or the new apparatus that may advance the identification of genetic diseases. Resolute efforts could greatly reduce the ill-effects of lead poisoning in American cities, and, quite possibly, do more to enhance the quality of the lives of American children than the attacks on all the single-locus genetic diseases combined. If pragmatists are to resist the idealist argument, they cannot do so simply by reiterating their general distrust of social interventions. They must produce economic and social analyses that reveal that the apparent benefits of specific programs are unreal, or are offset by excessive costs. In short, they must engage in the same kinds of detailed data-collection and calculation that is needed to assess the levels of social support that applications of molecular genetics will require.

Although idealists and pragmatists may find some common ground in the thought of redistributing resources for parallel programs that seek specific, local, ways of reversing the depressing effects both genes and environments have on the quality of lives, significant moral differences are likely to linger. Because of the scale of the social problems, there will be disagreements about how far societies are obliged to go, about how to balance concern for those who are worst off with avoidance of society-wide losses in quality of life. The debate between idealists and pragmatists is multi-faceted—perhaps more many-sided than is commonly supposed. But I believe the idealist analogy cannot simply be dismissed; and, if the analogy does nothing else, it should force each of us, privileged citizens of affluent societies, to reflect on the extent of our obligations to promote the quality of lives that are currently bleak, to think about our own chosen place on the moral spectrum.

SECTION FOUR

Let me now turn to the second part of the idealist critique. Champions of molecular medicine commonly believe they can achieve its benefits without attending to the broader social inequalities. The difficulties of avoiding serious harm, of preventing the reinforcement of prejudice and inequality, already surfaced in my discussion of the short-term problems of the HGP. Some of those difficulties can be resolved—at least in principle: we could ensure adequate counselling for all, provide universal health coverage, enact strict regulations to block discrimination in employment.

Yet are these important social programs likely to be implemented and sustained in a society whose prevailing concerns are with efficiency rather than equality, whose members position themselves on the moral spectrum at a far remove from the egalitarian pole? We live at a time in which the dominant political rhetoric opposes the idea that affluent members of our society owe anything to children currently born into poverty. Utopian eugenics requires the construction of expensive forms of social support. Consider the important conditions that secured it against the charge of repeating the moral blunders of the past.

(1) All members of the society must have access to information that will enable them to make free and informed reproductive decisions.
(2) Personal choices must not be limited by background social prejudices.

(3) Those born with genetic conditions that can be ameliorated by social support must be guaranteed those forms of support.

Translating these ideals into practice requires bringing millions of Americans, whose connections with the system of medical care are currently tenuous, into a position in which they can draw on extensive genetic counselling, and receive costly forms of support (special diets, wheelchairs, special classes for developmentally disabled children, and, quite possibly, an increasing number of drugs patented by biotechnology firms anxious to recoup initial investments). Will a society that grudges the children of East St. Louis unpolluted schools take on these obligations?

As I was writing the current draft of this essay, this question took on renewed force. The governor of California (Pete Wilson) and California legislative leaders agreed on a proposed budget, providing more funds for public schools and for the state university system. At the same time, welfare funding has been cut. Some salient points are made clear in a newspaper report:

> Welfare for families will be cut $141 million statewide, pushing the monthly grant for a mother with two children down 4.9% from $595 to $566 in high-rent areas, including Los Angeles, Orange and Ventura Counties. In lower-cost areas . . . the grant level will drop to about $538 for a mother with two children.
>
> Aid to the poor who are elderly, disabled and blind will be cut by similar percentages—pushing the grant for an individual from $614 a month to $584 in high-rent areas, and $566 a month in low-rent areas.
>
> As legislative leaders tried to sell their colleagues on the deal, several disabled people in wheelchairs blocked the front door to Wilson's outer office, vowing to stay until they were arrested. Twenty-three demonstrators were cited for disturbing the peace and released, but none was jailed.
>
> "I'm having trouble making ends meet now", said Daryl Wisdom, 47, who is in a wheelchair because of polio. Wisdom's pharmacy bill is $200 a month, he said. (*Los Angeles Times* 27 July 1995: A18)

The apparent cruelty of these budgetary cuts might be understood as a necessary measure. In hard economic times, perhaps California must sacrifice the interests of some to attain other worthy ends— pumping money into an ailing educational system by drawing from the impoverished. But a later paragraph in the same report belies the idea of economic necessity.

Although there is no general tax cut, high earners stand to benefit. The two upper-income tax brackets for people whose annual taxable income exceeds $100,000 will expire at the end of the year. As a result, the state will lose about $300 million in revenues. (Ibid.)

It is not hard to do the arithmetic. If the tax brackets had been renewed, then, instead of cutting funds to the poor, the elderly, the disabled, and the blind, those people could actually have benefited from a modest increase (at least 3%). The 1994 Tax Schedule locates the top bracket at $429,858 (for couples filing jointly) and $214,929 (for a single person); the second bracket begins at $214,928 (for couples filing jointly) and $107,464 (for a single person). Under the new measure couples whose joint taxable income exceeds $61,000 will all pay at the same top rate. Mr. Wisdom, on the other hand, will receive $384 each month after he has paid his pharmacy bill. We should ponder the high ideals of utopian eugenics in the light of these political realities.

Utopian eugenics is a fragile enterprise. Critics of the promise of molecular biology are wrong to fear that we shall *inevitably* be swept into repeating past eugenic excesses—there are theoretical possibilities, exemplified by utopian eugenics, for a far more benign program. It would be small consolation if that were simply a theoretical point, if the possibilities could not be actualized. Ironically, we can now see that *both* idealist critiques of applications of molecular medicine and pragmatist defenses mix optimism and pessimism. Idealists believe that we can take steps to eradicate the inequalities in affluent democracies, but are skeptical about the possibility of preventing new injustices, spawned by applications of molecular knowledge, if the broader social issues are not addressed. Pragmatists maintain that piecemeal efforts are sufficient to ward off the danger, and that we can sustain a benign form of eugenics, but they doubt our ability to remedy more general social ills. What if both the pessimistic assessments are correct?

CONCLUSION

We don't need the HGP to inform us that there are important social problems that face our society, and that concern for the welfare of others cries out for a more egalitarian (and compassionate) approach to those who are born into poverty. Reflection on the HGP doesn't somehow, suddenly, make an argument in social philosophy that wasn't available before. Rather the HGP is a lightning rod, a program that draws fire because of the presence of background instability. Its critics have, I believe, muddied the main issues by using rhetorically powerful devices: they have charged that the HGP overestimates the idea that genes are powerful, or that they are destiny, and that the program is a return to eugenics. Intelligent supporters of molecular medicine recognize that these charges are quite unfounded. We don't have to become genetic determinists or revive the eugenic errors of our predecessors to appreciate the real benefits that advances in human molecular genetics may bring.

I have been trying to identify what I think is most important in the critics' case. Because of urgent social problems that arose before the HGP was proposed, problems that we have had a moral obligation to address quite independently of the possibilities opened up by molecular genetics, the beautiful world that the enthusiasts envisage may be a long way from reality. The main argument of this essay is that discussions of the HGP ought to lead into a much broader, and more difficult, set of issues—that leave genetics far behind—because these issues are deeply implicated in the social setting that will surround the new medicine.[21] We need, in short, explicit discussions about the moral spectrum and appropriate regions along it, supplemented with careful analyses of the potential costs and benefits of various programs, both social and medical. If the HGP leads citizens and policy-makers, as well as philosophers, to engage in serious discussion of these broad questions, then, whatever else it achieves, it will have done us an important, and unexpected, service.

---

21 In short, the big ethical and social issues surrounding the HGP are not *implications* of the HGP. They are huge social problems that American society has shamefully neglected for a decade and a half.

Commentary on "Utopian Eugenics"

Diane R. Paul

This is a rich paper with many strands—of which I will focus only on two. In general, Philip's claims seem to me both important and right. More specifically, I agree that the most productive way to approach the eugenics issue is to identify those features of past eugenics that make the enterprise now seem repugnant to us and ask whether these features are likely to inform future applications of genetic knowledge. (If the U.S., Britain, and Germany had followed the same path as France and Brazil—where eugenicists emphasized healthy babies and breast-feeding—we would not now be anxiously asking if medical eugenics is disguised eugenics).[1] I also agree that it is possible to envisage, in principle, a form of eugenics in which these odious features are absent, that whether this ideal is realizable depends on the context ("social surround") in which the applications of genetic knowledge are made, and that without significant social reforms (which include "the construction of expensive forms of social support"), the benefits promised by utopian eugenics will not be achieved in practice; indeed, that we risk creating just the kind eugenics we especially fear.

---

1 However, I do not agree with his claim that, given available knowledge, we must practice some kind of eugenics. It is certainly true that people who opt for ignorance also make a choice with consequences for future genotypes. But I do not see why we are logically forced to call this choice "eugenics."

263

I also consider it a great virtue of this paper to have moved beyond "in principle" arguments about genetic testing to consider the social and political setting that largely determines how tests will be used; to recognize that "the real difficulties are not philosophical but practical" and to specify what at least some of these difficulties are. But I think Philip could (and should) go further—broadening his concept of social context and drawing the conclusion that seems to me to follow from his analysis. In the end, a paper that starts bravely suffers, in my view, from a kind of failure of nerve. In short, I want to push Philip on two points: What it *means* to consider the "social surround" and what the *implications* are of taking it seriously.

My first point is that the "social surround" is not simply economic; it includes the knowledge and values that citizens bring to decisions about who should employ genetic tests and for what purposes. To predict how tests will actually be used, we need to recognize not only that they will be introduced in a society marked by large and increasing economic inequalities and few restraints on firms with commercial interests in testing—important as these factors are. We must also recognize that widespread ignorance about what genetic tests can and can not do has been no bar to forming strong opinions about their value. Most people know little about testing, and what they think they know is often wrong. Nonetheless, they tend to believe that testing is good—even when there is no effective way to act on the results.

Philip writes that a benign eugenics requires that decisions be made freely, reflectively, and in light of the best scientific information. However, in effect he reduces this complex of issues to the problem of "access to information." Access is not the real problem. We are swamped with information. Part of the problem consists in the way that information is structured. But we should also acknowledge that many people do not want to know more; indeed, they may actively ignore information (sometimes for good reasons). Moreover, even if people were generally well-informed, knowledge would not necessarily lead to reflective decisions. To cash out these claims, consider the following social science research on knowledge about and attitudes toward genetic testing.

Attitude surveys indicate that opinions about prenatal testing are both favorable and confused. For example, about two-thirds of the respondents to one survey said they would want to undergo such tests (or would want their partner to); they believe that tests will do more good than harm. However, they understand very little about them. A majority of respondents thought that current tests could predict whether or not a person will suffer a heart attack and that genetic screening could "correct the genetic defects that cause such inherited diseases as Down's syndrome, Tay-Sachs disease, and hemophilia" (Singer 1991, 240–41). A recent survey commissioned by the March of Dimes discloses similar confusions. Thus sixty-eight percent of the respndents said they knew "relatively little" or "almost nothing" about genetic testing. Nonetheless, most were of the opinion that someone else (including insurers) had a right to an individual's genetic information. While eighty-six percent of the respondents said they knew little or nothing about gene therapy, eighty-nine percent approved of using it to treat genetic diseases and forty-two percent to improve children's intelligence (March of Dimes 1992).

In California, health care providers are legally mandated to offer the maternal serum alpha feto-protein (MSAFP) test to all pregnant women as part of routine prenatal care.[2] The procedure is described to the women, who are also given a booklet prepared by the State. Seventy-five percent of women interviewed by Nancy Press and Carol Browner said they read and understood the booklet (Press and Browner 1994). Few actually did. More than a third thought the test was required. Consistent with other studies, Press and Browner found their respondents to be confused about the test's purpose. Indeed, many considered it therapeutic; they agreed to be tested in order to insure that their "baby would be the healthiest that it could be." The women had a deep faith in science and thought the more one knows the better (even though all were Catholic and none would have chosen abortion). In their view, scientific information is always valuable and should not be refused.

Press and Browner noted that the booklet was badly designed, but they also concluded that the women did not want to hear more. They were favorably disposed toward testing because they believed that prenatal care would help produce a good birth outcome. Yet in respect to prenatal testing, this optimistic faith is in tension with its point: to discover serious anomalies, which may lead to selective abortion. Health care providers are uncomfortable discussing this possibility (in forty intake interviews, pregnancy termination was

---

2 This screening test is used to identify women at elevated risk of carrying a fetus with neural tube and some other serious birth defects.

mentioned twice; in written materials; code words were used) and women are equally uncomfortable contemplating it. So together the women and the providers create a "collective fiction": that testing is good medical care that has nothing to do with abortion (ibid.; Faden et al. 1985).

At the 1994 student Pugwash conference, Dorothy Wertz, Dorothy presented preliminary results from an ongoing project on ethical views of genetic counseling clients (based on questionnaires administered to first-time visitors to genetics clinics). In his essay, Philip comments that "there is little good to be done in telling people that they are at high risk for a disease where there is nothing that can be done to alleviate that risk (or nothing to be done that would not have been recommended whether or not the people had tested positive)"(Wertz 1997). But the survey results concerning adult-onset disorders indicate that forty-seven percent of respondents think that parents should tell a child the results of an Alzheimer's test immediately; eighty-four percent would have their thirteen-year-old daughter tested for a breast cancer gene. Concerning rights not to know, only thirty-nine percent agreed with the statement, "After taking a test, people have a right not to know results." Fifty-eight percent agreed that the "doctor should tell patient's relatives their genetic risks, even if relatives do not want to know" and forty-three percent that "people should be told test results, even against their wishes, if there is risk for mental illness"(ibid.).

Most people speak a language of autonomy and the right to choose. But they are little concerned with their actualization. Concerning rights to be tested, sixty-one percent of those interviewed for the Wertz survey agreed that "patients are entitled to whatever services they ask for, if they can pay out of pocket." Everyone has the right to what they can afford. As Ruth Chadwick has noted, autonomy has come to be identified with self-reliance; "standing on your own two feet." (That is why the same people can vote to cut services while upholding freedom of choice) (Chadwick 1993). On the prevailing view, if the market limits choices, no rights are violated. That view is also part of the "social surround." It helps explain why the actions Philip considers necessary to avoid social prejudice—ensuring adequate counseling, providing universal health coverage, and enacting strict regulations to block employment discrimination—are unlikely to be adopted.

Indeed, we are moving—rapidly—in the opposite direction. The situation in respect to phenylketonuria (PKU) provides a good example; while more money is being devoted to the development of gene therapy, less is available for the programs of social and psychological support that enable individuals to remain on or return to the required low-phenylalanine diet. Philip acknowledges this general trend. He asks whether social programs required for a benign eugenics are "likely to be implemented and sustained in a society whose prevailing concerns are with efficiency rather than equality and whether "a society that grudges the children of East St. Louis unpolluted schools [will] take on these obligations?" (Kitcher this volume, 260). I assume that these questions are rhetorical. He also implies that advocates of "molecular medicine, who believe that they can achieve its benefits without attending to the broader social inequalities," are dead wrong. The implication seems obvious. But Philip's paper leads to a conclusion he stubbornly declines to draw.

In the end, Philip leaves open the question whether the geneticists' benign intentions won't inevitably turn sour; whether they won't end up producing the kind of eugenics we fear. Indeed, he says that "critics of the promise of molecular biology are wrong to fear that we shall *inevitably* be swept into repeating past eugenic excesses" since we can imagine a different outcome (ibid., 261). But whether results are inevitable—or at least extremely likely—is not decided by the fact that we can imagine a different theoretical possibility. It is decided by concrete social structures.

I think the uses we fear are inevitable, and would be, even if the motives of HGP defenders were (and were to remain) wholly benign. I am in fact skeptical of Philip's claim that defenders of the HGP are committed to the utopian enterprise he describes. How do we know what they think? I assume that if their motives *weren't* benign, project advocates would try to keep this knowledge to themselves. But in any case, their intentions are not what matters. As Langdon Winner, Langdon has compellingly argued, there are instances where the process of technical development is so thoroughly biased in a particular direction that it produces harm when no one consciously intends it (Winner 1980). The problem is that the technological deck is stacked in advance to favor some social interests; e.g., the mechanical tomato harvester, developed at the University of California, which was able to harvest tomatoes in a single pass through a row, cutting the plants from the ground, shaking loose the fruit, and sorting the tomatoes electronically. The researchers didn't intend that 32,000 jobs would be lost and small growers forced out of farming. On the contrary. But they were part of a process that made those results inevitable (ibid.). So were the atomic scientists who worked on a weapon they thought would be used to defeat Nazi Germany and who continued their work in order to demonstrate why nuclear power should be brought under international control. These scientists never understood the nature of the social forces that would actually determine the uses made of their research. Neither do many genome scientists. But we should. The HGP will doubtless produce many medical and scientific benefits. But we can also predict with some confidence that it will bring in its train considerable social harm—whether or not anyone wills this.

Part Three

Is a Strong Genetic Program Possible?: Introductory Comments

As discussed briefly in the introductory essay, one of the recurrent philosophical issues surrounding discussions of the philosophical and social dimensions of human genetics is the degree to which it is possible to "explain," in the strong sense of this term, the properties of human beings by reference to the structural properties of their genetic makeup. As the underlying structural "code" defined by the base sequence in human DNA is more fully disclosed by HGP research, this issue becomes pressing, with social, legal and philosophical ramifications following upon the conclusions drawn about the strength of explanation supplied by genetic understanding.

On one level genetic explanation bears upon the medical implications of contemporary genetic research, particularly as involves the genetic causation of diseases and abnormalities. Presumably a causal understanding of the structural genetics related to known genetic diseases offers some possibility of technological interventions that can circumvent or even counteract the predictable effects of genetic abnormalities. This can be either by direct administration of biochemical treatments that counteract the phenotypic expression of the disease by interventions someplace in the pathway from DNA sequence to phenotype, as in the well-known case of PKU therapy, discussed in the previous section by Philip Kitcher and Diane Paul. Or it can take the form of more radical "gene" therapy that seeks to alter the genotype itself. Both forms of therapy presume a strong causal connection between underlying "genes" and their phenotypic manifestations or expressions. A reductive understanding of genetic disease may also help determine pre-dispositions or susceptibility some may have to certain kinds of diseases, or it may assist medical science in understanding the reasons why certain individuals may be resistant to epidemic diseases, such as AIDS, or even resistant to forms of cancer.

On another level, the evidence for genetic causation of manifest traits raises a more philosophical set of issues. If there are strong causal linkages between the structural properties of the genome and physically-expressed traits, can similar causal reasoning be extended beyond the more obvious case of clinical abnormalities or gross traits of the phenotype to more complex mental and behavioral properties? To answer this question, we must attend to at least three issues.

First, it is necessary to be able to give some kind of unitary characterization of an expressed mental or behavioral character sufficiently precise that it can be related to underlying genetic structure in a way that is truly "homologous" to the genetic causation of physical properties, and not simply "analogous" to it. To pose this issue in terms of a question, is the *same kind* of strong causation that can be presumed for physical properties with complex genetic causes, such as Sickle Cell anemia, also operative in the case of complex behavioral traits, such as homosexuality, in which the phenotypic expression is in the domain of behavioral choices rather than in visible structural effects?

A second issue, related to this problem, is the degree to which the relationship between genetics and behavior stronger can be claimed to be stronger than mere *correlation*? The history of human genetics is replete with instances of gross behavioral properties that were taken by medical scientists, geneticists, and social reformers to be evidence for the operation of strong genetic causation, only to be seen later as embodying a large number of questionable assumptions. Characteristic of the missteps of the eugenics movement of the 1920s and 30s was the common assumption that it was possible to infer from observed behavioral properties, such as "feeblemindedness," to underlying genetic causes with the same warrant that one could reason to the complex genetic causes of such properties as coat color in guinea-pigs (Kevles 1985, 2000 in press).

A third difficulty concerns the issue of the complex relationship between base-sequence structure and phenotypic expression, a problem that is multiplied when one is dealing with multi-gene effects. For example, the *functions* of the rhodopsin protein involved with vision in two of the model multicellular organisms being studied in the HGP, *Drosophila* and humans, might be very similar. However, their actual amino acid sequences may be only in the order of 25% exact identity, and the actual DNA base sequences involved in coding for these proteins is likely even less.[1] To what extent we can speak of the "same" genes being operative in producing this protein is obviously not a simple matter to decide. Consequently, the claim that genes "cause" complex behavioral patterns in human beings must at some point confront the enormous complexity of the process of genetic expression. The completion of the HGP will supply great technological power for determining the base-sequence of human DNA rapidly. But this is considerably distant from being able to apply this information to the intricate problem of developmental gene expression and its relation to behavioral patterns.

Efforts at reviving this "strong" program in behavioral genetics by a group of contemporary geneticists display the importance of the concept of the "gene" and its use in genetic discourse, a concept with a complex history. In this section of the book, Evelyn Fox Keller leads off with a valuable historical discussion of the gene concept in light of current knowledge of the complexity of developmental genetics, and raises some of the significant questions concerning its continued viability. The commentary by Jean Gayon pursues these issues and also seeks to link Keller's discussion more closely with the HGP.

Following this is an extended discussion of the empirical warrant for the strong genetic program by Kenneth Schaffner. He will analyze the question first by examining the researches on simple organisms, particularly focusing on the round-worm *Caenorhabditis elegans*, one of the five original target genome project organisms. The complete sequencing of this organism's genome, announced in December of 1998 (Genome Sequencing Consortium 1998), represents the first complete genetic mapping of a multicellular organism. He will then extend these issues into the domain of behavioral genetics by looking closely at the controversial research of David Hamer on the genetic causes of human homosexuality. Edward Manier's commentary en-

---

1 I thank my colleague, David Hyde of the Notre Dame genetics unit, for very illuminating conversations on this issue. Responsibility for these claims is solely my own.

gages Schaffner in a critical discussion of these issues. Aspects of these discussions also refer back to the comments raised in Timothy Murphy's commentary on Arthur Caplan's contribution to this volume.

The questions raised in this section about genetic explanation of human traits set the groundwork for the reflections on reductionism and theology that are pursued in the final section of the book.

1 0

Is There an Organism in This Text?[1]

Evelyn Fox Keller

It is a well-established fact that language is not only our servant, when we wish to express—or even to conceal—our thoughts, but that it may also be our master, overpowering us by means of the notions attached to the current words. This fact is the reason why it is desirable to create a new terminology.... Therefore, I have proposed the terms "gene" and "genotype"....The "gene" is nothing but a very applicable little word, easily combined with others, and hence it may be useful as an expression for the unit factors". . . in the gametes, demonstrated by modern Mendelian researches.... (Johannsen 1911, 132)

Ten years ago, Richard Burian observed, "There is a fact of the matter about the structure of DNA, but there is no single fact of the matter about what the gene is" (Burian 1985, 37). In the interim, things have only gotten worse. At the very least, we must admit to some irony here: attempts to define the "gene" have all but collapsed at the very moment in which belief in the causal powers of genes has reached a cultural and scientific zenith. More than temporal coincidence is in fact at work here, for both the ideological force of genes and their conceptual disarray are intimately tied to current developments in molecular biology—most notably, to the success of the Human Genome Project.

---

1 The title is a play on Stanley Fish's famous question: "Is There a Text in the Class?" (Fish 1980), later transmuted by Mary Jacobus to "Is There a Woman in This Text?" (Jacobus 1982).

In large part, the definitional disarray of the gene, by now observed by numerous authors in the history and philosophy of biology (e.g., by Burian 1985; Falk 1986; 1990; 1995; Rheinberger 1995; Buerton 1995; Kitcher 1992; Carlson 1991; Atlan and Koppel 1995)—and even by some working biologists (e.g., Lewin 1983; Watson et al. 1987; Fogel 1990)—has become visible as a direct consequence of information about nucleotide sequences that has been emerging from molecular biology for the last quarter of a century, and especially, since the initiation of the Human Genome Project. Sequence information has led to the identification of repeated genes, split genes, overlapping genes, cryptic DNA, anti-sense transcription, nested genes, transposition, multiple promoters that allow for alternative sites of and variable criteria for the initiation of transcription—all of which have hopelessly confounded hopes for a structural definition of the gene. Similarly, the identification of alternative splicing and other forms of editing of the primary transcript, as well as of regulatory mechanisms operating on the level of protein synthesis and function, have confounded attempts at a clear-cut functional definition of the gene, and most recently, research on processes of methylation and gene imprinting even disturb accepted definitions of the gene as unit of transmission. Thus, in addition to all the older problems having to do with the many to many relation between gene and chromosome, today, the gene can no longer be said to be an identifiable sequence of DNA, a mappable locus on the chromosome, a clearcut functional unit, or even a stable entity transmitted through the generations. Perhaps, as Petter Portin suggests,

> Our knowledge of the structure and function of the genetic material has outgrown the terminology traditionally used to describe it. It is arguable that the old germ gene, essential at an earlier stage of the analysis, is no longer useful . . . . (Portin 1993, 208)

Many authors would seem to agree. Philip Kitcher, e.g., notes in concluding his own review of current uses of the term that "it is hard to see what would be lost by dropping talk of genes from molecular biology and simply discussing the properties of various interesting regions of nucleic acid" (Kitcher 1992, 130). In a similar vein, Hans-Joerg Rheinberger asks, "Why stick to a 'unit' structure of the hereditary material when there appears to be no general and universal genetic unit of the type suggested by classical genetics?" (Rheinberger 1995, 13).

Why, indeed? What *do* we need the notion of a *gene* for? Talk of *genes* is easy, catchy, and it has manifest promotional value, but the question is, does it—at least today—have conceptual value? Kitcher suggests no; he would be content to speak instead of "interesting regions of nucleic acid," and Rheinberger suggests that "it might be more reasonable today to speak of genomes than of genes. . ." (Rheinberger 1995, 11). Yet, for all my sympathies with their arguments, and for all my concern about the undue cachet *genes* have now acquired—both in popular and in scientific culture—I am not sure I am willing to so readily dispense with the term. At least not without a better understanding of how that term (and concept) has functioned historically, and how it is functioning today. The suggestion that we dispense with *genes* in favor of regions of nucleic acid or a genomic text makes me just a little bit wary, especially in the context of the alternative metaphors now being promulgated (e.g., the "book of life"). I would argue that the *gene* does continue to have conceptual value, but now that value is to be found in its very disarray.

It may be worth noting, at least parenthetically, that there is in fact considerable tension among biologists today over this very question, separating those who identify themselves primarily as geneticists from those who identify themselves primarily as molecular biologists. To a certain degree, the differences between these two groups are reminiscent of an earlier division, namely that between embryologists and geneticists. Surprisingly, and perhaps paradoxically, *genes* function in these debates to remind modern geneticists of what it is that makes a region of nucleic acid "interesting," or of what constitutes "meaningful structure" in the genome. In a word, *genes* have come to remind at least some biologists of the organism, pointing to its peculiarly biological as distinct from its strictly biochemical properties.

This curious shift in function of the concept of *gene*—from its historical role in eclipsing the organism to its current (at least occasional) role in signalling organismic dynamics—is precisely the effect of the conceptual unravelling we are now witnessing. Of course, Rheinberger and others remind us that the concept of the *gene* was never either unitary, comprehensive, or "clean." Nor, for that matter, has the concept of *gene* been stable over the years. But one feature distinguishes the present from the past: throughout its many variations and transformations over the course of its lifetime, it had always been possible in the past to contain whatever definitional difficulties surrounding the concept of *gene*; one might even say that it had been functional, now both experimentally and professionally, to keep its internal incoherence under wraps. What is distinctive today (not to say ironic) is that molecular biology—with considerable help from the Human Genome Project—has provided the conditions for breaking this historic silence. The sequence information which has now become available puts a demonstrable wedge between the various attributes of a *gene* (e.g., its physical location, material composition, structure, function, and causal efficacy).

If, for example, a gene is taken as that sequence of nucleotides transcribed and translated into a particular polypeptide chain, this particular entity may not exist as DNA at all, but only as mRNA, and may not come into existence until after the mRNA has been spliced. As such, this *gene* can no longer can be said to reside on the chromosome, and perhaps not even in the nucleus.[2] By contrast, Singer and Berg define a gene as "a combination of DNA segments that together constitute an expressible unit, expression leading to the formation of one or more specific functional gene products" (Singer and Berg 1991, 440). Their definition keeps the gene on the chromosome, but leaves open the question of its exact constitution. The key word here is "expressible," allowing, as it does, for an after-the-fact designation of the particular combination of DNA segments actually employed.

Historically, the self-identity of genic structure, location, composition, and function had always been taken as a given. Now, that self-identity has been disrupted: the functional gene has been released from its traditional material location on the chromosome, be it as a bead on a string or a stretch of DNA, and accordingly, from the unit of transmission, i.e., the entity that is presumed to guarantee intergenerational memory. Indeed, the functional gene may have no fixity at all: its existence is often both transitory and contingent, depending critically on the functional dynamics of the entire organism. Considered as a functional unit, the gene is no longer a static entity, set above and apart from the processes that specify cellular and intercellular organization, but itself a part and parcel of these processes, defined and brought into existence by the action of a complex self-

---

2 The spliced mRNA, although constructed in the nucleus, is immediately transported into the cytoplasm.

regulating dynamical system in which and for which the inherited DNA provides the crucial raw material. One way of posing the particular difficulty that is here made explicit might be to paraphrase Howard Pattee's question from 1969 and ask, "How does a sequence become a gene?"[3]

A second—and perhaps more important—difficulty revealed by the new research on gene structure and function pertains to the question of causal efficacy. In both classical and early molecular genetics, the *gene* was assumed to be not only a fixed and unitary locus of structure and function, but also of causal agency. T. H. Morgan, for example, regarded the assumption that genes are the causal agents of development as so basic and so self-evident that an understanding of heredity did not require its elaboration.[4] The attribution of causal agency is implicit in what I have elsewhere called the "discourse of gene action" (Keller 1995). I argue that the notion of "gene action" has functioned historically to endow the material gene simultaneously with properties of life and mind. As a unit of transmission, the *gene* was credited with permanence; as an autonomous entity capable of reproducing itself, it was credited with vitality; as ontologically prior to life, with primacy; as the locus of action, with agency; and as capable of directing or controlling development, with mentality. Part physicist's atom and part Platonic soul, the *gene* was assumed capable simultaneously of animating the organism, of directing, and of enacting its construction; as Schrödinger put it, it was taken to be "law-code and executive power–architect's plan and builder's craft–in one" (Schrödinger 1944, 23), or, in more recent parlance, as "the cell's brain" (Baltimore 1984, 150). Indeed, it was the attribution of vitality and mentality that invited the trope of the gene as organism so frequently invoked by its early advocates, and that, at the same time, invited the charge of an implicit preformationism brought (albeit somewhat less frequently) by some of its critics.

Those days are all but over. Today, the mental and vital powers of the classical and early modern gene are rapidly being dispersed throughout the organism. Contemporary metaphors are more likely to

---

3 Pattee's question was, "How Does a Molecule Become a Message?" (Pattee 1969).

4 As he wrote in 1928, "In this sense, the theory of the gene is justified without attempting to explain the nature of the causal processes that connect the gene and the characters" (Morgan 1928, 25).

locate the cell's brain in the regulatory dynamics of an intra- and inter-cellular signalling system than in the gene itself. Eric Davidson, for example, locates it in a "complicated assemblage of proteins known as a transcription complex" (quoted in Beardsley 1991, 87); others would insist on including those complexes involved in post-transcription regulatory processes. But whatever one's experimental focus, in much of the laboratory practice (even if not in the more public rhetoric) of contemporary molecular biology, the privileged status of genes (however defined) as the pineal gland of the biological organism has undergone major erosion. And as genes lose vital, causal, and even operational force in biological thought—as the locus of control is seen to shift to the complex biochemical dynamics (protein-protein and protein-nucleic acid interactions) of cells in constant communication with each other—so too does "the discourse of gene action" lose *its* cognitive and linguistic force in the formulation of questions and in the design of experiments.[5] In a recent review, *Scientific American* glosses the changes I am describing as the news "that organisms control most of their genes" (ibid.). We might think of this turn of events as marking a return of the repressed. Or, to put it only slightly differently, we might say that the "organism" has been disinterred from its earlier submersion inside that material entity referred to as the *gene*.

Several different kinds of questions arise from these considerations: What held these various attributes together in the past? To what extent ought we try to reread some of the classical controversies in genetics with particular regard to the multiple meanings the term 'gene' has in the past been taken to cover? And, of course, how are we to account for such a turn about today? Earlier, I credited the Human Genome Project with providing (however unintentionally) much of the impetus for this transformation, arguing that despite all expectations, the sequence information that has emerged from that project has worked to subvert its founding"master molecule" dogma. That would seem to lead to a conclusion of the form, "nature will out." But surely, nature itself is neither sufficiently pristine nor sufficiently powerful to shoulder full responsibility for such an upheaval—no more so today than it could have borne responsibility for older ways of figuring genes and organisms. We might like to think

---

5 Though far from perfect, "gene activation" is generally seen as closer to the mark.

the claim that "organisms control their genes" is closer to the truth of the matter than the assertion that "genes control organisms," and that this shift in discourse reflects the natural progress of science, but we must remember that both are, as they say, "just" ways of talking. And as ways of talking, they inevitably reflect far more than the experience of scientists generated by their experimental practices (see e.g., Keller 1995). Furthermore, at least some of the difficulties that have now been made explicit might have been evident (and indeed were evident to some) from the beginning of genetics. An obvious task which thus faces historians is that of better understanding how these different ways of talking have worked in the past, and how they work in the present—in short, how scientific language, experimental practice, and social expectations have worked (and continue to work) in concert and mutual reinforcement.

But another question also arises, though it may not be a question for historians. And that is, what new ways of thinking about biological questions do the new results and the new metaphors permit? And what changes in experimental practice do they invite? Rather than try to answer these questions, I will attempt to sketch out some of the preliminary groundwork that might be useful in thinking about such questions. In the remainder of this paper, I offer something of a Cook's tour of the shifting metaphors relating genes and organisms over the course of the century—both in earlier "preformationist" conceptions of *genes* and in the "epigeneticist" conceptions currently gaining favor. In conclusion, I turn (even more briefly) to some opportunities for radical reformulation now before us.

TOWARD A HISTORICAL THESAURUS

Perhaps the first formulation of the classical gene—and the first formulation of the gene as a miniature organism—was given by Hugo DeVries in 1889. Seeking to retrieve what he saw as the crucial component of Darwin's theory of pangenesis, he introduced the term *pangens* to refer to "the minute material particles in the germ cells which Darwin had hypothesized to represent" all the hereditary characters of the organism, and to distinguish these particles from the more questionable entities Darwin had claimed could be transported from full-grown organs back to the germ cells. Like Weismann, he located these pangens in the nucleus. But unlike Weismann, he located these self-reproducing particles in the nuclei of all cells: against Weismann's distinction between germ and somatic cells, he offered a parallel distinction between nucleus and cytoplasm, in which the pangens in the nucleus would maintain their structural integrity, but those entering the cytoplasm could be converted into cellular specific organelles. DeVries also differed from Weismann in his conception of what these particles are. Where Weismann had suggested a substance "with a definite chemical, and above all, molecular constitution" (Weismann 1885, 170), DeVries argued that if these particles were to represent the *Anlagen* of the organism, they would need to be something more than mere chemical molecules. In a spirit somewhat reminiscent of Buffon's notion of *living organic molecules*, he wrote,

> These minute granules are not the chemical molecules; they are much larger than these and are more correctly to be compared with the smallest known organisms. (DeVries [1889] 1910, 4)

The crucial term is, of course, "represent." What does it mean to *represent* the properties of the adult organism? The problem DeVries faced in 1889—that which led him to invoke the figure of pangen as an organism—was nothing less than the problem of integrating the demands of developmental determination with those of transmission, of reconciling the regulative capacity of the organism with the material units of heredity (see Ravin 1977). In short, the same problem that would plague genetics for the next century. How could a mere chemical molecule carry such regulative powers?[6]

After the the rediscovery of Mendel's laws, Johannsen's coining of the term *gene*, the mapping work of Morgan's group, and the accumulating evidence for the location of the units of transmission on chromosomes, the ground for the materiality of the *gene* became vastly more solid. Looking back, later generations of geneticists might have said that DeVries had been tripped up by a confusion between "characters" and *genes*.[7] It is not characters themselves which are transmitted, but only the antecedent material entities. As early as 1910, Morgan was able to write,

---

6 For a particularly interesting analysis of the way E. B. Wilson employed a diagrammatic representation to manage this problem, see Griesemer and Wimsatt 1989.

7 Especially after the distinction Johannsen introduced in 1911 between genotype and phenotype.

When we speak of the transmission of characters from parent to offspring, we are speaking metaphorically; for we now realize that it is not characters that are transmitted to the child from the body of the parent, but that the parent carries over the material common to both parents and offspring. (Morgan 1910, 449)

This material "common to both parents and offspring" did not actually "represent" the characters, it merely caused their appearance.

But DeVries' problem would not go away so easily. The question still begged, and Arnold Ravin has put it clearly:

[To] what did [genes] owe their powers of affecting the outcome of developmental processes? Would it be possible to reconcile, within the context of a material hereditary factor, the properties of transmission and potency in developmental determination? (Ravin 1977, 19)

Ravin argues that the most important attempt at just such a reconciliation in the early years was provided by Leonard Thompson Troland's "enzymatic theory of life" (Troland 1914). Troland, a graduate student in psychology (at Harvard) with a background in biochemistry, set out to find "*a material principle of regulation having an essentially chemical nature and function*" (ibid.). He latched onto the enzyme as a substance which by its mere presence in a mixture is able to facilitate and direct the chemical changes taking place" (ibid., 98). An enzyme, he argued, would have both the autocatalytic and heterocatalytic properties required; it could account not only for growth and reproduction, the defining features of life, but also for life's origin:

That control of self and environment, which the vitalist rightly regards as [life's] defining feature, appertains to life only by virtue of the store of *enzymes* which is hidden away in the protoplasm, and when one of these enzymes first appeared, bare of all body, in the aboriginal seas it followed as a consequence of its characteristic regulative nature that the phenomena of life came too. (Ibid., 104)

Enzymes, he argued, could fit the bill simultaneously for Weismann's "hereditary determinants" (ibid.,128) and for the supposedly nonphysical factor ("E for entelechy") which Driesch had posited to account for the "autonomy of life," i.e., for "the specifically regulative function of the organism"(ibid.,133). Driesch's "entelechy," as Troland wrote,

> is the ground of every complex purposeful adaptation; in a word, it accomplishes just those tasks which we have assigned to the enzymes of the germ-plasm . . . . Is it not, then, legitimate to parody the symbolism of Driesch, and to write: E = *enzyme*? (Ibid.)

In conclusion, Troland argued that we should regard "the enzyme— and not the mystical 'entelechy'—as the pilot of life's journey" (ibid.).

Three years later, in an account of his theory that was more directly aimed at biologists, he identified enzymes yet more explicitly with genes:

> On the supposition that the actual Mendelian factors are enzymes, nearly all [the biological enigmas] instantly vanish. (Troland 1917, 328)

Troland's aim was not simply to rebut Driesch's vitalism, but also, and perhaps even more to the point, to close the gap between DeVries' "organism" and Weismann's chemical molecule. He sought to demonstrate the possibility of a fully mechanistic account of life, one that "links these great biological phenomena . . . directly with molecular physics" (ibid., 327). In effect, his aim was to show that the "organism" could itself be a chemical molecule, albeit of a particular kind (i.e., an enzyme). Thus, while C. M. Child objected that,

> The hypothetical units [the genes] are themselves organisms with all the essential charactersitics of the organisms that we know . . . . In other words, the problems of development, growth and reproduction, and inheritance exists for each of them, and the assumption of their existence brings us not a step nearer the solution of any of these problems…. (Child 1915, 11–12)

Troland's argument was not so much that these units are not organisms, but that organisms could themselves be molecules.

His intervention in these debates is notable on several counts: its ambitiousness (particularly given that he was only a graduate student, at least until 1915); its consonance with modern theories (he even suggested that it "now seems probable [that] the genetic enzymes must be identified with the nucleic acids" (Troland 1917, 342); and perhaps especially, its influence. Despite the fact that Troland was both a novice and an outsider (he was neither a geneticist nor even a biologist), and despite the extremely hypothetical nature of his theory and its almost complete lack of experimental support, Troland provided a physical-chemical analogy to fill a conspicuous gap in the gene theory, and one that could be used to directly counter remaining proponents of vitalism. Probably the most direct influence of this work was on H. J. Muller, explicitly so in Muller's 1922 paper, "Variation Due to Change in the Individual Gene," (Muller 1922) and at least implicitly, in his 1926 paper, "The Gene as the Basis of Life" (Muller 1929).

Muller was somewhat skeptical of the notion of genes as enzymes, a notion he ultimately abandoned, but he kept Troland's characterization of the fundamental properties of genes as their autocatalytic and heterocatalytic capabilities, with particular emphasis on the specificity of the autocatalytic property and on the gene's temporal and ontological priority.[8] He wrote:

> Genes (simple in structure) would, according to this line of reasoning, have formed the foundation of the first living matter. By virtue of their property . . . of mutating without losing their growth power they have evolved even into more complicted forms with such by-products—protoplasm, soma, etc.—as furthered their continuance. Thus they would form the basis of life. (Muller 1929, 920)

It is noteworthy that what Muller was able to take from Troland was simply the claim that there could be physical-chemical entities with the necessary property, stripped of Troland's more specific proposal of what kind of entity might qualify. Where Troland had felt the need to ground such a claim (theoretically if not experimentally), it appears that Muller felt no such need. Indeed, this may be the most critical point: geneticists like Muller seemed to need neither experimental nor theoretical grounding for the multiple powers they attributed to genes. What they needed was a reassuring way of talking, and this they (or at least Muller) did take from Troland.

Of course, not everyone was so reassured. As J. B. S. Haldane pointed out in 1928, one could equally well cast genes in a quite different role:

---

8 By 1947, in order to better distinguish his own entities from enzymes, Muller changed these terms to "autosynthetic" and "heterosynthetic."

In the present state of our ignorance we may regard the gene either as a tiny organism which can divide in the environment provided by the rest of the cell; or as a bit of machinery which the living cell copies at each division. The truth is probably somewhere between these two hypotheses. (Haldane [1928] 1932, 147)

Three years later, his father, John Scott Haldane, expressed his own sympathy for the charges some had leveled (e.g., by E. S. Russell 1930) that genetics was still caught in a kind of preformationism in its assertion that genes are "so constituted physically and chemically that . . . [they] gave rise to all the amazingly specific details of the structure and activity observed in the adult organism:"

> Dr. Russell points out that this is in reality only a variant of the box-within-box" theory, an extremely complicated molecular structure capable of producing the adult form being substituted for the original miniature adult. (J. S. Haldane 1931, 147)

As we know, however, such concerns—either father's or son's—were not the ones that prevailed. What prevailed was a particular way of talking that collapsed all vital functions onto the unit of transmission, and not Haldane's (or any other) alternative. This conflation, despite changes in vocabulary, persisted well into the early modern period.[9] Most notably, it persists both in Schrödinger's essay, *What is Life?*, and in the rhetoric of early Molecular Biology.

It is not clear whether or not Schrödinger was familiar with Muller's writings, but the resemblance between his essay and Muller's own essays does not require that we suppose he was: it is sufficient to recognize that both were drawing on the same traditions. Schrödinger's most significant contribution to this tradition was, on

---

9 Various attempts over the years to introduce alternative (more interactionist) formulations were of course made. One such attempt can be seen in the papers of a Symposium on Causality organized by developmental geneticist Donald Poulson and Heinz Herrman in 1948 (Poulsen Papers, Yale University Department of Biology Archives 1948). Poulson in particular argued that the evidence from genetics demands a more complex understanding of the causal relations between genes and effects: "First, the efficacy of a gene depends . . . on the second allele in the gene pair; the gene action is influenced by other genes in the same or in other chromosomes, and eventually it is subject to factors originating in the inner environment of the organism itself or in the external environment." Poulson Papers, fol. 7.

the one hand, his particular rephrasing, and, on the other, the authoritative imprimature he was able to lend to the collapse between transmission and development (i.e., between autosynthesis and heterosynthesis). Invoking both quantum mechanics and the Second Law of Thermodynamics, he authorized a formulation of the inscription of mentality and vitality in the chromosome structures in the terms of physics (see also, Doyle 1994; Keller 1992; 1995). Just a single quote should suffice:

> It is these chromosomes . . . that contain in some kind of code-script the entire pattern of the individual's future development and of its functioning in the mature state. Every complete set of chromosomes contains the full code . . . .
> But the term code-script is, of course too narrow. The chromosome structures are at the same time instrumental in bringing about the development they foreshadow. They are law-code and executive power—or, to use another simile, they are architect's plan and builder's craft—in one. (Schrödinger 1944, 22–23)

Even Schrödinger, notwithstanding his formidable theoretical powers, was in fact unable to solve this problem—i.e., he was unable to provide an account of just what kind of entity might simultaneously embody the gene's law-code and executive power." The best he could do was to take recourse in a "device" that would function as a Maxwellian Demon—a device, as he put it, that "really consists in continually sucking orderliness from its environment" (ibid., 79); in short, a device that could enact the Cartesian imperative with which his essay begins: "Cogito ergo sum."[10]

Within little more than a decade, a newer and better set of metaphors became available, which molecular biologists were able to put to remarkably effective use. A great deal has been written by historians of biology about the importation of computer science metaphors into biology—especially, those of program and information, but here my point is somwhat different from the arguments that have most frequently been made in this literature.[11] It concerns both

---

10 See Keller 1995, chap. Two, for further discussion.
11 By far the most insightful account of this subject is more of a philosophical than a historical treatise—namely, that provided by Susan Oyama (1985). Much of my point here might be thought of as underscoring her earlier and more extended argument. See also Lily Kay, *Who Wrote the Book of Life? A History of the Genetic Code* (Stanford University Press, 2000), and Kay this volume.

the gap between the technical definition of *information* in computer science and the use of that term made by biologists, and the obfuscation of this difficulty brought about by the subsequent collapse of *information* with the notion of *program*.

As early as 1960, André Lwoff pointed out the inadequacy of the technical notion of information for encompassing "the idea of quality, the specific value [that] is included in the biological concept of genetic information"(Lwoff 1965, 93–94).[12] But Lwoff did not address the critical question to which his observation points: How *does* the gene acquire its "idea of quality," its "specific value"? In short, how does the gene become a meaningful entity?

In the 1960's, molecular biologists could not resort to the older notion of gene as organism, nor, for that matter, was Schrödinger's notion of a device" much more helpful. What *was* helpful was the notion of *program*. That concept was invoked to fill the breach that had earlier been filled by "organism" (or, in Troland's efforts, by "enzyme").

A handy guide to how the notion of *information* as *program* (or *instructions*) was used in the early days of molecular biology to manage the question of how the gene acquires its "specific value" is provided in the opening pages of François Jacob's, *The Logic of Life*. Reiterating Schrödinger's dual attribution with only minimal rephrasing, he writes,

> What are transmitted from generation to generation are the instructions specifying the molecular structures: the architectural plans of the future organism. They are also the means of executing these plans and of coordinating the activities of the system. In the chromosomes received from its parents, each egg therefore contains its entire future: the stages of its development, the shape and the properties of the living being which will emerge. The organism thus becomes the realization of a programme prescribed by its heredity. The intention of a psyche has been replaced by the translation of a message. (Jacob 1974, 1–2)

Susan Oyama (1985) and others have shown how the appropriation of computer science metaphors worked to collapse form and intent, control and controlled, subject and object. In Oyama's terms, the collapse of *information* and *program* worked to reimport an "argument

---

12 It was the recognition of this difficulty that led to the collapse of earlier efforts to quantify the informational content of DNA (see, e.g., Quastler 1953).

by design,"—in effect, to reinscribe the homunculus in the gene. But it would be a mistake to see this collapse as evidence merely of a misunderstanding of computer science terminology, for the fact is that the same conflation was (and continues to be) widely employed in that discipline. To computer scientists, a program, insofar as it can be represented in digital code, *is* information, and to Norbert Wiener, communication *is* control. Perhaps, then, there is no problem. Might we not say (as many have) that computers provide just the kind of theoretical rationale that Troland had earlier (albeit mistakenly) sought in enzymes for the vital duality of function so deeply and so reiteratively built into its definition? Perhaps Schrödinger was closer to a cogent account than I have given him credit for—while not actually an organism, might not a gene be thought of as very special kind of device, a device which, in contemporary thought, is itself often imagined to be "alive"?

My answer to these questions is a decisive no. The crucial difference between computers and organisms, as Henri Atlan and others have emphasized, is that, for computers, function and goals are externally prescribed—in fact, by the human beings who program the computers and interpret its results, whereas for organisms, function and goals are generated by the actual organisms themselves. Atlan describes this peculiar feature of natural systems as the "Self Creation of Meaning:"

> [I]n living organisms . . . , the meaning of information has to do with a physiological function. The same holds true for a machine where the meaning of a part or a program instruction has to do with it function, with what it is good for within the overall functional organization of the system. However, the origin of the function is very different in a natural organization, such as a biological system, from what it is in a man-made machine. In the latter, the task to be performed is specified from outside and . . . the opposite is the case when we analyze a natural organization. Here we cannot assume that its finality has been set up by some consciousness either from the outside or from inside. (Atlan 1987, 564)

Indeed, it is just because the task of a classical program is pre-given from the outside, and can therefore be taken for granted, that computer scientists as such do not need to deal with its function (or meaning): the technical problem they face in the design of computers is the problem of transmission. It is thus for practical purposes that the meaning of a program in computer science can be (and is) bracketed and that, therefore, a program can be treated as pure information. For organisms, however, no such bracketing is possible: meaning, or function, is an intrinsic part of what we call the program. Hence, Atlan argues, programs and information (or data) need to be distinguished for natural systems, and we are back to asking the fundamental question: What kind of system is capable of generating its own program (goals, meaning, task, or meaningful data)? In other words, what is a "natural machine"?

Atlan and his colleague Moshe Koppel believe they can answer this question. They offer a measure of "meaningful complexity" to distinguish program from data,[13] which, they claim, can arise (complexity from noise) as an emergent property of self-organizing automata networks (Atlan and Koppel 1989). Mapping the current information about genetic processes onto their own experience with computer simulation, Atlan suggests a view of

> genetic determinations [as those which] result from the sequential structure of the DNA, functioning . . . not as a program, but as data which is recorded, processed, and used in a dynamic process playing the role of a program. This process is produced by the set of biochemical reactions coupled with the cell metabolism. (Atlan 1998, 4)

In the end, however, he opts for a picture in which DNA might be viewed as functioning alternately as program and as data, depending on time-scale:

> Through these *two metaphors*, we can conceive the image or model of an *evolving network in which two dynamics would be superimposed, on different time-scales.* The first-order dynamic would depend on the structure of the metabolic network and on the data it receives in the form of active genes. However, the second-order, slower, dynamic of stable states of the network would modify the activity of certain genes . . . thus producing a change in the structure of the network. . . .(Ibid., 5)

Atlan rejects the notion of a classical computer as a natural system, but he has another kind of machine which he believes *will* do the trick, at least given enough time to evolve. What kind of machine is

---

13 Closely related both to Chaitin's measure of "logical depth" (1977; see also Bennett 1986) and to Gell-Mann's notion of "effective complexity" (1995).

this? It is "a multi-layered parallel computer network." Maybe. But even if not, this newest of machines does provide a metaphor with which one can talk about genes in ways that are consonant with the findings of the new molecular biology. In particular, it enables Atlan to argue that "what is genetic is not in the gene" (ibid., 1).

CONCLUSION

But it is not Atlan's conclusions that interest me here so much as the questions he makes explicit. Indeed, I cannot even promise that the work on which they are based bears close technical scrutiny. More important is the fact that the particular questions he raises—about the emergence of function, about the generation of biological meaning, about what a meaning-making machine might be—are the central questions that biology faces today—or perhaps I should say, that it is possible for biology to face today. Indeed, my own question, "Is there an organism in this text?," is just a variant of the larger and more general problem Atlan is posing.

Questions about what makes a particular structure an organism are among the oldest and most central questions of biology. What is new today is the reopening of a space in which to raise these questions, in which to generate models in the hope of finding possible answers, and to design experiments or simulations to test such models. And Atlan and Koppel are not alone: in one form or another, this has become the central focus for a number of researchers, perhaps signalling the reemergence of a theoretical biology.[14] For its particular role in helping to create this space, however inadvertently, and however self-defeating to its original logo, I want to suggest that the Human Genome Project deserves at least some credit.

---

14 For other perspectives, see e.g., Gunter Wagner; Stuart Kauffman; Stephen Wolfram.

The Human Genome Program, Cognitive and Practical Aspects: A Commentary on "Is There an Organism in This Text?"

Jean Gayon

There is an obvious irony in Professor Keller's presentation. Except for a few allusions at the beginning and at the end of her paper, she carefully avoids saying anything explicit about the Human Genome Project. Instead, she offers us a reflection upon the concept of the gene in contemporary biology. Let me first summarize this elegant reflection.

Professor Keller shows that the gene concept has recently undergone a major change, both in terms of its operational content and with respect to related philosophical categories and metaphors. Starting with the operational aspect, she reminds us that recent molecular biology precludes any single and unambiguous definition of the gene: we can no longer see the gene as an identifiable sequence of DNA, endowed with the privileges of being simultaneously a stable structural, functional and transmission unit. In particular, "the functional gene may have no fixity at all: its existence is often both transitory and contingent, depending critically on the functional dynamics of the entire organism" (Keller this volume, 276). Many biologists and philosophers would certainly agree on this (see for instance Burian

1985; Gros 1986; Rheinberger 1995), and some of them have indeed proposed to give up the term "gene" as a much too ambiguous term (Kitcher 1992; Burian 1995; Rheinberger 1995).

However, this is not Professor Keller's mood. The second step of her argument consists in pointing out an important shift in the philosophical interpretation of genes. The big issue is that of causation. Older conceptions of the gene tended to emphasize the causal powers of the gene: self-reproduction, and determination—both in terms of program and control—of the organism. Today, biologists tend to reverse this schema: it is not the genes which control the organism, it is the organism which controls the genes. In an earlier publication, the author nicely captured this idea, saying that the "discourse of gene action" tends to be replaced by the "discourse of gene activation" (Keller 1995, 26). Indeed, this idea underlies her whole paper. Keller argues that the discourse of gene action has endowed the gene with the properties of life and mind, whereas modern molecular biology develops a more decentralized view of the gene-organism relation: "today, the mental and vital powers of the . . . gene are rapidly being dispersed throughout the organism" (Keller this volume, 277). Such declarations show that Keller's philosophical approach to the concept of the gene is primarily driven by the analysis of the metaphors which underlie the geneticist's causal claims. This is indeed the third aspect of the text: Keller provides a fascinating historical account of various metaphors invested in past and present concepts of the gene. I will return later to some of these metaphors. It is enough for now to observe that this kind of analysis convincingly demonstrates that there has been and there now is more in the concept of the gene than its strictly operational ingredients. As I show later, this is particularly important in the context of the present volume on the Human Genome Project.

However, as noted at the beginning of this comment, Professor Keller avoids speaking of the Human Genome, except in a very few sibylline allusions, to which I will return later. In consequence, we are left with two possible interpretations of her paper. (1) The paper is not seriously concerned with the Human Genome Project, or only in a quite indirect manner. In this case, it must be appreciated for what it seems to be: a philosophical evaluation of the history of the gene concept. (2) This paper has important implications for the Human Genome Project. Then, these implications must be explained.

Or in other words, there is a missing paragraph, which must be restored. In fact, these two interpretations are not exclusive. I will explore each of them successively.

Let us first assume that Keller's paper is a philosophical evaluation of the history of the gene concept. Obviously, her method consists in taking seriously some metaphors which have been intimately associated with the scientific use of the concept of the gene. I will make two kinds of remarks here, one of which bears on the content of the metaphors, and the other on the relation between metaphors and methodological issues in the history of genetics.

Let us begin with metaphors. A striking feature of Keller's paper is the emphasis she lays upon the attribution of the properties of life and mind to the genetic material. I was puzzled by this assertion, which at first sight I found somewhat excessive. But after reflection, I dare say that I find it extremely enlightening. I would even propose to expand it somewhat. The story told by Keller deserves to be inserted in a wider context, the history of concepts of heredity.

A few years ago, I myself wrote a paper in which I claimed that on the broad scale of the nineteenth and twentieth centuries, heredity has been successively conceptualized in relation with three major metaphors, corresponding to the Aristotelian doctrine of the three parts of the soul: vegetative, animal, and intellective (Gayon 1991). The demonstration runs as follows. The first theoretical formulations of heredity in the nineteenth century were expressed by two large rival schools. On the one hand, people like Galton, Weismann, and others advocated that heredity was to be thought of in terms of an independent line of cells or tissues (the so-called "germline"). The vegetal metaphor is obvious: "germs," and Galton's "stirp" (from the latin *stirpes*, "root") are vegetal metaphors (Weismann 1883; Galton 1875). The other school was that of the Neo-Lamarckianism: and hereditys, many of whom insisted on heredity as a particular aspect of "nutrition" or "assimilation" (Bernard 1867; Gayon 1991). Since Antiquity, nutrition has been the characteristic property of the vegetative soul. Therefore the vegetal metaphor is equally obvious here. What about traces of the animal soul in modern theories of heredity? Between 1870 and 1930, an incredible amount of literature over the "mnemonic theory of life" was produced (Hering [1870] 1920; Haeckel 1876; Coutagne 1903; Semon 1911; 1912; Rignano 1923;

Bertalanffy 1927; 1928).[1] Memory can be seen as a property of animals. Another manifestation of animal metaphors can be recognized in the discourses emphasizing heredity, not as a substance, but as the *behavior* of a dynamical equilibrium. D'Arcy Thompson was a major advocate of such a conception, but it was also quite popular in the heroic phase of molecular biology, when people commonly opposed two concepts of genetical mechanisms: "master molecules" and "dynamic organization" (Thompson 1942, 248–49; Nanney 1957, 134–35). Finally, concerning the metaphor of heredity as intellect, we are so familiar with it today that I do not need to expatiate on it. Our modern discourse on heredity is pervaded by analogies with language and computers.

I do not want to suggest here that biologists tried consciously to shape their successive concepts of heredity in relation to the Aristotelian doctrine of life and soul. But the fact is that they did so without knowing it. In that perspective, Professor Keller's discussion of the metaphoric projection of the properties of life and mind upon the material gene fits well with the general history of the concept of heredity. Indeed, her analysis adds a crucial dimension to it in the sense that it makes understandable the shift we are presently observing in molecular biology. Conceptually speaking, we are probably attending the death of heredity as *the* privileged vital actor. Modern molecular biology forces us to decentralize the most common vital and mental metaphors.

Keller's analysis of the history of the gene concept in terms of metaphors deserves another observation. To be fair, Keller is not interested in metaphors for their own sake, but only insofar as they interfere seriously with scientific research programs, or provide a powerful instrument for the analysis of the relation between internal and social history of science. In that respect, a major interest of Keller's paper is the link she suggests between what she calls "the discourse of gene action" and some methodological debates within genetics, such as the problem of the definition of the gene, and that of its causal efficacy. On this occasion, I would like to underline an interesting contrast between past and present genetics. Speaking of the interwar years, Keller says that "geneticists like Muller seemed to need neither experimental nor theoretical grounding for the multiple

---

1 For a description of this tradition, see Bowler 1983; Gould 1977.

powers they attributed to genes" (Keller this volume, 283). It is important here to recall that classical Mendelian genetics admitted a typically operationist interpretation of the gene concept. Morgan, for instance, declared on the occasion of his Nobel lecture in 1934: "at the level at which the genetic experiments lie it does not make the slightest difference whether the gene is a hypothetical unit, or whether the gene is a material particle" (Morgan 1965, 315). Therefore it seems that classical geneticists, although they had an operationist and non-realist interpretation of the gene, believed nevertheless in "gene action." Today molecular geneticists, after a thirty year period of enthusiastic scientific realism, come back more or less to an operationist interpretation of the gene: operational definition of concept.[2] This is not for lack of knowledge concerning the material entities and processes involved in genes. It is because biologists no longer believe in the existence of a non-ambiguous entity that can be called a "gene." The property of being a gene is instantiated in many ways, and through a number of different processes. In consequence, as Keller recalls at the beginning of her paper, contemporary biology can no longer offer a single definition of the gene. It is in this precise context that genes, to borrow Keller's nice formulation, tend to be viewed more and more as "activated" rather than actors, or as processes rather than substances (Keller 1994). In other terms, the better our causal account of what genes are and do, the less these entities are likely to be interpreted as "agents."

Enough now for Keller's penetrating insights regarding the transformation of the gene concept. I now turn to a more puzzling aspect of the paper. As I already mentioned, Professor Keller says hardly anything about the link between her analysis and issues—scientific, ethical or philosophical—raised by the Genome Project. Looking very carefully, there are in fact a few mysterious allusions. All of them boil down to saying that the Human Genome Project has provided "much of the impetus for [the] transformation" of the gene concept advocated by the author. However, even this rhetorical concession is somewhat obscure: Keller adds that if the Genome Project has contributed to this conceptual transformation, it has done so "unintentionally" or "inadvertently." What should we understand here? There is seemingly a missing paragraph. Perhaps Keller's mes-

---

2 On the historical and philosophical aspects of this question, see Beurton 1995; Gayon 1995a; 1995b.

sage is so evident that it can be easily deduced, or else that it would be too long to explain. Another conjecture might well be that the author, who had already published on the Human Genome Project, encountered some unknown difficulty in relating these issues to one another. Whatever that may be, I will try to provide the most probable reconstruction of the "missing paragraph." Of course, it must be clear that it will be a pure conjecture, the responsibility of which is fully mine.

The only cue we have is Keller's suggestion that the Genome Project has "unintentionally," "inadvertently," and "dramatically" provided "much of the impetus for [the] transformation" of the gene concept and for the rehabilitation of "organization" in biology. I understand these declarations in the following way. The Human Genome Project has its origin in a certain ideology, widespread among contemporary biologists and influential enough in society to obtain substantial funding. I suppose that Keller would call this ideology the "discourse of gene action." According to this discourse, since genes are the basis of life and the ultimate determinants of anything biological that happens to living creatures, if we know all of them, we will increase our biological knowledge and power as we have never done before.[3] Now Keller's idea is probably that the development of the human genome program has produced, and will seemingly produce, various effects which do not fit well with this ideological impulse.

What are these "inadvertent" or "unintentional" effects, not to mention their "dramatic" aspects? I think it is essential here to distinguish two kinds of effects: cognitive effects and practical effects.

Until now, the cognitive effects of the Genome Project have been controversial. Consider for instance the debate about whether one should sequence only the coding regions, using cDNA as much as possible,[4] or whether it is preferable to sequence much larger regions. The molecular biologists who argue in favor of sequencing more than cDNA say that, by doing so, biologists gain considerable information about many regulatory processes. For instance, they will learn about the processes which make it possible for a single sequence to be processed in various ways according to some developmental stage, and how it relates to the particular location of some tissue within the organism (Hood 1992). In some sense, most, if not all,

---

3 A good example of this can be found in Watson 1992.
4 "Copy DNA," obtained from RNA transcripts with the aid of reverse transcriptase.

molecular biologists would certainly admit this kind of argument. In Keller's own philosophical terminology, the discourse of such biologists would belong to the "gene activation" type, rather than to the "gene action" type. In time, the most probable cognitive effect of the Genome Project will be to foster a huge amount of work on unknown organizational properties of organisms that we are just beginning to suspect. The scientific future of the project belongs certainly more to physiology than to anatomy, whatever may be said about the difficulties of the present task of "mapping." Therefore, to sum up, on purely cognitive grounds, the kind of discourse which initially supported the Genome Project will most certainly be subverted from the inside. This is in fact a rather obvious consequence of the present state of knowledge in molecular biology, independently of the Human Genome Project. The Genome Project did not create this cognitive situation. It amplified it. Enough now for Keller's allusions to the "unintentional" aspect of HGP.

Let us now turn to the practical effects of the development of the Human Genome Project. I suppose that Keller would call it the "dramatic," if not the "tragic" aspect of it. Thousands of people have made the same observation: in the short run, therapeutic prospects are weak, because there are many steps between the location of a gene and the knowledge of its function, and sometimes even more between the knowledge of its function and the invention of a new therapy (see for instance Lewontin 1992). Therefore, at least in the proximate future, the major application of our knowledge of the human genome will be prenatal and presymptomatic diagnosis. This means that for a good deal of time, the main practical prospect of the Genome Project will be to dramatically increase, or reveal, the impotency of therapeutics. This is not to say that pre-symptomatic diagnosis is completely useless from a medical point of view. The prediction of diseases may indeed be prove very useful in suggesting precautions which, if observed, could lessen the risk of manifestation of the illness, or even cure it.[5] But, no doubt, prenatal diagnosis has revived the eugenic question. I do not want to enter this question, which Evelyn Fox Keller has herself nicely treated in another context (Keller 1992), and which is dealt with by Martin Pernick, Philip Kitcher, and Arthur Caplan in this

---

5 I am thinking of the examples of Phenylkeptonuria and Retinoblastoma pigmentosa explored at this conference by Theodore Dryja and by Philip Kitcher and Diane Paul in this volume.

volume. I will content myself with pointing out the relation between this kind of issue and Keller's reflection upon the "discourse of gene action." Present debates on predictive medicine and so-called "new eugenics" are deeply impregnated by the discourse of gene action. In this area, most people go on reasoning as if genes were indeed the "basis of life," the ultimate living particles responsible for life and health.

This brings me to my conclusion. Earlier, I suggested that, if there was a "missing paragraph" in Professor Keller's presentation, there might be a difficulty somewhere. The difficulty, if there is one, is a very objective one, a real challenge for modern societies. The difficulty consists in a conflict between the two sides of the Genome Project that I have evoked, its cognitive effects and its practical effects.

From a purely theoretical point of view, the extensive study of the genome will almost certainly force biologists to return to questions of organization and to give up the naive view of genes as ultimate vital agents. Once they will have mapped the genome, or once some imaginative companies will have done this for them, biologists will not be able to go on believing in the unilateral powers of genes. Biologists will have to work hard on the causal network in which genes are involved. And this will lead them to abandon narrow-minded genetic determinism, even though genetic determinism—or the "discourse of gene action"—provided the initial ideological impulse for the Genome Project.

However, from a practical point of view, we do not know whether modern societies will give up so easily the "discourse of gene action." We do not know whether they will renounce the vitalist and preformationist view of the gene and heredity that prevailed throughout the nineteenth and twentieth centuries, and we do not know whether they will modify their expectations, individual and collective, relative to the knowledge of the Human Genome. As Henri Atlan declared recently, genetic preformationism: and gene concept involves some sort of magical thinking: the cause seems to contain the effect; the genome seems to "contain" the properties of life (Atlan 1994). Hence the metaphor of "culprit genes," genes which deserve elimination, and possibly the elimination of their owners. I do not want however to take a stance on abortion or *in vitro* sorting of embryos. This is discussed in limited ways by Diane Paul, Arthur Caplan, and Philip

Kitcher in this volume. I only wish to insist on the philosopher's responsibility in these circumstances. What present science tells us is that genes are nothing but molecules, sometimes—to quote again Keller's formula—transitory and contingent molecules. Such molecules cannot be culprits, nor can they be the first and last word on life and health.[6] We can be grateful to Evelyn Keller for reformulating this idea in a philosophically suggestive way.

---

6 DNA cannot be "guilty." But to borrow the title of a recent technical book, it can be a "drug" (Kahn 1993). "DNA as a drug" might after all be an efficient antidote to the discourse of "gene action" and "culprit genes." Please note that this meaning of 'DNA' is not a metaphor.

11

Reductionism and Determinism in Human Genetics: Lessons from Simple Organisms[*]

Kenneth F. Schaffner

Many molecular biologists now frankly admit that the ultimate object of their interest is not simply the system with which they work. It is not simply *lambda*, the T4 phages, or *E. coli*. It is not even *C. elegans*,

---

[*] This paper was presented at the University of Notre Dame's Conference "Controlling Our Destinies: Historical, Philosophical, Social, and Ethical Perspectives on the Human Genome Project," 5–8 October 1995. The research leading to this article has been partially supported by the National Science Foundation's Studies in Science, Technology, and Society Program and by a National Institutes of Health Grant R13 HG00703 to the University of Maryland. No endorsement by the NSF or the NIH for the conclusions of this article should be inferred from this support. I would like to express my gratitude to Drs. Steve Klein and Norman Krasnegor for inviting me to a special August 1993 workshop, convened under the auspices of the NIH's NICHD, at which work in progress by Drs. Bargmann and Chalfie on *C. elegans*, Drs. Hall and Tully on *Drosophila*, Dr. Hamer's studies on humans, and Dr. Plomin's general methodological approaches, were presented and discussed. I would also like to thank Drs. Cori Bargmann, Martin Chalfie, and Shawn Lockery both for providing me with information about their research and for comments on an earlier form of this paper. Also thanks to Dean Hamer and Irving Gottesman for references to the behavioral genetics literature. This paper draws in part on some other recent essays of mine, including "Genetic Explanations of Behavior: Of Worms, Flies, and Men," prepared as a background paper for a conference sponsored by the University of Maryland, "Complexity and Research Strategies in Behavioral and Psychiatric Genetics," prepared for a University of Texas Medical Branch at Galveston conference (1997a), and "Genes, Behavior, and Developmental Emergentism: One Process, Indivisible?" *Philosophy of Science*, 65 (1998): 209–52.

Drosophila, or *Aplysia*. It is human biology. And some biological researchers are so bold as to see their ultimate interest as the function of
the human mind. (Kandel 1987, vii)

Over the course of the past century, a considerable amount of
philosophical and scientific inquiry has been directed at the issues of
reduction and reductionism. Reduction, in the context of the Human
Genome Project, is closely associated with the conceptually distinct issue of determinism, and in particular with the notion of 'genetic determinism.' By 'genetic determinism' I mean that *inherited* genes are
taken to be the principal drivers of an organism's behavior. In this
paper, I am largely going to bypass an analysis of various quasiformal
models of theoretical reduction, though I will *very* briefly situate the
paper in the context of some of that work. Rather, I am going to con-
centrate on a set of examples that illustrate some very recent and ongoing attempts to provide genetic explanations of behavior, and will
assess those research programs largely in the light of their own
strategies. Enormous strides have been made in molecular biology,
including the molecular neurosciences, but I will argue that many of
the results have been seriously oversimplified and overinterpreted.
Important advances have been made and will be made in these areas,
but we should understand the nature of the difficulties and limitations,
lest inappropriate inferences be drawn from, and harmful policies be
based on, these discoveries.

My approach in this paper will be to examine the extent to which
there are purely, or even primarily, "genetic" *explanations* of behavior. The relation of 'explanation' to 'reduction' is reasonably clear—
or at one time was. In his classic analysis of reduction, Ernest Nagel
stressed the theoretical sense of that notion, and characterized it as
"the *explanation* of a theory or a set of experimental laws established
in one area of inquiry, by a theory usually though not invariably
formulated for some other domain" (Nagel 1961, 338; my emphasis).
In this paper, however, I am not going to emphasize either a notion of
"theory," or the concept of a "*set* of experimental laws," since these
are notions that are better suited to characterize historically completed reductions in the physical sciences, which involve "clarified"
sciences and employ theories that can be formulated at unified levels
of aggregation. For reductions involving explanations that are
"partial," interlevel, and evolving, it seems better to turn to the notion

of a causal explanation that is formulated largely, but not completely, at one level of aggregation, and which partially explains some phenomena described primarily in "higher level" terminology. The following two-by-two table (Table One) situates the Box One type of *causal explanation* I am going to be discussing below, and indicates where it fits with respect to more formal models of theoretical reduction:

Explanatory Approach→ / State of Completion of the Reduction ↓	Causal Explanation (CE)	Formal Reduction Model (FR)
Partial/ Patchy / Fragmentary/ Interlevel	**Box 1:** CE approach usually employed; interlevel causal language is more natural than FR connections.	**Box 2:** In the complex FR Model the connections are bushy and complex when presented formally, but FR does identify points of identity, as well as the generalizations operative in mechanisms.
Clarified Science/ Unilevel at Both Levels of Aggregation.	**Box 3:** Either approach could be used here, but where theories are collections of prototypes,[1] the bias toward axiomatization or explicit generalization built into the FR approach will make it less simple than CE.	**Box 4:** Here we find a Simple FR Model – the best match between Nagelian-type reduction and scientific practice.

TABLE ONE

Causal and Formal Approaches to Reduction.

---

1 For a discussion of this notion of theories as collections of prototypes, see my Schaffner 1993, esp. 98, and also chap. Nine of that book for more details on the information in this table.

This paper, then, examines genetic explanations of behavior as are available, given the still incomplete state of behavioral genetics in relating genetics to behavior patterns. Thus, in some ways, the analysis is more akin to what Kitcher (1984; 1989) has termed an "explanatory extension" than to a full-scale reduction. The focus will initially be on "simple" organisms, about which a rapidly growing body of information exists concerning their genetics and its relation to behavior. Within the paper, the generalizations obtained from simple organisms are explored for the possibilities of "scaling up" to include aspects of at least some human behaviors, and I will then discuss briefly some methodologies currently in practice, and some that are emerging, to study genetic explanations of human behavior. I will not in this paper, however, attempt to embed my discussion of genetic explanation, and its limitations as I see it, within the context of an analysis involving the various models of scientific explanation that have occupied the attention of philosophers of science, especially over the past fifty years.[2]

NATURE *AND* NURTURE

A 1992 review by Kupferman in Kandel, Schwartz, and Jessel's *Principles of Neural Science*—a book that is a bible of sorts in neuroscience—begins by noting that behavior in all organisms is shaped by the interaction of genes and environment. The relative importance of the two factors varies, but even the most stereotyped behavior can be modified by the environment, and most plastic behavior, such as language, is influenced by innate factors (Kupferman 1992, 987).[3] Kupferman then focuses on *aspects* of behavior that might be inherited, and on the processes of interaction between genes and environment affecting behaviors. Thus the point about the incompleteness of any exclusively genetic basis for the explanation of behavior is taken as a general premise in the scientific community, though there are more radical views that question the ability to

---

2 I have done this elsewhere in my article "Genetic Explanations..." (in press).

3 Kupferman stresses that not only do genetics and environment always interact to produce behavior, but in addition there is "no sharp distinction between learned and innate behaviors. There is instead a continuous gradation. . . "(989). This section relies extensively on Kupferman's review to present a mainstream neuroscientific position, but more radical critiques of this view are to be found in the biological, ethological, and psychological literature; see below.

meaningfully even make the "nature-nurture" distinction (Oyama 1985; Johnston 1987; 1988; Gottleib 1992; Gray 1992; Lewontin 1995) that will be considered further below.

Scientists examining inherited aspects of behavior now frequently refer to what the ethologists called instinctive behaviors as *species-specific behaviors,* since these are inherited as characteristic of a species (Kupferman 1992, 989). Ethologists such as Lorenz and Tinbergen introduced two theoretical concepts to describe such behaviors: the cause (or releaser) of the behavior was termed the *sign stimulus,* and the stereotypical response of the organism was called a *fixed-action pattern,* often abbreviated as a FAP. A FAP can be quite complex, and in simple organisms the firing of a single *command neuron* can trigger activity in over a thousand different neurons in different neuronal subsystems. Such command neurons have been found in a variety of organisms including the crayfish and in *Aplysia* (ibid., 990–91). Quite recently Frost and Katz (1996) reported finding a single neuron in the marine mollusc *Tritonia diomedea* that triggers a long-lasting motor program governing the organism's escape swim. The input to command neurons are from a type of sensory neurons that serve as quite specific feature detectors. It should be emphasized that though these types of behaviors are highly stereotypical, environmental factors and learning history can modify them to some extent.

A simple model of the implementation of a FAP is suggested by Kupferman, who shows an essentially linear flow chart leading from sensory input(s) to a sensory analyzer, to a command system, to a motor pattern generator, and resulting in motor output. There are some examples of a sign stimulus and a FAP in humans. Ahren's work in the 1950s on sign stimuli that elicit smiling in young infants illustrates this.[4]

Genes do not control even species-specific behaviors directly, and the synthesis of even a single neuron (say a command neuron) requires the coordinated action of *many genes.* I shall call this principle the rule of *many-genes one-neuron.* This principle is one of about

---

4 See figure in Kupferman's review (1992), based on Ahren's study, in which he notes that stimuli based on two large dots in an otherwise featureless face are more effective in eliciting smiling behavior in young infants than either one large dot or a fully-featured face. As the infant matures, the double-dot patterns become less effective, and a face image with features more effective.

eight complicating rules to be discussed in the next section. It should also be noted that the conceptual orientation underlying notions, such as fixed action pattern, has been questioned by a number of ethologists, who contend that it represents inappropriately "dichotomous" thinking (see especially Johnston 1987; 1988) that should be replaced by an "interactionist" perspective often associated with the work of Lehrmann (1970).

In the present essay, I will not explicitly discuss this "interactionist" position, which also has been termed a "developmentalist" and even "constructionist" view, but have done so elsewhere in my (1998), where I delineate and criticize the assumptions of the "Developmentalist Challenge" to the "nature-nurture" (and "learned-innate") distinctions. Suffice it to say that *some* elements of the interactionist or developmentalist view are represented in the eight rules discussed next.

CAENORHABDITIS ELEGANS *AS A MODEL ORGANISM, BOTH FOR BIOLOGY AND PHILOSOPHY*

Much of what we know about the biological basis of behavior and learning is based on studies involving simpler organisms. This comparative approach to understanding behavior and learning has been called the "simple systems" approach, and has been widely adopted in learning and memory studies in "psychology, physiology, biochemistry, genetics, neurobiology, and molecular biology" (Gannon and Rankin 1995, 205). Here I extend that approach to philosophy. In this section I summarize information about two such simple organisms, *C. elegans*, a small round worm, and *Drosophila melanogaster*, the fruit fly.

C. elegans: *An Introduction*

The cover of the 1995 annual issue of *Science* magazine on the Human Genome Project (see Figure Seven, Dreger this volume, 165), displayed the outline of a human and a worm. That worm is *Caenorhabditis elegans* or *C. elegans* for short, and its attractive features are the subject of the "centerfold" for that issue of *Science* as well. *C. elegans* is what is known as a "model organism," which means that it is an ideal organism for learning about fundamental biological processes that obtain both in the worm system as well as for other organisms, including humans. A quotation from the centerfold points this out, and states:

> More than 40% of *C. elegans* genes have significant similarity togenes from other organisms. These similarities range from sequences shared by all organisms to those found only in Metazoa. To illustrate the potential utility of *C. elegans* as a model system, [a computer program] BLASTX was used to determine that 32 of the 44 human disease genes identified by positional cloning had significant matches to worm genes (P<0.05). The table at the top right provides some examples. In some cases, such as the recently-discovered early-onset Alzheimer's disease genes, the *C. elegans* gene represents the only significant database match. (Jasny 1995, centerfold)

C. elegans does, however, display some important differences from mammals and in particular humans. Greenspan, Kandel, and Jessel (1995) point out that the small size and quite simple forms of behavior make studying its electrophysiology very difficult. In addition, there is a lack of "any significant anatomical homology with distantly related organisms such as mammals" (ibid., 556).

The relationship between genes, the nervous system, and behavior is, however, probably best understood in *C. elegans*. Though it would be dangerous to extrapolate uncritically from this organism, a close analysis of the genes-behavior relationship reveals some important generalizations that, from a philosophical perspective, provide a framework that suggests caution in interpreting behavior as primarily directed by any organisms genes, especially in any "one-gene–one behavior type" sense. In order to make these points, and consider these philosophical implications, I will briefly summarize some of the methodology and findings in the area of the worm's behavioral genetics.

Though the organism has been closely studied by biologists since the 1870s (see von Ehrenstein and Schierenberg 1980, for references), it was the vision of Sydney Brenner (1988) that has made *C. elegans* the model organism that it is today. *C. elegans* is a worm about 1 mm long that can be found in soil in many parts of the world.

It feeds on bacteria and has two sexes: hermaphroditic (self-fertilizing) and male.[5]

The organism has been studied to the point where there is an enormous amount of detail known about its genes, cells, organs, and behavior. The adult hermaphrodite has 959 somatic nuclei and the male 1,031 nuclei. The haploid genome contains about 100 million nucleotide base pairs, organized into five autosomal and one sex chromosome (hermaphrodites are XX, males XO), comprising about 13,000 genes. The organism can move itself forward and backward by undulatory movements, and responds to touch and a number of chemical stimuli, of both attractive and repulsive forms. More complex behaviors include egg laying and mating between hermaphrodites and males (Wood 1988, 14). The nervous system is the largest organ, being comprised, in the hermaphrodite, of 302 neurons, subdividable into 118 subclasses, along with 56 glial and associated support cells. The neurons are essentially identical from one individual in a strain to another (Sulston et al. 1983; White et al. 1986), and form approximately 5,000 synapses, 600 gap junctions, and 2,000 neuromuscular junctions (White et al. 1986). The synapses are typically "highly reproducible" from one animal to another, but are not identical.[6] The *C. elegans* community is an large and international one, numbering about 1,000 researchers, and displays extraordinarily cooperativity. A current snapshot of its extensive resources is available on the World Wide Web at Leon Avery's homepage (see Avery 1996).

In 1988, Wood, echoing Brenner's earlier vision, wrote that:

> The simplicity of the *C. elegans* nervous system and the detail with which it has been described offer the opportunity to address fundamental questions of both function and development. With regard to function, it may be possible to correlate the entire behavioral repertoire with the known neuroanatomy. (Wood 1988, 14)

---

5 See figure in Wood (1988) that shows a photograph and a labeled diagram of each sex.

6 Bargmann quotes figures from Durbin 1987: "For any synapse between two neurons in any one animal, there was a 75% chance that a similar synapse would be found in the second animal. . .[and] if two neurons were connected by more than two synapses, the chances they would be interconnected in the other animal increased greatly (92% identity)" (Bargmann 1993, 49).

Difficulties and Complexities with C. elegans *Behavioral Genetics*

There are some serious limitations that have made Wood's optimistic vision difficult to bring to closure easily. I have already mentioned that the small size of *C. elegans* makes studying the neurons difficult, a point noted by two prominent investigators in the field who wrote in 1988 that "because of the small size of the animal, it is at present impossible to study the electrophysiological or biochemical properties of individual neurons" (Chalfie and White 1988, 338). Only very recently have patch clamping and intracellular recordings from *C. elegans* neurons begun to be feasible (see Raizen and Avery 1994; Thomas 1994, 1698; Avery, Raizen, and Lockery 1995; Lockery 1996). In her 1993 review article, Cori Bargmann writes that "heroic efforts" have resulted in the construction of a wiring diagram for *C. elegans* that has "aided in the interpretation of almost all *C. elegans* neurobiological experiments." But Bargmann goes on to say that:

> However, neuronal functions cannot yet be predicted purely from the neuroanatomy. The electron micrographs do not indicate whether a synapse is excitatory, inhibitory, or modulatory. Nor do the morphologically defined synapses necessarily represent the complete set of physiologically relevant neuronal connections in this highly compact nervous system. (Bargmann 1993, 49–50)

She adds accordingly that the neuroanatomy needs to be integrated with other information to determine "how neurons act together to generate coherent behaviors," studies that utilize laser ablations (of individual neurons), genetic analysis, pharmacology, and behavioral analysis (ibid. 50).

In this essay I will not have the opportunity to present many details of the various painstakingly careful studies that have been done comparing behavioral mutants behaviors with neuronal ablation effects, in attempting to identify genetic and learning components of *C. elegans* behaviors. Interested readers should consult the Wood (1988) volume as well as Bargmann's (1993) review article; several of these are also discussed in my (1998). Suffice it to note that the work on the worm that seeks to relate genes to behavior follows a typical methodology. A general statement of this methodology containing some of the complicating rules regarding the relation of genes to behavior is found in Avery, Bargmann, and Horvitz (1993) who write that:

> One way to identify genes that act in the nervous system is by isolating mutants with defective behavior. However the intrinsic complexity of the nervous system can make the analysis of behavioral mutants difficult. For example, since behaviors are generated by groups of neurons that act in concert, a single genetic defect can affect multiple neurons, a single neuron can affect multiple behaviors and multiple neurons can affect the same behavior. In practice these complexities mean that understanding the effects of a behavioral mutation *depends on understanding the neurons that generate and regulate the behavior* [my emphasis]. (Avery et al. 1993, 455)

These "rules" can be further elaborated and have some additional generalizations about the relations of genes and behavior added to them.[7] A summary of these from my (1998) is presented in Table Two below.

In my view, these rules, based on empirical investigations in the simplest model organism possessing a nervous system that has been studied in the most detail, should serve as the *default assumptions* for further studies of the relations of genes and behavior in more complex organisms. These eight rules are generalizations involving principles of genetic pleiotropy, genetic interaction, neuronal multifunctionality, plasticity, and environmental effects, and like virtually any generalization in biology, they are likely to have exceptions, or near-exceptions, but I think these will be rare.[8]

Taken together, what these rules tell us is that the relation between genes and behavior types will be "many-many" (on this point compare Lewontin 1995, 27). There is also a *prima facie* stochastic component present that shows up in some developmental variation for the synapse/wiring-diagram, even in simple systems in which the complete lineage of cells has been identified and is traceable. Waddington (1957), Stent (1981), and Lewontin (1995) have stressed the negative

---

7 I refer to these general principles relating genes and behavior as "rules" because they are like spelling rules: they hold quite generally but there are exceptions that do not disprove the general rule.

8 I consider one type of exception involving an almost "one gene–one behavior type" association in my 1998.

implications of this stochastic element, that they term "developmental noise," for any predictive/explanatory relations between genes and

---

1. *Many genes→ one neuron*

2. *Many neurons (acting as a circuit)→ one behavior*

3. *One gene→ many neurons (pleiotropy)*

4. *One neuron→,many behaviors (multi-functional neurons)*

5. *Stochastic [embryogenetic] development→ different neural connections*[9]*

6. *Different environments/histories→ different behaviors* (learning/plasticity)*

7. *One gene→ another gene behavior (gene interactions→ including epistasis and combinatorial effects)*

8. *Environment→ gene expression behavior (long-term environmental influence.)*

TABLE TWO

Some Rules Relating Genes (Through Neurons)
to Behavior in *C. elegans.*

traits, including behavioral traits. *C. elegans* researchers, however, seem to be able to deal with this variation and obtain reproducible and general circuits that govern specific types of behavior. The environment can and does affect gene expression, most specifically through temperature-sensitive mutants and via organism-pheromone interactions, as in the production of long-lived larvae, termed the "dauer state," during conditions of crowding and starving.[10]

The lessons gleaned above from *C. elegans*, and embodied in the eight "rules" proposed above, also seem to apply to other biological organisms, including the fruit fly, *Drosophila*, as well as more complex vertebrate organisms such as mice and humans. Though I do not

---

9 In prima facie genetically identical (mature) organisms, the starred items can be read as "affect(s)," "cause(s)," or "lead(s) to."

10 Details can be found in my 1998.

have space to review the extensive *Drosophila* work done over the years by a number of investigators from Benzer on (see Hall 1994; R.Greenspan 1995a; 1995b), I want to cite comments of one of the major investigators in the field, Ralph Greenspan. Greenspan writes that his work leads to the conclusion that, ". . . behaviors arise from the interactions of vast networks of genes, most of which take part in many different aspects of an organism's biology"(1995b, 78). To this theme of networks involving multifunctional neurons, Greenspan also adds that evidence from *Drosophila's* courtship behavior indicates that both male and female fruit flies "have the ability to modulate their activity in response to one another's reactions," writing:

> In other words, they can learn. Just as the ability to carry out courtship is directed by genes, so too is the ability to learn during the experience. Studies of this phenomena lend further support to the likelihood that behavior is regulated by a myriad of interacting genes, each of which handles diverse responsibilities in the body. (Ibid. 75–76)

If this "network" type of genetic explanation holds for most behaviors, including even more complex organisms than worms and fruit flies, such as mice and humans, it raises barriers both to any simplistic type of genetic explanation, as well as to the prospects of easily achievable medical and psychiatric pharmacological interventions into behaviors. These eight "rules" may be some of the reasons why it has been so difficult to find single-gene or even oligogenetic (involving a *few* genes) explanations in the area of human behavior. There have been a number of attempts to do so, for simple genetic explanations are seen as a kind of "holy grail" for biological psychiatry (and psychology), and I consider how this might work in the following concluding section after we examine one example of human behavioral genetics in some detail.

SEXUAL ORIENTATION IN (SOME) HUMAN MALES AS A EXAMPLE OF CURRENT APPROACHES TO HUMAN BEHAVIORAL GENETICS

A search for genetic explanations of behavior leads to a question that arises from the caveats about complexity discussed in the previous section. The question is whether there may be behavioral "simplifications" of a sort that can be discovered—perhaps a kind of "final common pathway" in the complex network of gene-neuronal net-behavior interactions. Certainly that seems to be what researchers approaching the genetic basis of human behaviors in the area of psychiatric disorders hope to find, so that they may intervene at specific pharmacological points, and treat those affected by these illnesses.[11] In this section, I move beyond so-called simple organisms to consider how behavioral genetics is beginning to study *Homo sapiens*—ourselves—and examine a putative genetic cause of sexual orientation in men, first reported by Hamer and his colleagues several years ago. In the following section I will provide an overview of several different but potentially mutually reinforcing study designs used in human behavioral genetics, and will place Hamer's continuing research program within that framework.

In July of 1993, Dean Hamer caused a stir in the behavioral genetics community, as well as a much broader reaction, with the publication of his group's paper in *Science* on "A Linkage Between DNA Markers on the X-chromosome and Male Sexual Orientation" (Hamer et al. 1993a). The study has been criticized from both scientific (Risch et al. 1993; Byne 1994) and social points of view (Rose 1995; CRG 1996), but in general it has been well-received scientifically as a first step in clarifying the possible genetic influence on human sexual orientation. Hamer has summarized his work and interrelated it with Simon LeVay's anatomical investigations of the basis of sexual orientation (LeVay and Hamer 1994). Also, collaborating with journalist Peter Copeland, Hamer has amplified on his research and its implications in a recent book (Hamer and Copeland 1994). More recently, Hamer's work was replicated and extended by his group. The work is controversial.[12] However the 1995 *Nature Genetics* analysis that added Fulker and Kruglyak to the research team, as well as a QTL extension of the original allele sharing methodology to be discussed later, represents a powerful and convincing set of results (Hu et al. 1995).

Hamer's studies are good examples of an increasingly methodologically sound, though frequently overly interpreted, study of putative

---

11 See Gershon and Cloninger 1994, for a recent summary of the field.
12 See the 30 June 1995 issue of *Science* (1841) for some questions by the Office of Research Integrity about the 1993 study, that are now apparently resolved (see *Science* 1997 [28 February]: 275: 1251.) Replication studies of Hamer's result continue, however, to generate controversy (see below).

genetic influence on human behavior. There are very few such studies that have been conducted at the chromosomal/DNA-level that have survived reanalysis, the cases of bipolar disorder on chromosome 11 and schizophrenia on chromosome 5 being well-publicized failures (see below). Most human behavioral genetic studies, including those involving mental disorders, are of the more general epidemiological type that I will consider in the following section. No "gene" for homosexuality has been found, only a region termed q28 on the X chromosome, and this only for thirty-three pairs of the forty pairs of subjects studied. Byne (1994) has argued that the study does not show that all sixty-six men from these thirty-three pairs shared the *same* Xq28 sequence; rather "each member of the 33 concordant pairs shared the Xq28 region only with his brother—not any of the other 32 pairs" (Byne 1994, 55). However, an examination of the number of markers used in the analysis does not seem to support significant variation.[13] But it is the case, as Byne states, that "no single, specific, Xq28 sequence (a putative "gay gene") was identified in all 66 men"(ibid., 55). Hamer et al. (1993a) estimate that this region is about 4M base pairs in length and believe it is likely to contain *hundreds* of genes. Nonetheless the study is often interpreted as a partial, if preliminary, genetic explanation of a human behavioral trait. This interpretation can be found in both scientific publications, as well as in the popular media, including gay-oriented publications (Burr 1995).[14] Hamer's work is also congruent with studies on *C. elegans* and *Drosophila* in which the genetic factors affecting mating behaviors have been studied (Liu and Sternberg 1995; Hall 1994; R. Greenspan 1995b), and thus is worth describing in some additional detail as an example of a potential partial genetic explanation of behavior.

In their 1993 article, Hamer's group reported pursuing a two-stage type of analysis, the first being a pedigree study of seventy-six homosexual volunteers, in which they found 13.5% of the gay men's brothers to be homosexual, in comparison with an estimated 2% general population rate (Hamer et al. 1993a). The phenotype was determined by using Kinsey scores (based on self-identification of sexual orienta-

---

13 I discuss the differences between "identity by descent" determination and "identity by state" inference later in this article.

14 See *Science's* "Research News" account with the headline "Evidence for Homosexuality Gene" (Pool 1993).

tion, attraction, fantasy, and behavior) that gave a dichotomous distribution for self-identifying gay men versus their heterosexual counterparts. Kinsey scores assigned on the basis of estimates from interviewees were also obtained for noninterviewed relatives. In looking beyond the immediate family, Hamer found more gay relatives on the maternal than the paternal side, suggesting that this type of influence might be X-linked, since it is the only chromosome inherited exclusively from the mother. Stage two of the study recruited forty pairs of homosexual brothers and examined the X-chromosome DNA using twenty-two markers looking for a linkage group common to most of the sib-pairs. Of the forty pairs examined, thirty-three shared a group of five markers in the Xq28 region. Hamer's group performed several statistical analyses examining the probability that an association between homosexual orientation and inheriting genes in this five marker region could occur by chance. Generally LOD scores of 4.0 were obtained, and interpreted as being "statistically significant at a confidence level of > 99%," concluding that "it appears that Xq28 contains a gene that contributes to homosexual orientation in males" (Hamer et al. 1993a, 325). Hamer's analysis was criticized on statistical grounds by Risch et al. (1993) but responded to (I think reasonably) by Hamer and his group (1993b). Subsequently Hamer's study was replicated by a Dr. Cherny in a new sample of thirty-three pairs of homosexual brothers (Cherny in Holden 1995, 1571) and also received some support in a study by Turner (1995). It was then, however, reported in the 30 June 1995 issue of *Science* (1841) that Hamer's study was being investigated by the Office of Research Integrity at DHHS, possibly because of controversy about the manner in which research subjects were selected; Hamer was subsequently exonerated (see note 12 above). This report also cites the inability of still another researcher, Dr. Ebers, to confirm Hamer's findings. In the November 1995 issue of *Nature Genetics*, Hamer's group, expanded to include additional investigators with special expertise in the analysis of complex trait genetics, reported a replication of their original finding as well as an extension of their results (Hu et al. 1995). Another group has also obtained replication data, but thus far it is unpublished (Hamer, personal communication, June 1998). I shall discuss this more recent work in some detail below, but suffice it to say now that the Xq28 region was again reassociated with a disposition to male homosexuality, but not to female homosexuality, in subpopulations of the gay and lesbian populations studied.

The study designs—as reported in the 1993 version and even more so in the 1995 replication—are methodologically and statistically sound. There are, however, a number of unresolved questions about these studies that have been raised in the literature. Both studies only report a partial association between a behavior pattern (more accurately an orientation) and a chromosomal region; they do not identify a gene or genes and report a DNA sequence. Thus, not surprisingly, it is not known what protein the gene (assuming it is just one gene) codes for, where such a protein might act and what it might do, how the gene in homosexual men might differ from heterosexual men, whether it invariably leads to homosexual orientation or not, or what role this gene might play in women (Pool 1993). Further, Hamer himself notes, citing M-C. King's additional questions, that we do not know what fraction of all gay men carry an allele in this region that might influence sexual orientation, how many different alleles there might be, and what other factors, including other genes, familial environment, and culture, might affect sexual orientation (Hamer and Copeland 1994, 145). In their brief collaboration, LeVay and Hamer stated that they see "two broad lines" of evidence pointing to a biological component for male homosexuality (LeVay and Hamer 1994). But LeVay and Hamer follow only weakly a "coevolutionary" approach that would carefully match the genetics with neuroanatomy and behavior, as is pursued in studies of *C. elegans* and also by R. Greenspan and his colleagues in *Drosophila*. One possibility LeVay and Hamer do speculate on is that there may be genetic differences in the way that individual brains respond to circulating androgen in fetal development, but an inquiry to determine if the gene or genes in Hamer's Xq28 region was the gene for the androgen receptor turned out negative: the locus of the androgen receptor was at Xq11—far from the Xq28 region (Macke et al. 1993).

Thus no "gay gene" has been found, and no molecular or even neurological pathway/circuit explanation for sexual orientation for humans is likely to be forthcoming in the near future. Hamer's group's work is significant, however, even if it frequently tends to be overinterpreted in the scientific literature and in the lay press, since it represents a methodologically well-designed genetic study of a complex behavior "trait"—a point to which I will return again below.

RECENT APPROACHES TO THE METHODOLOGY OF HUMAN BEHAVIORAL GENETICS

The purpose of this section is to put the Hamer studies discussed in the previous section into a broader methodological context, and pave the way for the concluding more philosophical section that follows. I will do this by briefly outlining four types of study designs that are used by geneticists to identify the genetic contributions to "complex traits."

Why I focus on *complex traits* requires a bit of a digression before we plunge back into the issue of study designs.

Single-gene Disorders

It is widely acknowledged that advances in the molecular genetics of a variety of *single-gene* somatic disorders have become mind boggling in the past half-dozen or so years. The prototype is probably sickle-cell anemia that dates to the 1949–1957 years when it was characterized by Pauling and Ingraham as a molecular disease due to a single mutation. In 1983, the gene for Huntington's disease was localized to chromosome 4, though the specific locus of this gene was difficult to determine, and had to wait until 1993. Genes for polycystic kidney disease, retinoblastoma, and cystic fibrosis were mapped in 1985, and this latter gene was localized and cloned in 1989. In the past three years, genes for two types of inherited colon cancer and two forms of breast cancer, have been localized and sequenced, and there are many other single genes associated with physical illnesses that have been found.

The genetics of mental disorders, however, has not fared as well, though there have been some important, though controversial, advances in the behavioral realm. The finding that there was a gene associated with schizophrenia on chromosome 5 turned out to be fallacious (Kennedy et al. 1988), and a report identifying a gene for manic-depressive disorder on chromosome 11 had to be withdrawn (Kelsoe et al. 1989). The only single-gene–behavioral trait example of which I am aware involves a point mutation in the structural gene for monoamine oxidase A that was associated with a somewhat vaguely defined impulse disorder/aggression phenotype reported by Brunner et al. (1993a; 1993b). Though this work represents methodologically sound science, Brunner himself does not characterize this mutation as an "aggression gene" (Brunner 1996), but does see this mutation as producing "a very general disturbance of brain metabolism." Moreover, Brunner adds that "behavior should and does arise at the highest level of cortical organization, where individual genes are only distantly reflected in anatomical structure. . . (ibid., 160). Recent discoveries in personality genetics include an identification of a gene, DRD4, for "novelty-seeking," but it should be noted that the group reporting a confirmation of that gene noted that this was only a partial—and not a single-gene kind of explanation. They wrote that "D4DR accounts for roughly 10% of the genetic variance, as might be expected if there are ten or so genes for this complex, normally distributed trait. These results indicate that Novelty Seeking is partially but not completely mediated by genes, and that the D4DR polymorphism accounts for some but not all of the genetic effect (Benjamin et al. 1996, 83).[15]

There are thus several problems with this focus on single-gene disorders, particularly in the behavioral/mental disorder realm. The Huntington's Disease (HD) paradigm of a (prima facie) single dominant 100% penetrant gene may be more the exception rather than paradigmatic, for the search for such single genes, as already indicated for bipolar disorder and schizophrenia, has *not* been successful.[16] Recent research on Alzheimer's Disease indicates there are probably three genes (on chromosomes 21, 14, and 1) associated with the familial early-onset form of this disease, and another gene on chromosome 19 that is partially linked to the more prevalent late-onset form of AD, but which may interact with the gene on chromosome 21 (Levy-Lahad et al. 1995). Investigators in the area of psychiatric genetics, and in behavioral genetics more broadly, have, accordingly, begun to think seriously about *multigenic* causes for behavioral disorders. In a 1990 essay Robert Plomin argued on rather theoretical grounds that there will be a very limited number of behavioral traits that will segregate as Mendelian single loci. Plomin

---

15 The novelty-seeking effects of these two alleles of D4DR continue to be controversial, with one additional replication but three non-replications. See Pogue-Geile et al. (1998) for a discussion and references.

16 A quite recent article in the *American Journal of Human Genetics* indicates that the Huntington's Disease mutation "may not always be fully penetrant" (Rubinsztein et al. 1996, 16).

pointed out that (1) most behaviors are not inherited in a dichotomous (either/or) fashion; (2) most behavioral traits appear to be influenced by many genes, each with small effects; and (3) behavior is substantially influenced by nongenetic factors (Plomin 1990, 183–84). This nongenetic or environmental factor influence is underscored by data presented in a review article "The Genetic Basis of Complex Human Behaviors" (Plomin, Owen, and McGuffin 1994), that displays probandwise concordances for identical and fraternal twins for several behaviors, including some classical psychiatric disorders, as well as a range of some common medical disorders. One point to note about that data is that genetics is *not* the entire explanation for these disorders—there is a large environmental component that will figure as well in any complete explanation of these behaviors and disorders.

Single-gene Disorders May Actually be a Myth

There is some evidence that the basic paradigm of what Plomin, Owen, and McGuffin call the OGOD approach–standing for "one gene-one disorder"—is in the process of replacement by a multi-genetic approach, even in those cases earlier thought to represent simple single-gene disorders. In this regard the history of recent genetic studies of cystic fibrosis is important, since close investigations by Cutting and his colleagues have shown that mutations that were thought to produce clear phenotypes of the disorder do not always do so. The title of their 1993 article tells it all: "A mutation in CFTR [the cystic fibrosis transmembrane conductor gene] produces different phenotypes depending on chromosomal background" (Kiesewetter et al. 1993). Furthermore, as Lander and Shork have noted, even the sickle-cell anemia prototype is in actuality a "complex" disease (Lander and Schork 1994, 2037). These findings should not be a surprise given the default rules presented above based on the *C. elegans* research.

Genetic Epidemiology as a Backdrop to Molecular Studies

It is important to note that in the absence of any results that localize traditional DSM–IV mental disorders, or other well characterized stereotypical (or otherwise) human behavior, to specific single genes (with the possible exceptions mentioned earlier), genetic-epidemiological findings from twin, family and adoption studies represent the soundest evidence for a genetic component of these disorders. The epidemiological studies, important as they are, however, do *not* provide answers to questions such as the mode of the genetic components transmission, the degree of biological heterogeneity, or the pathophysiological process underlying the behavioral manifestations (compare Gershon 1990, 373, on manic-depressive disorder). As R. Greenspan, Kandel, and Jessel (1995) similarly note, twin studies "do not tell us anything about how many or which genes are important, let alone how genes affect behavior" (557). Further studies involving linkage, as well as other types of approaches, will be required to elucidate these important dimensions, including the possibility of a biochemical means of ameliorating or curing these disorders. We have already seen this approach exploited in a very preliminary way in our account of Hamer's research, though on the latter point I hasten to add that Hamer construes homosexual orientation as a *normal variant*, and not a disorder to be treated or "cured." Using genetically-based information to alter sexual orientation, of either homosexual or heterosexual form, is viewed by Hamer et al. as "unethical" (see Hamer et al. 1993a, 326).

Lander and Schork's Four-Fold Way

In an attempt to zero in on the genes that underlie behavior patterns that thus are expected to be multigenically caused, behavioral geneticists have explored four genetic approaches for dealing with more complex modes of inheritance. These are (1) traditional linkage studies, but in a more complex guise with special strategies and refinements; (2) modernized allele sharing methods; (3) association studies; and (4) quantitative trait loci, or QTL analysis. Each of these designs has its strengths and weaknesses, and behavioral geneticists need to pay close attention to the assumptions that each method makes regarding the populations, and the definition of the behavior they plan to investigate, as they design their studies. This essay is not the place to pursue these details and the reader is encouraged to consult Lander and Schork's (1994) excellent overview of what they term the "fourfold way" for an account of the methods and references to the literature. It may be appropriate, however, to note that in contrast to the first three methods which are fairly traditional, albeit with refinements and extensions, the QTL approach is quite new and since it may be unfamiliar to many readers, it is worth saying a bit more about QTLs.[17]

QTL methods in their current form were developed by Lander and Botstein in 1989 and first applied to fruit characteristics in the tomato, and then were quickly extended to epilepsy in mice and hypertension in rats. The latter study led to an identification of angiotensin in human hypertension, and more recently QTLs have been also been used in connection with animal models of alcohol and drug abuse (Crabbe, Belknap, and Buck 1994). QTL methodology required both the development of high density marker genetic maps, as well as advances in the statistical methods, so that QTLs, each with a small effect, but collectively having a major effect, could be detected against background variation due to other loci and the environment. An application of QTLs in Hamer's research program will be considered in the next section.

Hamer's Studies Revisited

It may be useful at this point to indicate where the Hamer studies discussed above fit into this four-fold way schema, and also how QTLs have been added into this research program. Hamer's first study utilized a pure allele sharing design (recall that this is number two in the four-fold way). In this approach, one does *not* seek to confirm an hypothesis that specifies a chromosomal location for a gene causing the trait being investigated, as is the case in traditional linkage analysis. Rather one seeks to reject a hypothesis (or model) of random Mendelian segregation for the trait of interest by showing that "affected relatives inherit identical copies of the [chromosomal] region more often than expected by chance" (Lander and Schork 1994, 2039–40). As noted earlier, in Hamer's first study thirty-three out of the forty gay brothers shared alleles in the Xq28 region. This type of analysis ideally proceeds using Identity-By-Descent (IBD) data

---

17 There were some attempts to develop QTL methods among the earlier geneticists, but these had to await the discovery of a dense set of genetic markers provided by recombinant DNA technology known as RFLPs. See my 1993 (453–55), for a brief discussion of RFLPs, and Lander and Botstein (1989), for some references to these earlier attempts. A good general introduction to QTL methods can be found in chapter twenty-one of the new (fourth) edition of Falconer and Mackay (1996).

by identifying how frequently the set of shared alleles has been *inherited from* a common ancestor, in this case the mother. This requires testing of the mother, but an alternative strategy may be pursued known as Identity-By-State (IBS), in which the common set of alleles is *inferred* to be IBD. Lander and Schork indicate that this inference is "usually safe" in those situations involving "a dense collection of highly polymorphic markers" (ibid., 2041). Hamer utilized both IBD- and IBS-based data in both of his studies. Again, as stressed in the previous section, this method will generally identify a chromosomal *region* of interest, not a specific locus.

In Hamer's 1995 replication, the analysis of the X-linked region was further extended to include a QTL methodology, in an attempt to "refine the localization of the putative X-linked male sexual orientation related locus. . . " (Hu et al. 1995, 250). This approach was based on an approach to linkage using sib-pairs developed initially by Haseman and Elston (1972), and then extended by Fulker and Cardon (1994) to estimate the location and effect size of a QTL. Hu et al.'s data was analyzed in terms of individual Kinsey subscales for gay male and lesbian female sexual behavior, fantasy, self-identification, and sexual orientation.[18]

Thus the Hamer studies demonstrate the synergistic use of several of the methods grouped in the "four-fold way." In their 1994 review article, Plomin and his coauthors similarly suggest that QTL, coupled with method (3) of the fourfold way—allelic association strategies— will be able to identify the more complex genetic causes of behavior, and moreover will also serve as a "guide" to "molecular genetic research by identifying the most hereditable domains of behavior and the most heritable dimensions and disorders within domains" (Plomin et al. 1994, 1738). Very recently, Risch and Merikangas (1996) have argued that association analysis, using candidate genes, will turn out to be the method of choice (over linkage studies) for complex disorders.

---

18 See figure in Hu et al. 1995, 250. Note that a review of the data from the gay males scales tracks together and peaks at locus C (which is marker DXS52 in the Xq28 region) with a corrected P value of <0.02, but that data from lesbian females does not attain significance nor show any peaking (though the subscales do track together as would be expected).

SUMMARY AND CONCLUSION: HOW MIGHT A STRONG GENETIC PROGRAM EVER BE POSSIBLE?

The search for *pure single-gene* genetic explanations of behavior might be thought of as the strongest of "strong" genetic programs, but this concept is easily seen to be incoherent: any behavior requires an organism acting in an environment. The organism is the product of many genes that are required to build the nervous system integrated within the organism that is productive of behavior. Still, the Huntington's Disease (HD) paradigm, in which we appear to have a single gene that is productive of HD in all current environments, exercises a powerful hold on geneticists' thinking (not to mention popular and media thought). This paradigm, suitably generalized, can be conceived of as articulating a not-incoherent *strong genetic research program* in the behavioral science area, and can be contrasted with weaker research programs that only seek to characterize genetic components of behavior, or genetic influences on behavior. If my review of the field of behavioral genetics is accurate and my examples representative, explanations that are *primarily* genetic—a class of explanations that would also perhaps support such a strong program—will probably be few in number, and are likely to be of narrow applicability, as in the discussion of FAPs earlier in this paper. There are a few comments, however, that should be added at this point that may somewhat ameliorate my criticism of strong genetic explanations.

It seems to me that even within a complex system of the genetically influenced neural networks as evidenced in *C. elegans* and *Drosophila*, there are two ways in which causal simplification may occur that may result in something close to a single-gene explanation of behavior. One simplification, that can also perhaps provide points of potential intervention, occurs when a "common pathway" emerges. This is usually referred to as a *"final* common pathway" in medical and physiological etiology, in which many different parallel-acting weak causal factors (often termed "risk factors") can coalesce in a funneling toward a common set of outcomes. An example is the pathogenetic mechanism by which the tuberculosis bacterium acts in a susceptible host after parallel risk factors predispose the host to infection (Fletcher, Fletcher, and Wagner 1982, 190). But we need to be attentive to the possibility of common pathways emerging at *any* stage (early, intermediate, and final) in the temporal evolution of a complex network involving multiple causes and complex "crosstalk."[19] Determining the effects of factors in complex networks is methodologically difficult and typically requires complicated research designs with special attention to controls.[20] The existence of a common pathway with a specific set of metabolites might permit intervention by manipulation of the metabolites in such a common pathway.[21] One way that a common pathway in the behavioral area could arise is as a consequence of a mutation in a neurotransmitter receptor on a cell or set of cells in a key circuit.

Another type of simplification that can emerge in a complex network of interactions is the appearance at any given stage of a *dominant* factor. Such a dominant factor exerts major effects downstream from it, even though the effects still may be weakly conditioned by other interacting factors.[22] Manipulation of such a dominant factor may thus have major effects on the future course of the complex system, though such effects can be quite specific and affect only a small number of event types. Sengupta et al.'s (1994) analysis of the *odr-7* gene (see my 1998), which appears to determine a nuclear hormone receptor in *C. elegans*, suggests that it may be such a dominant factor. Such factors are major leverage points that can permit interventions, as well as simpler explanations, which focus on such factors. Whether such dominant factors exist, as well as whether any common pathways exist, is an *empirical* question to be solved by laboratory investigation of specific systems. This is, in point of fact, where the power of

---

19 The term "crosstalk" for complex regulatory interactions is used by Egan and Weinberg in their description of the *ras* signaling network (1993, 783).

20 See my (1992) and my (1993, esp. 142–52) for a discussion of this type of problem.

21 It might be that focus *only* on common pathways could lead to an overly simplistic, reactive, and reductionistic approach to health care, and to a downgrading of more complex "risk factor" types of influences. For cautionary comments along these lines see Rose 1995.

22 It is possible that some of the work on temperament might reflect such a dominant factor/gene, or it may be that this is such a broad "phenotype" that generalizations in this area reflect many different factors. (See Kagan 1994, for an account of this research area. I thank Ed Manier for drawing my attention to this work). Support for the position argued by Manier in his oral comments presented at the Notre Dame Conference (and developed in written form following this essay) has received some empirical support in the two studies that report finding a "Novelty Seeking" gene briefly discussed above (Benjamin et al. 1996); also see my 1999.

model organisms is likely to become most evident.[23] Carrying out the type of investigation described by Sengupta et al. (1994) in an organism several orders more complex than *C. elegans* becomes considerably more difficult. One might hazard a guess that the difficulty may increase exponentially with the numbers of genes and neurons. The prospects of recognizing highly specific single-gene and single neuron effects in complex organisms is likely to be accomplished only if highly homologous and strongly conserved genes can be identified in much simpler model organisms. Such identifications can give us powerfully directive hints where to look for such genes in more complex organisms, and may help begin to characterize dominant factors or common pathways.[24] As in connection with the behaviors of even simple organisms such as *C. elegans* and *Drosophila*, however, the answer thus far appears to be that dominant factors and common pathways will be rare.[25]

A third type of simplification that may occur is what might be termed "emergent" simplifications, perhaps of the type that Cloninger and his colleagues claim exists in personality genetics.[26] It may well be that this is what Manier has in mind in his approach to temperament genetics in his commentary following. This notion of emergence is not an in-principle emergence, but rather a pragmatic one, similar to the type that has been discussed by Simon (1981) in connection with complex systems, and by Wimsatt (1976) and Bechtel and Richardson (1993). It is like what we encounter when a simple gas law is used to describe a very complex system of gas molecules.

It is difficult to predict the fortunes of behavioral genetics, and my somewhat pessimistic comments outlined in this paper may turn out to be wrong, and important single- (or pauci-) gene explanations of behavior may be discovered for both simple organisms and for humans. My own view is that progress will be made by following three routes,

---

23 I thank Sally Moody for the suggestion that this point needs emphasis here.

24 A good example of the utility of model organisms is the discovery of the DNA repair gene in humans, termed hMSH2, that is strikingly similar to the MutS gene in *E. coli* and to the MSH2 gene in the eukaryotic yeast *Saccharomyces cerevesiae* (see Schaffner and Wachbroit 1994, for a discussion).

25 Bargmann takes a more optimistic view and believes not only that dominant factors will become evident as research proceeds, but that "dominant genes will be quite common in behavior once we succeed in breaking behavior down into small precisely defined components" (personal communication, August 1995).

26 For a discussion see Schaffner 1999.

hopefully in a synergistic manner. The first route or research strategy is the more classical search for simple gene-neuron behavior interactions, looking for common pathways and/or dominating factors. These may, however, turn out to be quite atypical, even in simple organisms, though some results obtained by Chalfie's and by Bargmann's laboratories (see my 1998) do seem to confirm the usefulness and importance of this type of strategy. The second route is to develop biologically-informed connectionist models of the neuronal circuits.[27]

This is the route that Lockery (1996) is following for *C. elegans*, and I suspect this connectionist methodology will turn out to be the method of choice, as he (and others) are able to transcend the technical problems of single neuron recording in this organism. The third strategy is to look for what might be termed emergent simplifications perhaps of the type that Cloninger and his colleagues claim exist in personality genetics. The eight rules discussed earlier in connection with *C. elegans* and *Drosophila*, involving pleiotropic, interactional, multifunctional, plasticity, and environmental effects, are themes that should not be barriers to approaches utilizing either a connectionist methodology or one seeking levels where lower-level detail is filtered out in higher level generalizations. Those rules, however, appear to argue strongly against finding many "one gene-one behavior" explanations in simple, and especially in still more complex, organisms, such as humans.

Finally I want to add in closing that though this paper has been critical of *genetic* determinism, it has not offered any arguments to undercut a combined (and historically-based) *genetic-environmental* determinism. The ethical and legal implications of such a determinism are significant, and there is a vast philosophical literature stretching back at least to Aristotle on the implications of determinism for our concepts of self and responsibility. These are issues I have not attempted to address in the present paper, but have been pursued elsewhere by others (Brock 1992; Patricia Greenspan 1993).

---

27 See Gardner 1993 for some recent examples of such an approach.

How to do Human Behavioral Genetics. . . ?
A Commentary on "Reductionism and Determinism in Human Genetics"

Edward Manier

> Enormous strides have been made in molecular biology, including the molecular neurosciences, but I will argue that many of the results have seriously oversimplified and overinterpreted. (Kenneth Schaffner this volume, 302)

Kenneth Schaffner and I agree that the short term utility of information flowing from the Human Genome Project to human behavioral genetics to molecular psychiatry has been overestimated and oversold. This could impede expert policy formation in the area of mental health care and the accuracy and scope of the general public's understanding of mental disorders, available therapies, and the relative efficiencies of various strategies for delivering mental health care.

But has any speaker at the conference asked, "If the Human Genome Project has been oversold, who are the salespersons?" Is this another of those tragic scenarios where we find the enemy in the mirror? Do science journalists and editors oversimplify and sensationalize? *Do elite scientists imitate the political elite in manipulating coverage of their activities?*

Schaffner's answer is that if the HGP has been oversold, it is because things have been made to seem simpler, easier than they really are, as if a succession of "molecules of the year" will eventually include the genes and molecular therapies for major mental disorders, if we have enough faith and patience. This is part of the answer.

A complete answer would take account of the complexity of the system that delivers mental health care in the U.S (Grob 1994; Mechanic 1994). If real harm results from the malaise Schaffner documents, it will be to the system for delivering mental health care both to chronic in-patients, and to temporary out-patients. Unfortunately, Schaffner also gives less than adequate attention to the problems of identifying genes whose putative phenotypes have been so poorly described that little or no work has been done on their ontogeny.[1]

Schaffner's basic expository strategy offers a remedy for the problem he has described by showing that even in the simplest of the simple systems being used to study molecular neurobiology and the genetic control of the nervous system, things are much more complex than they seem. To make this point he offers a thorough description of strengths and the limits of a specific organism, *Caenorhabditis elegans,* a simple worm.

The story yields "rules" illuminating the need for caution and skepticism in examining claims including the phrase "the gene for *B*," if *B* is some simple unit of behavior. The rules have been staples in gene based accounts of evolution and development produced by T. Dobzhansky and his students for all most half a century. Almost all of them are well entrenched in the relevant scientific communities. Only Rule Two (Table Two, 311), which could be misread as "one neural circuit, one behavior" will raise the eyebrows of those not "friends of the modularity hypothesis." The dispute between the friends of modularity and the friends of distributed neural nets, functioning in parallel, however, has little relevance for Schaffner's argument.

Otherwise, Schaffner's Table Two reviews basic principles of the molecular biology of cell differentiation: 1) some of the simplest organic entities or processes are *polygenic,* i.e., assembly of the products of many genes may be required to produce a molecule with a de-

---

1 See the discussion of J. Kagan's work, *infra,* for a full account of a phenotype and its ontogeny.

tectable function in ontogeny or physiology, 1a) gene products may modulate the expression of other genes (*epistasis*); 2) many genes are *pleiotropic*, i.e., a gene may make a protein implicated in many diverse physiological functions; 3) ontogeny is *epigenetic*, i.e., the developing brain contains more information than is coded in the genes underlying its development (coincidences of spatial proximity result in salient diffusion gradients affecting patterns of neuronal growth and formation of synaptic connections); 4) *learning exploits the plasticity of the nervous system*, e.g., the "weights" or "connection strengths" of various synapses are open to modification by experience (interaction with salient aspects of its physical and social environments) throughout the organism's life span; and 4a) organisms with different "learning histories" may respond to similar patterns of stimulation in quite distinct ways.

It would be hard to "oversell" the HGP to anyone who fully understands those rather straightforward tenets of molecular neurobiology. The so-called "rules" make it perfectly clear that locutions including the phrase "the gene for B" must be labeled "handle with extreme caution." Schaffner's rules deserve the widest possible promulgation, because and in spite of the fact that they are utterly non-controversial, never disputed, and so well integrated into laboratory practice that we all risk overlooking them from time to time. They are tacit, implicit propositions which are so non-remarkable, non-sensational, that they are rarely expressed. Perhaps *never* expressed in the media. Too bad.

Ken's next topic concerns the problem that arises in molecular genetics when the topic of interest is identifying, cloning and sequencing segments of the human genome which are neither necessary nor sufficient for determining a course of development leading to a so-called "complex" (multigenic) trait. The most typical examples of such traits include those whose phenotypic expression varies in degree (such as height) and not simply in presence or absence (Mendel's "wrinkled, smooth", "green, white"). Complex traits are thought to require contributions from many different segments of the genome, perhaps widely dispersed, and having little in common except that each accounts for, say, less than 10–15% of the phenotypic variance of the same trait, say hand-eye coordination.

Perhaps sexual behavior is a complex (QTL) trait in all infra-human metazoa, perhaps not. It would surprising if sexual behavior in humans were under less complex control than in infra-human primates, where Harlow famously demonstrated its dependence upon normal parent/offspring bonding in neonates. There is an important lesson in the sexual ineptitude of Harlow's "unloved monkeys" which should not be too far from our minds whenever we consider "inherited predispositions to learn sexual preferences."

The lesson is that learning may be tacit as well as explicit. The lesson is that a literally unknown variety of ontogenetic stressors may predispose a particular organism to atypical approach/avoidance behaviors in every relevant aspect of the species specific repertoire, and so, *a fortiori*, in that aspect most highly charged with emotions from every segment of the rainbow.

However expert the group of elite neuroscientists and quantitative geneticists working with Dean Hamer and Simon LeVay on the question of a biological predisposition to homosexuality, they will have to address the fact that existing systems of classification of the relevant phenotype are antiquated, based on self-report of the variety of a subject's erotic fantasies, and that there is utterly no current consensus concerning the etiology or ontogeny of male homosexuality (Cole and Cole 1993, 632–49).

Whatever may be established about correlations between various QTL and adult human male homosexuality, the epigenetic path between the genome and the phenotype is simply unknown at the present time (Bem 1996, 320–55). Human sexual behavior, whether straight or gay, is not sufficiently well understood to lend itself to a strategy of simply "opening the black box" of the human genome. I would fry other fish, nominate other more thoroughly and carefully studied human behaviors as exemplary of the problems of human behavioral genetics and of possible paths to be pursued.

Human behavioral genetics is a hybrid technoscience buffeted by all the social forces surrounding a nation's concerns for its mental health.

Please shift your gaze to a methodologically proximate issue. I suggest the first task is precise description of the human *behavioral phenotypes* of interest in psychiatry and abnormal psychology and analysis of the various physiological and ontogenetic mechanisms underlying those phenotypes. The scientists responsible for exact identification and analysis of these phenotypes specialize in a "human science," psychology or one of the social sciences, e.g., developmental psychopathology.

The reducibility of behavioral, cognitive or affective processes to biological models is, at root, an open question. Philosophical extremists (of both camps), friends of supervenience, and moderates, all sketch plausible scenarios, more or less frequently instantiated in biopsychology. There is no persuasive analysis of the general case.

To the extent that reduction is of interest in behavioral medicine and psychiatry, present circumstances require case by case assessment of its utility. The necessary first step in assessing any such case is precise description and analysis of the behavioral (cognitive, affective) phenotype of a specific mental disorder. The absence of rigorous information on that point makes assessment of the utility of any given reduction (and, *a fortiori*, of the whole set) a highly dubious undertaking, often blind *and* empty.

If the phenotype is not precisely characterized, both in terms of its component behaviors and their neurophysiological substrata, the quest for the underlying genes will be based on crude molecular biological guesswork by brilliant, well trained and highly expert neuroscientists. The path from genes to human behavioral phenotypes always passes through proteins, cell adhesion molecules, various epigenetic processes such as programmed cell death, and the molecular bases of learning (G proteins, cyclic adenosine monophosphate). But one simply must get the behavior right or all the high tech neuroscience will produce useless results forever and ever.

The brick by brick inferential path is from behavior to its underlying physiological basis, to the neural and endocrine architecture underlying such function, to the inter-cellular messengers and intracellular second messengers that mediate neuronal function, finally activating or deactivating genes coding varying neuronal functions.

To illustrate these *dicta*, I turn to recent reviews and reflections by Jerome Kagan, a pioneer in longitudinal, multidimensional description and analysis of the aspect of human temperament he calls "reaction to the unfamiliar" (Kagan 1997a, 139–43).[2] This senior psy-

---

2 My understanding of Jerome Kagan's work owes everything to the patient generosity with which he answered my questions, supplied a steady flow of preprints and permitted me to observe procedures in his laboratory during a sabbatical year, 1993-94. John Dowling, Professor of Neuroscience and Master of

chologist, in addition to his sustained record of meticulous experiments, has written frequently on topics familiar to investigators in history and philosophy of science, the field whose reflection upon the Human Genome Project is the unique feature of this book.

Kagan's publications are notable for their persistent critical reflection concerning the methodological and conceptual structures which distinguish the study of child development and psychology in general. He has recently directed critical analysis to four "habits" of his colleagues:

1. Reliance on constructs of processes not specifying the class of agent or context of action;

2. Reluctance to use profiles or types comprised of differently weighted characteristics;

3. Aversion to treating subjects with extreme values as representing special categories;

4. Skepticism of empirical truths inconsistent with political or social goals; (Kagan 1997b, 321–34)[3]

This critique undercuts the Platonism of cognitive science as well as the exaggerated atheoretical empiricism of most psychiatric diagnostic and statistical manuals (*DSM–IV* 1994), but aims its heaviest charges at a sort of anemic folk psychological functionalism in psychology.[4] Kagan, although his avoidance of this reductionist metaphor is quite

(contd.)
Leverett House, Harvard College, graciously and generously permitted me to serve as Visiting Resident Tutor during this period. I also benefited from the hospitality of the McLean Branch of Massachusetts General Hospital and Joseph Coyle, M.D., Director of the Consolidated Departments of Psychiatry, Harvard Medical School, as well as the History of Science Department at Harvard under the direction of Peter Galison. Good conversations with Ross Baldessarini, Terry Deacon, Anne Harrington, Allan Hobson, Gil Noam, and Marty Teicher assured that I benefited from the year to the best of my ability. My disabilities are my responsibility.

3 *Note to the reader*: this and other references to Kagan's recent publications will take the place of full bibliographic apparatus in this commentary. Kagan is surely a leader in his field, but not the only leader. For a rather complete bibliography of the field, circa 1993, see Kagan 1994.

4 The major symptom of this anemia is failure to move beyond folk psychology to descriptions of the biological etiology and substrata of traits and processes of interest.

significant, insists upon "opening the black box" of temperament, probing all its physiological and ontogenetic components. He does this by experimental analysis of the contextually situated initiatives of young agents in discrepant or unfamiliar contexts.

This is not a flirtation with reductionism. Kagan deploys a multidimensional (psychological, physiological, environmental—social and physical—and genetic) investigative scheme for describing and explaining child behavior.

Kagan places himself squarely in the middle of the bridge between biological and psychological studies of behavior, suggesting that the bristly suit of armor which stands in one corner of his office (nearly the right size, always well dusted and sparkling) may have seen recent use. Jerome Kagan is a prototypical and ideal biopsychologist.

He published *Birth to Maturity, a Study in Psychological Development* (=*BM*) (Kagan 1962), an analysis of the Fels Institute's thicket of interview and test results tracking the psychological maturation of a large cohort, born in the Cincinnati, OH, area between 1929–31, for a quarter century. Called to Harvard, he used the data underlying *BM* to locate a small number of tractable and salient research problems.

Always apparent and finally unavoidable was a subcategory comprised of a small group of individuals who had strong symptoms of shy, quiet, avoidant behaviors as children and who retained these features into adult life. During the last twenty-five years Kagan has pursued this finding within the context of a set of longitudinal, multidimensional experiments, or standardized behavioral probes, now extended to include markers from inter-uterine signals (high, steady heart rate prior to birth) through adolescence.

The behaviors of interest are age appropriate reactions to unfamiliar or discrepant stimuli, objects, or situations. Such entities are difficult to recognize, classify or assign a calming emotional valence. About 20% of the infants in Kagan's sample exhibit responses to discrepant stimuli (odors, noises, moving visual stimuli, voices) with species-specific indicators of distress, e.g., thrashing, crying or both.

Such behaviors are homologous to comparable species-specific responses to the discrepant in other vertebrates: e.g., "freezing," pilo-erection, arched back, unwavering visual attention (staring), usually attended by the typical physiological profile of autonomic (sympathetic) arousal: elevated heart rate, narrowed cardiac inter-beat interval, suppression of respiration, elevated galvanic skin response, asymmetric cooling of distal extremities and face, tightened vocal cords, elevated levels of cortisol and circulating glucocorticoids. As children develop new skills and strengths and repertoires, both the inhibited (avoidant, high reactive) responses to discrepant situations and the configuration of situations counting as discrepant, gradually change.

For children 2–4 years of age, the situations include a strange experimental room (mother present), a strange technician attaching electronic leads measuring heart rate and inter-beat interval, respiratory rate and blood pressure, and probes including discrepant ("monster") toys, disembodied (electronic) voices, the entrance and departure of a third adult, masked but friendly, inviting the child to play. Measured behavioral responses include latency to first move away from the mother, or to the toys or other objects in the room; time spent apart from the mother, latency of time to first vocalization, and latency of time to initiating standard play or social behavior with the third adult. With school age children the situations include opportunities to interact with a group of unfamiliar same age children and with an unfamiliar interviewer. Behaviors then include latency to first spontaneous vocalization and first approach, frequency and duration of vocalizations and other spontaneous interaction with peers or interviewer during test period.

Many aspects of the normal social environment of Caucasian American middle class children *discourage* timid, tense, avoidant, quiet behaviors and reward relaxed, gregarious, smiling and socially engaged repertoires. Fear and timidity exhibit significant plasticity under varying constellations of environmental and endogenous factors. Nevertheless, a subset (n*) of originally timid children,[5] retain their timidity into adolescence, suggesting a strong predisposition (*diathesis*) to such behavior, and possible continuity with, or relatively reliable prediction of, adolescent or adult behavior either violating social norms in certain contexts (agent-environment dyads) or manifesting chronic anxiety or obsessive-compulsive behaviors.

---

5 n* < 5% of n > 400, and n* ~= 20% n´, n´ ~= 20% n. "n´" is a set, n>n´> n*, where n is the set of those displaying inhibition to the unfamiliar (IU) at t, n´ exhibits age appropriate IU at t + a, and n* exhibits age appropriate IU at t+a+b.

Kagan and several colleagues have published data from twin studies indicating a relatively high and statistically significant level of *heritability* for timid behaviors (DiLalla et al. 1994, 405–412; Robinson et al. 1992, 1030–37). The heritability of a specific phenotype is a measure of percentage of the variance of the phenotype in a given population correlated with the genetic variance in the same population. Alternative sources of phenotypic variance are the variances in shared and non-shared environments faced by the members of the population. Twins, whether reared together or apart, are easily divided into the two well-known classes of monozygotic and dizygotic, the former with identical genes and the latter no more closely related than full sibs, i.e., sharing 50% of the genes causing phenotypic variance in human beings. The widely used convention for establishing heritability in twin studies is stupefyingly simple:

$$\text{Heritability} = 2(r_{\text{MZ}} - r_{\text{DZ}})$$

where r is the frequency with which each member of a twin pair is "concordant for" a specific trait, (i.e., each twin—MZ or DZ—exhibits the trait).

Kagan and his colleagues found that for twins in the statistically normalized population constituting the "MacArthur Longitudinal Twin Study" (MALTS) (Emde et al. 1992, 1437–55), studied by the behavioral and physiological measures and in the experimental situations described above, r_{MZ} for timidity = .82, while r_{DZ} = .47. Hence heritability = 2 (.82–.47) or 0.7 at 24 months, the age at which these measurements were taken.[6]

In that population of twin pairs, at that time, with timidity measured in that way, 70% of the phenotypic variance in the population could be accounted for by the sort of genetic variance that obtains between MZ and DZ twins, while only 30% of the phenotypic variance in timidity (at 24 months) was explained by variation in shared

---

6 Very interesting twin studies in Sweden have recently established that although values of heritability fall when traits are mutable as a result of environmental influence, that—in the long run of an ordinary human life—heritability tends to rise, for traits studied in aging twins. The assumption is that environmental variation, given time and the vicissitudes of life, evens out and genetic variance counts for an increasing fraction of phenotypic variance in an aging population (Finkel et al. 1996, 84–99).

and non-shared aspects of the environment. Twins in the same family may experience non-shared environmental features of considerable behavioral salience, if their own behavioral predispositions conflict with or conform to comparable predispositions in other family members. The variables of age, sex and birth order may also contribute to non-shared environmental niches within the same family.

A relatively recent study, unavailable at the time of this writing, identifies one environmental variable inevitably shared by the members of every twin pair, date of conception, may account for as much as one quarter of shyness prevalence in young children (2–7 yrs.) during a one year survey in the U. S. and New Zealand.

> Data indicate that maternal exposure to short daylength during pregnancy, especially the midpoint of gestation, predicts an increased risk of shy behavior in children. . . .This phenomenon might be mediated by changing concentrations of melatonin, serotonin, or other neurotransmitters or corticoids known to covary with seasonal variations in daylength. (Gortmaker et al. 1997, 107–14)

While it is too soon to regard this finding as confirmed, such a powerful parameter, acting in exactly the same way on each member of every twin pair, MZ or DZ, would reduce the value of (r_{MZ} - r_{DZ}) and, consequently, the heritability of any trait it influenced. Striking implications of this new finding include its clear illustration of the efficacy of exogenous modulation of specific neurochemicals themselves implicated in the control of nerve growth factor and neurotrophins, with consequent implications undercutting genetic determinism in the control of development and emphasizing the interaction of genes and environmental factors in producing the substrata of this behavioral phenotype.

The Kagan lab has been collecting other data which also point to a putative site of gene action, early in development, with effects comprising significant elements of the typical profile of the inhibited child: light blue eyes, narrow face, susceptibility to atopic allergies, and a reactive cardiovascular system. These features are all derivatives of the neural crest, a necklace of cells forming around the nascent neural tube and giving rise to sensory ganglia, bones of the skull and face, the autonomic nervous system and the melanocytes (Kagan 1994, 161–69).

It is time to review Kagan's caveats concerning research in this area With the intent of illuminating the work we have just reviewed:

1. Reliance on constructs of processes not specifying the class of agent or context of action;

2. Reluctance to use profiles or types comprised of differently weighted characteristics;

3. Aversion to treating subjects with extreme values as representing special categories;

4. Skepticism of empirical truths inconsistent with political or social goals; (Kagan 1997b, 321–34)

Kagan's analysis of reactions to the unfamiliar is, first and foremost, psychophysiological. His experimental design tracking the ontogeny of this trait from late stages of pregnancy, through puberty and beyond is pioneering work. It deserves this accolade because of its meticulous interweaving of state of the art themes concerning human cognitive and affective development with the latest and most solid achievements in neurobiological modeling of the fear circuit (Ledoux 1995, 1049–1062; McEwen 1995, 1117–36; Hyman and Nestler 1996, 151–62), the autonomic nervous system (Davis 1992, 256–305; Berntson, Cacioppo, and Quigley 1991, 459–87), and the ontogenetic or epigenetic differentiation and maturation of these same brain structures.

Reaction to the unfamiliar as Kagan understands the phrase, is not a phenomenon with an exact equivalent in the language of folk psychology. It could not be identified by a questionnaire written in language subjects could understand. The man/woman in the street is unaware of medical histories including the formation of a necklace of cells (the neural crest) around the nascent neural tube in the third week of gestation. And, although anyone who knows his/her natal day can calculate the midpoint of their own gestation, few of us would volunteer information concerning the daylength during the second trimester of our own fetal life in response to questions about an aspect of our temperament. The point is not that questionnaires cannot be modified to incorporate the results of other forms of research, it is that they cannot be expected to *discover* ontogenetic mechanisms (Kinsey 1948; 1953; Hamer 1994; Cloninger 1996; 1997).

Questionnaires rarely achieve the rich idiomatic fluency and subtlety of colloquial speech. They necessarily use (or are guided by) folk psychological terms, e.g., "novelty seeking," as foci in developing questions accurately probing the various contexts in which, and the corresponding levels of intensity with which, respondents seek novelty. Kagan's method is not discursive, i.e., his subjects do not write memoirs or produce *memoir fragments* during his experiments. But then subjects completing a Temperament and Character Inventory (TCI) (Cloniger et al. 1977, 120–41; 1996, 247–72) are not subsequently asked to write short autobiographical sketches illuminating their answers to the questions they found most pertinent, or impertinent. In that sense, neither Kagan nor any experimentalist, completely satisfies all the implications of "specifying the class of agent or the context of action," (see Kagan's caveat #1 above). But Kagan's work is not only culturally specific with cross cultural components (Kagan 1994, 254–60), it also places agents in a specifiable class in well designed simulacra of culturally appropriate contexts.

In lieu of memoirs, Kagan draws our attention to a special subcategory of those whose reactions to the unfamiliar are colloquially characterized as shy, timid, or "up-tight." He focuses upon the "right tail" of the normal bell shaped distribution curve of this trait. This segment of the curve, representing those in whom "inhibition to the discrepant" is over determined are particularly susceptible to anxiety disorders, phobias, and obsessive-compulsive disorder. This completely satisfies caveat #3 cited previously. Kagan has consistently maintained that the subjects of greatest experimental interest are those in which every variable affecting the "fear" or "inhibition" circuit(s) is set at values intensifying avoidant behaviors. It is this extreme form of the phenotype which is the focus of his interest.

Kagan has sharpened our sense of the multidimensional character of the psychophysiological phenotype which distinguishes this group. He has further traced this phenotype back through fundamental ontogenetic processes identifying both surprisingly salient environmental influences during gestation, and even more surprisingly, an event, early in ontogeny, which may underlie the constellation of phenotypic features found regularly associated with "high reactivity" to the unfamiliar.

This effort has not resulted in specific weights for the different dimensions or stages of the ontogeny of the phenotype of fearfulness or inhibitions, but it has opened up the following possibility: Kagan starts from the first clear appearance of the phenotype at the age all children have the perceptual capacity to distinguish individual adults (~18 months), and then following the development of homologous behaviors and physiological indicators forward as children age, and backward to earlier stages of development. This has given him the opportunity to discover a crucial role for the neural crest in week three of gestation, and the neural crest may be under control of a single "hox" gene—genes which control the timing of development of specific regions of the embryo (Gerhard and Kirschner 1997, chap. Seven, 314–25). In embryology, as in traffic control, timing is everything. Mistimed development can disrupt normal function. In this case it might result in heightened sympathetic sensitivity. Kagan, not reluctant to use multi-dimensional profiles, or search for ways to validating the assignment of different weights to different dimensions, satisfies his second caveat.

Kagan's long held political goals are emphatically egalitarian. He is a staunch critic of the view that experience by age three irrevocably limits a child's developmental capacity. He has developed a pilot program to repair reading readiness in children at risk in the Cambridge, MA, primary grades. He has pursued a research program identifying individual variability in a significant dimension of human behavior capable of powerfully influencing chances for social success in the most dramatic ways. He satisfies the fourth caveat as well as anyone can.

The history of his work provides a map of the time consuming and difficult procedures necessary if we are ever to discover, clone and sequence genes underlying this special class of reactions to the unfamiliar.

Jerome Kagan provides a superb cautionary tale for science writers and journalists constantly bombarded with claims for the isolation and cloning of the neuroreceptors and genes underlying novelty seeking, human sexuality, violent criminal behavior, and intelligence

(Johnsson et al. 1997, 697–99; Plomin et al. 1997, 29–31).[7]

Who has oversold the significance of the Human Genome Project for our understanding of human behavior? First and foremost, I have argued, those molecular biologists, psychiatrists, and their commentators in the media and elsewhere, who have failed to note the indispensible significance of longitudinal analysis of the development of complex human behaviors.

---

7 LeVay 1995 is an excellent book with a fundamental flaw. Its identification of the suprachiasmatic nucleus as a center controlling the initation of male copulatory behavior is based on the reproductive behavior of rats. For a subtle critique of interpretive errors often associated with the use of animal models in such gender sensitive contexts, see Longino 1990.

Part Four

Reductionism, Determinism and Theological Humanism: Introductory Comments

As discussed in the introductory essay, several general questions surrounding the human genome project and the use of modern bio-technology concern the challenge of biological reductionism and the implications of this for a theological and philosophical understanding of the human person. These issues also bear on the significant ethical questions that arise from the genome project.

The discussions of the previous section of the book have broght to the fore some of the challenging issues raised by contemporary genetics and by the evidence for genetic causation of complex behavioral traits. We have also seen some of the complexities in what Evelyn Fox Keller has termed the "discourse of gene action." We have also seen discussion of some of the pitfalls in the assumptions behind the genetic determinism of mental and behavioral properties.

If it might be argued that the HGP presents no strikingly new issues in these debates that have not been previously raised in long-standing debates over materialism, evolutionary naturalism, and scientific determinism, what it does bring to the fore is new data about the more general genetic causes of human traits and demonstrates the power of reductionist molecular biology. This has

342 Section Four: Theological Issues

extended the image of biological determinism and the possibility of a complete reductionism to new levels, what Ted Peters has termed "puppet" determinism (Peters 1997, chp. 2). To the degree that the "strong" genetic program can claim success in accounting for behavioral traits long held to be in the domain of free rational human choice, the theological understanding of human freedom and moral autonomy, and a theistic interpretation of human nature faces new challenges.

The essays in this section bring together the reflections of five participants in the Notre Dame conference who were concerned with some of the science and theology questions of interest to the conference. Arthur Peacocke initially explores the meaning of reductionism in relation to theology by developing on the concept of biological emergentism and its relevance for understanding the relations between physical, biological, human, and theological levels of existence. These issues are then pursued in a more explicitly theological direction by Ernan McMullin, who examines the debate between reductionists and emergenists in contemporary philosophy of biology and its relevance for theological questions concerning ensoulment.

Kevin Fitzgerald then follows by situating these issues within an analysis of competing models of human nature, exploring the different ways in which new genetic and biological developments can be assimilated within competing models of human nature, developing particularly on insights of theologian Karl Rahner. A commentary by John Staudenmaier assists in drawing out the implications of these issues.

Moral issues in the HGP are then addressed from a theological perspective by moral theologian Richard McCormick, who analyzes the ethical issues with reference to the Catholic tradition.

The volume as a whole terminates with a reflective commentary by clinical geneticist John Opitz, who draws together several strands of discussion running throughout the volume, and who offers reflections on the new human genetics from the standpoint of a practicing human geneticist and humanist.

12

Relating Genetics to Theology on the Map of Scientific Knowledge[*]

Arthur Peacocke

There has been considerable confusion in the public mind about the aims of the Human Genome Project—is it to provide the data to enable genetically based diseases simply to be diagnosed and treated, or to enable society to reconstruct humanity in the image of some, presumably widely accepted, ideology? The first aim is unexceptionable and relates to scientifically well-established relations, in certain cases, between gene structure (=DNA sequence) and particular diseases or metabolic deficiencies. But the second, more general, aim is based on a presumption about the relation between genes and human behavior and dispositions, as well as (even more doubtfully) about what is universally desirable in these latter.

It is with respect to this second, wider aim, usually expressed hopefully and journalistically, that it is important to examine the

[*] This contribution to the 1995 Notre Dame conference is a modification and abbreviation of a chapter to appear in the book *Science and Theology: The New Consonance* (Peters 1998), to be published as the outcome of the Center for Theology and the Natural Sciences-Human Genome Project (HG00487) examining the theological and ethical implications of the Human Genome Initiative which has been undertaken at the CTNS at Berkeley, California, with Dr. Peters as the Principal Investigator. The fullest version, with a different focus and in a wider theological context, is to be found as chap. twelve of my *Theology for a Scientific Age* (Peacocke 1993).

relation of any knowledge about human genes firstly to scientific understandings of the many-levelled hierarchy of structure and function that constitute the individual member of the species *Homo sapiens*; and secondly to more general considerations of the relation of genetic information to the nature of human personhood with which theology is so deeply and directly concerned. This paper therefore attempts to relate genetics to theological reflections on humanity by locating more precisely genetics on the map of scientific knowledge of human beings.

I would hazard the guess that one of the most influential catchphrases of our times which has influenced popular perceptions of the implications of biology, especially genetics, is Richard Dawkins' dubbing of genes as "selfish" in the title of his widely-read book, *The Selfish Gene* (Dawkins 1976, 21). Within the general academic community a close runner-up, if not as an influence but rather as a goad, had appeared in the previous year in the first few defining pages of that seminal work of E. O. Wilson which launched the ship of sociobiology:

> Sociobiology is defined as the systematic study of the biological basis of all social behavior . . . . One of the functions of sociobiology . . . is to reformulate the foundations of the social sciences in a way that draws these subjects into the Modern Synthesis. (Wilson 1975, 4)

But in the scientific world, especially that of the molecular biology which developed with the discovery of the molecular basis of heredity in DNA, much more influential in shaping the stances of many scientists was an earlier remark of Francis Crick. One of the discoverers of the DNA structure, he had, some ten years before Wilson, thrown down the gauntlet by declaring that "the ultimate aim of the modern movement in biology is in fact to explain *all* biology in terms of physics and chemistry" (Crick 1966, 10). Such a challenge can, in fact, be mounted at many interfaces between the sciences other than that dividing biology and physics/chemistry. The ploy is called 'reductionism' or, more colloquially, 'nothing-buttery'—"discipline X (usually meaning yours) is really nothing but discipline Y (which happens to be mine)."

Before investigating further the whole question of reductionism, there is an even more sweeping claim that is sometimes implicit in the writings of certain scientists, namely, that, not only is (scientific) discipline X nothing but (scientific) discipline Y, but also that the *only* knowledge worthy of the name is *scientific* knowledge. All else is mere opinion, emotion, subjective, etc. This is the belief system called "scientism" (hence the adjective "scientistic"), that the only sure and valid knowledge is that which is found in the natural sciences and is to be obtained by its methods. It is the belief of only some scientists and of very few philosophers. Nevertheless scientism, together with reductionism, often underlie, as all-pervading assumptions, statements made by a number of influential biologists and geneticists which penetrate into the public consciousness of the Western world.

There has, of course, been a strong and often effectual response to these exaggerated claims—mounted, be it noted, often by scientists as well as by philosophers of science—but their proponents have nevertheless succeeded in conveying to many thinking people in our Western society that those who work on genetics, especially human genetics, share their (apparently)[1] reductionist and scientistic stances. It is this, I would surmise, which often engenders much suspicion of the whole Human Genome Project which is consequently believed to possess, over and beyond its avowed aims to counter human genetic disease deficiencies, a hidden agenda to control the future of humanity through manipulating its genes.

In order to allay such suspicions it is not enough simply to affirm the integrity and good intentions of the scientists and fund-providers involved. For the suspicion arises from the belief—encouraged by the "selfish gene" terminology and the philosophical stance of many sociobiologists—that these scientists think that it is the genes alone which are indeed the control centres of human behavior, and even of human thought. This involves an implicitly reductionist assumption and the whole question of reductionism as a philosophy of the relation between the sciences therefore needs clarifying.[2]

---

1 I say "apparently" because both R. Dawkins and E.O. Wilson, in the appropriate contexts, deny such intentions—in spite of their emphatic statements such as those quoted. Thus Wilson sees the relation between different scientific disciplines as one of conflict and confrontation in which every discipline has its 'anti-discipline' at the level below (in the sense of Figure One of this paper, described later) and is an 'anti-discipline' for that above (Wilson 1977). Yet the impression of a take-over bid by biology (as sociobiology) directed at sociology, anthropology and the behavioral sciences persists.

2 The broader issue of 'scientism' will not be pursued here for it is less directly relevant to the assumptions implicit in the role assigned to genes by some authors.

In order to examine the consequences for theology of genetic research it is, then, necessary to clarify the relation between knowledge about human beings gained from the various sciences. A closer look will, in fact, show a widening horizon of understanding as we move from the physical sciences to the life and social sciences and finally to the realm of human culture with its apprehensions of a transcendent dimension to human experience.

REDUCTIONISM, EMERGENCE, AND REALITY

To indicate the kind of issue at stake here, let me recount how the discovery of the structure of the genetic material, DNA, led me—as a physical biochemist studying, in the early 1950s, its behavior in solution—to *anti*-reductionist conclusions, unlike Crick. What was impressive about this development—and it is a clue to many important issues in the epistemology and relationships of the sciences—is that for the first time we were witnessing the existence of a complex macromolecule the *chemical structure* of which had the ability to convey *information*, a program of instructions to the next generation to be like its parent(s). Now, as a chemistry student, I had studied the structure of the purine and pyrimidine 'bases' which are part of the nucleotide units from which DNA is assembled. All that was pure chemistry, with no hint of any particular significance in their internal arrangement of atoms of carbon, nitrogen, phosphorus, etc. Yet here, in DNA, there had been discovered a double string of such units so linked together through the operation of the evolutionary process that each particular DNA macromolecule has the new capacity—when set in the matrix of the particular cytoplasm evolved with it—of being able to convey a program of hereditary instructional *information*, a capacity absent from the component individual nucleotides. Now the *concept* of 'information', originating in the mathematical theory of communication (C.E. Shannon), and indeed that of a *program*, had

---

(contd.)
Needless to say, it *is* a major issue in our culture and can colour many attitudes relevant to the application of scientific knowledge. However, it seems to me to be less widely assumed by scientists than the reductionism which, I believe, unwarrantedly underlies many discussions of genetics and its applications and is therefore the chief concern of the text at this point. For references to the literature on reductionism, see *inter alia*: Peacocke 1986, chaps. One,Two; Barbour 1990, 165-172.

never been part of the organic chemistry of nucleotides, even of polynucleotides.[3]

Hence in DNA we were witnessing a notable example of what many reflecting on the evolutionary process have called 'emergence'—the entirely *neutral* name for that general feature of natural processes wherein complex structures, especially in living organisms, develop distinctively new capabilities and functions at levels of greater complexity.[4] Such emergence is an undoubted, observed feature of the evolutionary process, especially of the biological. It is in this sense that the term 'emergence' is being used here and *not* in the sense that some actual entity has been *added* to the more complex system. There is no justification for making such assertions (as, for example, in the discredited vitalist postulate).

DNA itself proves to be a stimulus to wider reflections, both epistemological, on the relation between the knowledge which different sciences provide; and ontological, on the nature of the realities which the sciences putatively claim to disclose. Figure Nineteen[5] is intended to clarify what I am referring to and, in particular, the scope of genetics. It represents the relations between the different focal levels of interest and of analysis of the various sciences, especially as they pertain to human beings, rather like the different levels of resolution of a microscope.

The following four focal 'levels' can be distinguished:

(1) *the physical world*, whose domain can be construed, from one aspect, as that of all phenomena since everything is constituted of matter-energy in space-time, the focus of the physical sciences;

(2) *living organisms*, the focus of the biological sciences (with a special 'box' for the key neuro-sciences);

---

3 See the papers in this volume by Evelyn Fox Keller and Lily Kay.

4 This term needs not (*should* not) be taken to imply the operation of any influences, either external in the form of an "entelechy" or "life force'" or internal in the sense of a "top-down" causative influences. It is, in my usage, a purely descriptive term for the observed phenomenon of the appearance of new capabilities, functions, etc., at greater levels of complexity.

5 This figure is reproduced with slight changes from the second, expanded edition of my *Theology for a Scientific Age: Being and Becoming—Natural, Divine and Human* (Peacocke 1993) and is an elaboration by me of Figure 8.1 of Bechtel and Abrahamsen 1991.

FIGURE NINETEEN

The relation of disciplines. "Focal levels" correspond to foci of interest and so of analysis (see text). Focal level Four is meant to give only an indication of the content of human culture (cf. Popper's 'World 3' in Popper 1968).

Solid horizontal arrows represent part-to-whole relationships of structural and/or functional organization. Dashed boxes represent sub-disciplines in particular levels which can be coordinated with work at the next focal level in the scheme (the connections are indicated by vertical, dashed, double-headed arrows).

In each of the focal levels One-Three, examples are given of the *systems* studied which can be classified as being within these levels and also of their corresponding *sciences*. Focal level Two elaborates additionally the part-whole relation of levels of organization and analysis of in the nervous system (after Figure One of Churchland and Sejnowski 1988).

In focal level Two, the science of *genetics* has relevance to the whole range of the part-whole hierarchy of living systems and so, if included, would have to be written so as to extend across its entire width. CNS = central nervous system.

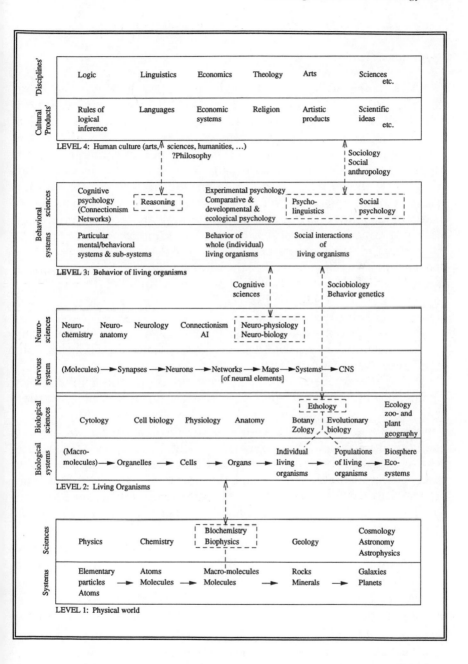

(3) *the behavior of living organisms*, the focus of the behavioral sciences;

(4) Human culture;

Within some of these four levels of interest at least portions of series of part-whole hierarchies of complexity can be found (the horizontal, solid arrows represent such sequences). This hierarchical
whole-part character of the relationships of the natural world is more
apparent in the relations *between* higher and lower levels in the
Figure. Moreover, within any particular analytical level of this
scheme of disciplines, there are often sub-disciplines that form a
bridge with an adjacent level by focusing on the same events or domains. This allows for, and shows the significance of, interdisciplinary interactions. These "bridges" are indicated in the Figure by
the vertical, dashed arrows between the focal levels of interest. Note
that genetics is relevant across the whole range of living organisms
(focal level Two) and, indeed, has an impact on level Three (the behaviour of living organisms) through the 'bridge' sciences of sociobiology and behaviour genetics.

The scheme represented in the Figure is an epistemological one
concerned with the foci of interest, and so of analysis, that naturally
arise from the, quite properly, *methodologically* reductionist techniques of the sciences—the necessary breaking down of complex
wholes into their smaller units for investigation. It serves to illustrate
"part-whole" hierarchies of complexity in which a science focusing on
a more complex "whole" is distinct from those focusing on the parts
that constitute them.[6] For, going up the Figure, one finds the need to
deploy distinctively new concepts and theories containing new referential terms in order to represent the observed capacities and functions and to describe accurately the structures, entities and processes
which occur at those more complex levels (inevitably "higher" on the
printed page).[7]

The concepts and theories, with their particular referential terms,
that constitute the content of the sciences focusing on the more complex levels are often (not always) logically not reducible to those operative in the sciences that focus on their components. This is an
epistemological affirmation concerned with the nature of our knowing

---

6 Such relations occur both horizontally, within the four broad categories of the
boxes, and vertically, though it is these latter which are most concern to us here.

7 No evaluative or normative meaning should be attached to the use of "higher"
in this context.

and knowledge. In particular, and with reference to our concerns in this inquiry, it seems unlikely that the contents of level Three and, even more so, that of level Four are likely to be reducible entirely to that of the science of genetics, which is located across the width of level Two.

Sometimes a variety of independent derivation, identification, or measurement procedures directed at a particular complex level find an invariance in the concepts and referential terms of the theories needed to account for the phenomena associated with them. William Wimsatt (1981) has called these "robust." I would argue that this indicates that what is yielded by the procedures appropriate to each level of investigation can then be said to have a putative, ontological status. They may even be said to be "real," if we use that term in the admittedly loose, but pragmatic, sense of what we cannot avoid taking account in our practical and experimental interactions with the system at the level in question. That is, when concepts and referential terms in theories applied to a more complex level, higher in the Figure, turn out to be "robust" in Wimsatt's sense, there is then a *prima facie* case that they actually refer to new realities distinctively emerging at that level of complexity.[8] There are, it would appear, (putatively) genuinely new realities to be observed in the behavior of living organisms (level Three) and in human culture (level Four) which are not subsumable under and reducible to the concepts which refer to the genetic realities observed at level Two. In particular, human culture includes religious experience and the theologies that are the intellectual reflections upon it, and upon all the other levels.

What I am saying here assumes some form of 'critical' realist philosophy of science. This continues to commend itself to me as the appropriate account of the scientific enterprise for many reasons (McMullin 1984; Peacocke 1984, chap. One; 1990, 11–14; Huysteen 1989, chap. Nine; Banner 1990; Barbour 1990, 41–45) it also seems to be the working philosophy of most practicing scientists who infer from their experiments to the best explanation, and thereby to

---

8 If a concept referring to a "higher" level of complexity *can* be reduced, one has no warrant for affirming that the higher level manifests a new *kind* of reality. Certainly the higher level will still have its distinctive properties as a whole (and the corresponding *qualia* for the observer) but these will not then be, if the concept is reducible, pointers to the higher level consisting in a new kind of reality, which is what the term *emergence* is intended to signify.

postulated provisional realities (candidates for reality), based on the normal criteria of reasonableness—fit with the data, internal coherence, comprehensiveness, fruitfulness and general cogency.

Such considerations now allow us to infer from the map of knowledge depicted in the Figure a kind of "scale" of being and becoming. Science has shown that the natural world is a hierarchy of levels of complexity, each operating at its own level, and each requiring its own methods of inquiry and developing its own conceptual framework, in which at least some of the terms can refer to new non-reducible realities distinctive of the level in question. Moving up the Figure from bottom to top also corresponds very closely to the sequence of the actual appearance in time of the entities, structures, processes, etc. on which the hierarchy of sciences respectively focus. That is, there is an emergence in time of new complex entities, etc., so that the natural world is now perceived as having evolved and as still evolving. The natural world manifests the emergence of new kinds of reality. For, on the basis of our considerations concerning the reality of what is referred to at different levels, one cannot say that atoms and molecules are *more* real than cells, or living organisms, or ecosystems.

Moreover there are *social* and *personal* realities, too. As one moves up the Figure, this recognition of the possibility of the emergence of new realities in the natural world gives a recognizable location within the map of knowledge for the emergence of the distinctively human, all that is signalled by the use of the word "person." The language of personal experience, especially that of personal relations (including, for theists, relations to God and so the language of theology), thereby acquires a new legitimacy as referring to realities which could be emergent in humanity and which are not prematurely to be reduced to the concepts applicable to the constituents of the evolved human body, in particular, its genes. The higher level concepts, in this case the personal ones, must be accorded a *prima facie* status of referring to realities until their respective terms and concepts have been shown unequivocally to be reducible totally to the sciences of the lower levels. So our particular concern in appraising the significance of genetic research, and concomitantly the implications and results of the Human Genome Project, must be to persist in asking whether or not a reduction of the personal to genetics is occurring with the elucidation of the human genome.

Structures, functions and processes present in human beings are, of course, to be found at focal levels One to Three of this scheme. No other part of the observed universe appears to include so many levels, and to range over so much within these levels as do human beings. Level Four is, of course, unique to them. Hence it is a preliminary and initial purpose of this paper to survey (in an admittedly recklessly brief manner) the gamut of the sciences which are relevant to understanding the human organism. Only then can the role of the gene— and so of the social and scientific priority, or otherwise, of genetics— be set in a context allowing judicious appraisal. This should then allow any excessive claims for genetics to be more accurately and suitably qualified. Even so, the level of operation of DNA, that of genetics, is certainly—there is no gainsaying—a key and influential one in the microcosm of the macrocosm which human nature transpires to be in this twentieth-century scientific perspective. Such a survey of the sciences pertaining to humanity could also serve a more positive and wider purpose. For it is imperative that investigators in theology and ethics seriously take into account the multi-layered complex entity that the sciences reveal human beings to be with causal influences going both up and down between the levels, as we shall see in the next section.

LEVELS OF HUMAN BEING

Focal Level One: The Physical Basis of the Human Being

From time immemorial human beings have known that they are made up of the same stuff as the rest of the world: "dust thou art, and unto dust shalt thou return" (*Genesis* 3:19). Today we would say that human bodies, like those of all other living organisms, are constituted of the same atoms as the rest of the inorganic and organic world and that, to varying extents, these atoms also exist throughout the universe. Many of them originated in supernovae explosions long before this planet formed.

A significant feature of the physical (atomic-molecular) level of natural reality pertinent to our existence is its capacity to form structures that can undergo replication in self-perpetuating processes and patterns. This is the focus of "molecular biology" which grew explosively from the discovery of the structure of DNA in 1953, and now forms the *bridge between levels One and Two.* These scientific disciplines have entirely exorcised any ghostly remnants of the "vitalism" that was mooted in the earlier half of the twentieth-century to account for the distinctive characteristics of living organisms and so of "living matter." But these developments equally do not warrant a reductionist interpretation of the functioning of biological systems, and in particular, a reduction to the DNA level. The pattern of "causal" relationships in biological evolution is interesting in this connection. We are dealing with a process in which a selective system "edits," as it were, the products of physico-chemical events (i.e., mutations, changes in DNA) over periods of time covering several reproductive generations.

Let us take an example from Donald Campbell (1974, 179–86) to illustrate this: the surfaces and muscle attachments of the jaws of a worker termite are mechanically highly efficient, entirely conforming with the best engineering and physical principles, and their operation depends on the combination of properties of the particular proteins of which the jaws are made. Selection has optimized these. So, from the perspective of the whole organism's activity and its being only one in a series of generations of termites, it is the efficacy of the proteins in constituting jaws, whose efficiency has been enhanced by natural selection. This efficacy apparently determines the sequences of the DNA units, even though when one looks at the development of a *single* organism, one observes only, with the molecular biologists, the biochemical processes whereby protein sequences, and so structures, are "read out" from the DNA sequences.

Hence there is a sense in which the network of relationships that constitute the evolutionary development and the behavior pattern of the whole organism is determining what particular DNA sequence is present at the controlling point in its genetic material in the evolved organism. Campbell called this "downward" or "top-down" causation insofar as specification of the higher levels of organization (the whole evolutionary system through time, in this instance) is necessary for explaining the lower level—in this case, the sequence in a DNA molecule.

Focal Level Two: Human Beings as Living Organisms

All the biological sciences depicted in this level in the Figure can, in one way or another, include within their scope some aspects of human beings. This is not surprising in view of the evolutionary origins of humanity, as manifested particularly in the fact that *ca.* 98% of human DNA is the same (homologous in nucleotide sequence) as that of the DNA of chimpanzees. The evolutionary process itself is characterized by propensities towards increase in complexity, information-processing and storage, consciousness, sensitivity to pain, and even self-consciousness (a necessary prerequisite for social development and the cultural transmission of knowledge throughout succeeding generations). Successive forms are likely to manifest more and more of these characteristics. However, the actual physical forms of the organisms in which these propensities are actualized and instantiated is contingent upon the history of the confluence of disparate chains of events. There can, it seems to me (*pace* Gould 1989, 51; see Peacocke 1993, 220ff), be overall direction and implementation of purpose through the interplay of chance and law, without a deterministic plan fixing all the details of the structure of that which emerges as possessing personal qualities. Hence the appearance of self-conscious persons capable, according to the Judeo-Christian tradition, of relating personally to God can still be regarded as an intention of God continuously creating through the evolutionary development. (It certainly must have been "on the cards" since it actually happened—with us!)

Remarkable and significant as is the emergence of self-conscious persons by natural processes from the original "hot big bang" from which the universe has expanded over the last ten–twenty billion years, this must not be allowed to obscure another fact about humanity, namely how relatively recent is its arrival in the universe. If one takes the age of the Earth as two days of forty-eight "hours" (one such "hour" = 100 million years), *homo sapiens* appears only at the last stroke of midnight of the second day. Our particular genetic constitution is a only a relatively recent arrival on the Earth and is closely related to that of our non-human forbears. Theists must not underplay the significance of all other living organisms to God as Creator—even though they are able to depict only in extrapolated imagination the kind of delight that God may be conceived to have in the fecund multiplicity and variety of created organisms.

There are some other features of this history that any contemporary theological account of human origins and the nature of human beings must also take into account. Evolution can operate only through the death of individuals: new forms of matter arise only through the dissolution of the old; new life only through death of the old. We as individuals would not be here at all, as members of the species *homo sapiens*, if our forerunners in the evolutionary process had not died. Biological death was present on the Earth long before human beings arrived on the scene and is the pre-requisite of our coming into existence through the processes of biological evolution whereby God, theists must assume, creates new species, including *homo sapiens*.

Furthermore the biological-historical evidence is that human nature has emerged only gradually by a continuous process from earlier "hominids" and that there are no sudden breaks of any substantial kind in the sequences noted by palaeontologists and anthropologists. There is *no* past period for which there is reason to affirm that human beings possessed moral perfection existing in a paradisal situation from which there has been only a subsequent decline. Reference to human behavior brings us within the scope of level three but, before this is examined, we need to look at:

Human Beings in the Perspectives of Sciences Bridging the Biological and the Behavioral. (Between Focal Levels Two and Three)

Sciences which bridge levels two and three include, on the one hand, cognitive science (or, "cognitive neuroscience") and, on the other hand, sociobiology (called by some "behavioral ecology") together with behavior genetics.

Cognitive science is concerned with relating meaningfully to behavior the different levels of analysis of information-processing (roughly, "cognition"). It thereby forms a bridge between the purely biological neurosciences and the sciences of behavior, and it is especially concerned with trying to understand how the mind-brain works, particularly in human beings. The detailed ways in which the various levels of analysis are being applied to shape investigations concerns us less here than the now wide-spread realization by cognitive scientists that, in order to understand the relation between the behavioral, at one pole, and the molecular, at the other, understanding of all levels of analysis, organization and processing is necessary.

This pressure to integrate the study of different levels is, it seems, generated by the very nature of the problems which cognitive scientists address. Moreover, what applies to the operation of the nervous system also applies to the operation of the brain as a whole (Churchland and Sejnowski 1988, 744). This is a clear example, in this instance from the cognitive sciences, of a general feature of biological systems. Because of their intricate complexity, especially that of nervous systems and *a fortiori* the human-brain-in-the-human-body, no one description at any one level can ever be adequate and therefore no one level has epistemological priority nor (on a critical-realist reckoning) ontological priority. The emergent properties and functions at the more complex levels of analysis, organization and processing are emergent *realities*, and those of consciousness and self-consciousness are notably so.

It has been proposed by the brain scientist Roger Sperry (1983, chap. 6 and in subsequent writings) that the way the brain acts on the body through the operation of the central nervous system is best conceived as an instance of that "top-down causation" (or, better, "whole-part influence") which we saw had earlier been postulated by Donald Campbell in connection with the evolution of DNA having a specific information content. Both are instances of the now widely-recognized feature of many complex systems (whose complexity can be structural and/or functional and/or temporal), in which the macroscopic state and character of the system a whole is an influence upon, effectively like a cause, upon what happens to the units of which it is constituted so that these latter behave in ways other than they would have were they not part of that system.

In Sperry's account, the total state of the human brain-as-a-whole, a state describable to our self-consciousness only in mentalistic language, is an influence upon, and so causally effective, on the firing of individual neurons, or groups of neurons, in such a way as to trigger, and actually be, the specific action intended in the consciousness which was that brain state. This amounts to a contemporary scientific analysis of what is involved in personal agency in non-reductionist terms. Indeed, the notion of whole-part influence can be regarded as a contemporary exposition of what is happening when human-brains-in human-bodies are agents. The mechanism through which they act is based on interconnected structures genetically coded in their formation in the embryo. However, the actions themselves cannot be accounted for by genetics alone, because total brain states are constraining factors in human actions and are only referable to, and perhaps describable by, mentalistic language, such as that of purposes or intentions. This places a fundamental limit on what we can expect genetics to explain.

None of this is inconsistent with that Christian anthropology stemming from that of the Bible, which regards human beings as psychosomatic unities displaying a many-faceted personhood uniting many properties, abilities, and potential relationships—rooted in materiality including, today, DNA. Nor need this material substructure of the human being, which arises *inter alia* from its genetic constitution, be regarded as a threat to the reality of subjectivity, of self-consciousness.

Sociobiology may be broadly defined as the systematic study of the biological, especially the genetic, basis of social behavior and, in relation to human beings, aims at exploring the relations between biological constraints and cultural change. It thereby encroaches, in the ambitions of at least some sociobiologists, on to level four. Clearly this whole development is of theological concern. By thus claiming to encompass in one theory human culture and the non-human biological world (especially in its genetical aspects), sociobiology inevitably challenges our thinking about what human beings are. The debate is not entirely a replay of the old nature-nurture dichotomy, for the subtlety and complexity of the strategies of gene perpetuation have undergone much revision and the many-leveled character of humanity is becoming increasingly apparent. The emphatically evolutionary outlook of sociobiology in fact raises no new questions for Christian theology that have not been raised by the general idea of evolution, both cosmic and biological. However, because of the predominantly reductionist tone in the writings of many sociobiologists, there has been a tendency to interpret human behavior functionally only as a strategy for the survival of genes. In its general thrust, the theological response to such suggestions must be that made to any purely deterministic and reductionistic accounts of human behavior. But in making such a riposte, theologians should nevertheless recognize, far more explicitly than they have done in the past, that human nature is exceedingly complex and dependent on its shaping by genetic inheri-

exceedingly complex and dependent on its shaping by genetic inheritance, however much that is overlaid by nurture and culture.

It has indeed been the purpose of *behavior genetics* since 1960, when it first came to be recognized as a distinct discipline, to examine "the inheritance of many different behaviors in organisms ranging from bacteria to man" (Hay 1985, 1). Behavior genetics is predominantly concerned with explaining individual differences within species. As a discipline, it represents a fusion of the interests of genetics and psychology moving between the two poles of a genetics of behavior and a genetically-aware psychology (ibid., 4; Vale 1973, 872). This new subdiscipline is currently being vigorously applied to human beings. The research proceeds and, like all scientific research, as it does so, it both clarifies and at the same time generates new problems. Even in their present form, such studies are producing evidence of the genetic underpinning of much in personal behavior and traits previously considered as entirely environmental and cultural.

Sociobiology and behavioral genetics cannot but influence our general assessment of human nature and, in particular, the degree of responsibility assigned to societies and individuals for their actions. The genetic boundaries which limit what we can do are, from a theistic viewpoint, what God has purposed shall provide the matrix within which freedom shall operate. Furthermore, theologians should acknowledge that it is this kind of *genetically-based* creature which God has actually created as a human being through the evolutionary process. However that genetic heritage itself cannot determine in advance the *content* of thinking and reasoning—even if it is the prerequisite of the possession of these capacities.

For example, to unravel the evolutionary and genetic origins of moral awareness is not to preempt its ultimate maturation in the moral sensitivity of self-aware, free, reasoning persons whose emergence in the created order God can properly be posited as intending. The vital question with which the religious quest, and so theology, is concerned now becomes: what do we human beings make of these possibilities? The biological endowment of human beings does not appear to be able to guarantee their contented adaptation to an environment which is inherently dynamic, for they have ever-changing and expanding horizons within which they live individually and socially, physically and culturally, emotionally, intellectually and spiritually. In particular, when one reflects on the balanced adaptation of other living organisms to their biological niche, the alienation of human beings from non-human nature and from each other appears as an anomaly within the organic world. Thus, it is not surprising to find Eaves and Gross (1990, 17), writing on behavioral genetics, pointing out what they call the "possible gulf between the ecosystem in which human evolution occurred and the global environment into which humanity is now projected;" and going on to suggest that the basically unethical, human favouring of genetic kin is a sign, at best, of tribal self-interest and "that humans bring into the world by virtue of their ancestry biological baggage which is ill adapted to the present world"(ibid.). For what constitutes the "world" for human beings transcends the purely biological.

Hence, as human beings widen their environmental horizons into ranges which are really more appropriate to level four, they experience this "gulf" between their biological past environment out of which they have evolved and that in which they conceive themselves as existing or, rather, that in which they aspire to exist. I am thinking of such experiences as: contemplation of our own death, our sense of finitude, suffering, the realization of our potentialities, steering our path from life to death. The mere existence of this "gulf" between our experience and our yearnings raises a problem for any purely biological account of human development. We may well ask, "Why has, how has, the process whereby there have so successfully evolved living organisms finely tuned to and adapted to their environments failed in the case of *homo sapiens* to ensure this fit between lived experience and the environing conditions of their lives?" It appears that the human brain has capacities which were originally evolved in response to earlier environmental challenges but the exercise of which now engenders a whole range of needs, desires, ambitions and aspirations which cannot all be harmoniously fulfilled—they are not compossible. The complexity and character of the human predicament clearly involves more subtle levels of human nature than are the focus of level two or of the "bridge" sciences to the next level. So we turn to those sciences concerned with human behavior.

Focal Level Three: The Sciences and Human Behavior

Some of the principal behavioral sciences and the systems on which they focus are indicated in level Three of the Figure. This includes various forms of psychology which is, in its usage since the eighteenth century, the study of the phenomena of mental life. Recently there has been a "cognitive," "consciousness," or "mentalist" shift of emphasis in psychology that moves its focus of interest towards the content and activities of ordinary consciousness (sometimes neutrally denoted as "self-modification").[9] Consciousness is now much more frequently regarded as a theoretical term that refers to realities whose existence is inferred from observation. How it is to be a thinking and feeling human being have again come on to the agenda of many of the behavioral sciences (level Three).

A re-habilitation thus appears to be occurring, from a scientific perspective, of the reality of reference of humanistic studies, in which theology should be included, if only because of its concern with religious experience. It also gives scientific credibility to what had never been doubted in theology—the pre-eminence of the concept of the personal in the hierarchy of our interpretations of the many-levelled structure of the world of which humanity is an evolved constituent. This has important implications for the relation of science and theology. Instead of a dichotomy between, on the one hand, a dualism of "body" and "mind" (a common misapprehension of the Christian view of humanity) and, on the other, a reductive materialism, a new integrated "view of reality" could emerge which, so Sperry hopes, "accepts mental and spiritual qualities as causal realities, but at the same time denies they can exist separately in an unembodied state apart from the functioning brain" (Sperry 1988, 609). Thus, the situation looks more encouraging for a fruitful dialogue between theology and the sciences of human behavior than it has been for many decades.

The Social Sciences (Between Focal Levels Three and Four)

The sciences variously designated as "social" form a bridge between the behavioral sciences and culture. The more the sciences are

---

9 Consistently with his ideas expounded earlier, Sperry affirms in (Sperry 1988, 608) that there is a new openness in the behavioral sciences, not only in a 'downwards' direction via cognitive science to the neurosciences, but also 'upwards' to all those studies and activities that regard human consciousness and its content as real and worthy of examination and interpretation. Such suggestions also allow us to understand better that much larger ontological transition from level three to level four, which one recognizes intuitively but finds more difficult to explicate scientifically than that between levels one, two, and three.

concerned with the mental life and behavior of human beings, the more they will impinge on the concerns of the Christian community—though it is worth noting that the social conditioning of religious beliefs which the social sciences disentangle and reveal does not, of itself, settle any questions as to the *truth* of these beliefs.

The evolutionary process, however, introduces another dimension into this complex relation between religious belief and social setting, namely that of "evolutionary epistemology"—the realization that cognition of its environment by a living organism has to be sufficiently trustworthy in its content to allow the organism to be viable under the pressures of natural selection. Cognition of its environment is "trustworthy" in this sense when the organism has to take account pragmatically of its content in its practical and experimental interactions with that environment in order to survive. Now, by setting up norms related to the existence of a "transcendental reality" other than human authority, the cumulative wisdom of the religious traditions has, it has been suggested (Campbell 1976, 176–208), contributed crucially to the process of human social organization, wider and more complex than that of any other living organism. In other words, humanity could only survive and flourish if it took account of social and personal values that transcended the urges of the individual "selfish" genes.

In the light of evolutionary epistemology, does this not imply that these social and personal values, enshrined in moral codes and imprinted in ethical attitudes, are part of the *realities* with which we humans have to deal and of which we have to take account or otherwise die out? Such a role for the religions of humanity in socio-cultural evolution points to the existence of values as constituting a reality system that human beings neglect to their actual peril. If this is so, religious beliefs are directly and causally pertinent to decisions relevant to the future of humanity, including its genetic future.

Focal Level Four: Human Culture and Its Products

Such perceptions bring us clearly to the domain of human culture, that of level four. The "cultural products" at focal level four are embodiments of human creativity in the arts and sciences and in human relations including, theists would add, relations to God.[10] Those patterns of discernible meaning within the natural nexus of events in the world, which are the means of communication between human beings and between God and humanity, are generated through their historical formation in continuous cultures which invest them with meaning enabling such communication. Thereby they have the unique power of inducting humanity into an encounter with the transcendence in the "other," whether in the form of another human person or, for theists, of God—the Beyond within our midst. George Steiner in his penetrating *Real Presences* has called such an encounter a "wager on transcendence."[11] We can expect all such encounters with "cultural products," all such "wagers on transcendence," to communicate only in their own way, in their own "language," with an immediacy at their own level which is not reducible to other languages.

This re-assertion of the conceptual and experienced autonomy of what is communicated in human culture is reinforced by the re-habilitation of the subjective, of inner experience, in cognitive science and in psychology—in fact, in the recovery of the personal, the recognition of the reality of personhood. We really do seem to be witnessing a major shift in our cultural and intellectual landscape which is opening up the dialogue between the human spiritual enterprise (broadly, "religion") and that of science in a way long barred by the dominance of a reductionist, mechanistic, materialism, thought erroneously to have been warranted by science itself. The human is undoubtedly biological, but what is distinctively human transcends that out of which and in which it has emerged.

This pressure for a wider perspective on humanity is being generated from within the sciences themselves (if not by all scientists, as such) in attempting to cope with the many levels of the Figure. Is it

10 I have argued elsewhere (Peacocke 1993, chap. Eleven) that relations to God are expressed by their own distinctive means through meaningful patterns created in what is received initially through our senses.

11 "[T]he wager on the meaning of meaning, on the potential of insight and response when one human voice addresses another, when we come face to face with the text and work of art or music, which is to say when we encounter the 'other' in its condition of freedom, is a wager on transcendence." (Steiner 1989, 4). He does not hesitate to point to its theological import: "The wager . . . predicates the presence of a realness...within language and form. It supposes a passage . . . from meaning to meaningfulness. The conjecture is that 'God' is, not because our grammar is outworn; but that grammar lives and generates worlds because there is the wager on God" (ibid.).

too much to hope that we see here the first glimmerings for some time of a genuine integration between the humanities, including theology, and the sciences? Are we seeing the beginnings of a breakdown of that dichotomy between the "two cultures" that was engendered by the previous absence of any epistemological map on which their respective endeavours could be meaningfully located?

CONCLUSIONS

A survey such as this highlights the superficiality of any talk which speaks of there being *particular* genes for *particular* kinds of human behavior.[12] Firstly, behavior is nearly always influenced by complex *sets* of genes, and the link between the immediate output of these genes and human behavior is extremely tenuous—not least because even the synapsal linkages in the brain are not entirely genetically determined, but depend on the input from learned experiences which create new ones, and which, in the human case, means the input from the experiences mediated by human culture.

Secondly, the genes operate within the hierarchy of nested levels shown in the Figure. The higher of these are sensitively dependent on their interactions with the environments specifically operative at those levels—not least that of human culture (level Four). Thirdly, a human person is not a static entity, but is always in process of becoming— one ought to speak of "human becomings," rather than "human beings."[13] This dynamic process has occurred historically, as humanity has evolved out of the inorganic material of planet Earth, rising from level one to four. It also occurs in the life of each individual. We start as DNA inherited from our parents immersed in the evolved cytoplasm of a fertilized egg, but gradually emerge as social beings/becomings, interacting personally with other members of our species, and eventually inducted into the world of culture and into the rich heritage of human concepts and values. These latter, in their multiform aspects and influences, shape our actions continuously

---

12 This statement is compellingly amplified and supported by the 'six rules (or generalizations)' which Kenneth Schaffner develops in his paper in this volume concerning the complex relations between genes, neurons, behaviors and environments/histories.

13 "We are not so much human beings as human becomings" (Morea 1990, 171).

within the limitations imposed by the genes which we possess. In those actions we have the irrefutable experience of being free to choose within limits (of which we may not necessarily be conscious), and this experience is a raw *datum* that even the reductionist cannot deny. This experience of freedom is acknowledged by all ethical systems, however differing in particular content, and by all systems of law in societies which attribute responsibility to individuals for their actions. These are observable features of all societies—even our own Western one— with their current loss of nerve concerning the content of law and ethics. They are not to be rubbished by the excessive claims of those geneticists who, expressing an unwarrantable reductionist philosophy, give the impression that we human beings are *determined* by our genes.

More generally, I conclude from this survey that, however much might possibly emerge in the future to remedy particular, localizable and identifiable biological deficiencies, it would be unwise to place too much hope in the ability of any directed genetic engineering to ameliorate the general human condition, especially the psychological and spiritual. It would be irresponsible of all those engaged in the Human Genome Project if they ever gave the impression that it could do so.

13

Biology and the Theology of the Human

Ernan McMullin

In 1981, at the height of the debate about "creation-science" in the U.S., the National Academy of Sciences stated in a resolution: "Religion and science are separate and mutually exclusive realms of human thought whose presentation in the same context leads to misunderstanding of both scientific theory and religious belief" (National Academy 1984, 6). "Separate and mutually exclusive"—This view of the science-religion relationship is an attractively simple one, one widely shared in our secular age. It implies that true conflict between the two cannot happen: if a debate does arise it can only be because one or the other has overstepped its bounds, epistemologically speaking. Long ago, in the first major confrontation between Christian theologians and practitioners of a natural science, Galileo quoted a *bon mot* he attributed to Cardinal Baronio: "The intention of the Holy Spirit is to teach us how one goes to heaven and not how heaven goes."[1] He argued, plausibly, that the Scriptures were never intended

---

1 Galileo 1957, 186. The same aphorism has twice been quoted, with apparent approval, by Pope John Paul II, both times in addresses to the Pontifical Academy of Sciences. In October 1981, speaking about the relation of Scripture and science, he made use of the aphorism to make the point that the Bible is not intended to be "a scientific treatise" but to illuminate the relation of man to the universe and to God (John Paul II 1981, 279). He quoted it again in a recent declaration terminating the work of the Galileo Commission he had set up in 1981. Elaborating on the aphorism, he goes on: "The Bible does not concern itself with the details of the physical world, the understanding of which is the competence of human experience and reasoning" (John Paul II 1992, 373).

to bear on such technical issues in the natural sciences as the true motion or rest of earth and sun; in this regard, the Biblical authors would assuredly have accommodated themselves to the capacities of their intended audience.

Though there would be pretty general agreement today on this clear-cut separation between theology and such sciences as astronomy, the matter is quite different in regard to one domain where the interests of the two sides clearly seem to converge. The sciences obviously have a great deal to say about human nature, but the Bible presupposes a very definite view of human nature also: Human beings are made in God's image, are morally responsible for their actions, are destined to eternal life with God. It is, of course, a matter of presupposition on the part of the Biblical authors, not of philosophical argument. Yet the commitment to certain very general views regarding human uniqueness, human unity, and human freedom is unmistakable.

TWO VIEWS

Pope John Paul II made this latter point very forcefully in a recent address intended for a meeting of the Pontifical Academy of Sciences that dealt with issues in those sciences that bear on evolution. It is proper for theologians to concern themselves with the question of evolution, he asserted, "for it involves the conception of man" (John Paul II 1996, 352). The Pope was perfectly willing to allow that the theory of organic evolution is supported by a growing body of evidence, impressive in its "convergence." (This was the feature of his address that was picked up by the press in the U.S., presumably because of the enduring controversy surrounding the efforts of "creation-science" advocates to oppose the theory of evolution in the name of Biblical orthodoxy.) He went on to insist, on philosophical as well as theological grounds, that this theory cannot account for the appearance of the human soul which, being spiritual, cannot originate from the resources of matter alone. The soul must, then, be "immediately created by God."

But does not this claim of an "ontological leap" to the level of the human create a severe tension between theology and the sciences, he asks? After all, physical continuity "seems to be the main thread of research into evolution in the fields of physics and chemistry." But closer consideration of the methods appropriate to "the various branches of knowledge makes it possible to reconcile two points of view which would seem irreconcilable." The "sciences of observation" are limited to what can be measured and correlated. "The moment of transition to the spiritual cannot be the object of this kind of observation," even though at the experimental level science can discover "signs of what is specific to the human being." Ultimately, however, it belongs to the competence of philosophy and theology to analyze the implications of such uniquely human features as self-awareness, freedom of choice, religious experience. It is at this level that the discontinuity of the human must be affirmed. Reductionist or materialist philosophies that deny this "ontological" discontinuity and "consider the mind as emerging from the forces of living matter or as a mere epiphenomenon of this matter are incompatible with the truth about man." They cannot "ground the dignity of the person." What is at issue, then, is "the true role of philosophy and beyond it, of theology." The Papal address simply restates traditional doctrine in regard to the human soul, with a fuller philosophical commentary than any other recent Roman pronouncement on the issue.

There are echoes here, ironically enough, of the position taken in the statement from the National Academy of Sciences. Apparent conflict between religion and science regarding the activities that distinguish human beings from other living things can be dealt with by ruling that the two are mutually exclusive, that they cannot in principle truly overlap in the claims they make about the world. Of course, the Papal statement backs this up in a way that the NAS would never allow by asserting that the sciences simply have no jurisdiction in the disputed domain. The claim is a familiar one, familiar in recent decades especially among those who belong in that group of philosophical traditions loosely called "Continental": existentialism, phenomenology, personalism. . . . One would look in vain for the most part in the works of leading figures in these traditions for any hint that the services of such sciences as evolutionary biology or anthropology or cognitive science could help to illuminate the philosophic quest for a better understanding of the human spirit. The group of philosophical traditions equally loosely grouped under the label of "analytic," associated especially with the English-speaking and Scandinavian countries, have been much more sympathetic to the claim that the sciences do indeed have something to say about the nature and origins of such human activities as thinking and willing.

Arthur Peacocke's essay in this collection reflects this latter sort of emphasis. In the map of scientific knowledge he provides the reader, the sciences all bear on the human in one way or another, and at the level of the *distinctively* human, such sciences as social psychology, anthropology, and psycho-linguistics, are given strong emphasis. Peacocke underlines, in particular, the way in which such "lower-level" sciences as molecular genetics, evolutionary biology, ethology, and behavior genetics, illuminate issues at the "higher" level of the human. The conclusion he draws from these sciences is that human beings "emerged" through the natural processes of evolution, so that from the theological standpoint evolution is the instrument by means of which God chose to bring the human race to be. "Furthermore, the biological-historical evidence is that human nature has emerged only gradually by a continuous process from earlier 'hominids' and that there are no sudden breaks of any substantial kind in the sequences noted by paleontologists and anthropologists" (Peacocke this volume, 356).

From Peacocke's perspective, then, there are no ontological dis-continuities of the kind the Papal statement finds necessary in order to safeguard human dignity. There is simply no evidence from the sciences that would support such an hypothesis, he maintains, and much that would count against it. How, then, does Peacocke, himself a de-vout Christian believer, secure the uniqueness of the human that ap-pears to be so central to Christian belief? This is where the notion of emergence comes to his aid. To say that humans "emerged" from the evolutionary processes leading up to their appearance is, for him, to say that the properties constituting the distinctively human level are "new," that they cannot be "reduced" to (explained in terms of) properties and configurations proper to the lower levels. A "new, non-reducible reality" gradually made its appearance with the advent of the human, something that "transcended" all that went before, even though it had evolved from it. The new reality is not a substance (mind-body dualism is, he says, "a common misapprehension of the Christian view of humanity"[ibid., 361]); it is a new level of being, characterized by "new capabilities and functions at levels of greater complexity," but not ontologically discontinuous with what preceded it and prepared the way for it.

Since Peacocke lays so much weight on non-reducibility as the criterion for this "new reality," it would clearly be important to establish that the distinctively human capacities are, in fact, non-reducible. Here his argument becomes muted: these capacities are not to be *"prematurely . . .reduced* to the concepts applicable to the constituents of the evolved human body" (ibid., 352, my emphasis); they must be accorded the *"prima facie* status" of new realities until they have been shown to be "unequivocally reducible." But this seems very weak. Elsewhere in his essay, he seems to assume that they have been shown to be irreducible, indeed, that the sciences of the human in general are coming round to this view in consequence of a "major shift in our cultural and intellectual landscape." But is this really so? What is needed here is at least brief reference to the lively debates going on around this very topic in what may be the most active area in philosophy at the present time, the philosophy of mind.

Here, then, are two Christian responses to the enormous advances in recent years in the connected sciences of genetics, evolutionary biology, and biochemistry. They could hardly be more different. It would be impossible within the scope of a brief comment to do justice to these differences. What I hope to do instead is more modest: to draw attention to troublesome ambiguities in some of the key concepts on which discussions of human uniqueness depend, to recall very briefly some of the difficulties philosophers have encountered in their attempts to define the relation of the human powers of mind to the material capacities of body, and finally to ask what the theological significance of all this is.

REDUCTION

In the context of the sciences, reduction is primarily an epistemological affair. To reduce a theory, T_2, by means of theory T_1, is to show that the explanation given, the concepts employed, the laws drawn on, the mechanisms postulated,[2] by T_1 can satisfactorily perform the equivalent tasks of T_2, thus "reducing" the sciences needed in that context from two to one. Examples would be Maxwell's explanation of optical phenomena by means of the electromagnetic field, thus reducing classical optical theory to electromagnetic theory, or the explanation of thermal phenomena through

---

2 These characterizations are not quite equivalent, and the differences between them give rise to somewhat different concepts of reduction. For our purposes here, these distinctions can be left aside.

the concepts and laws of mechanics, thus reducing thermodynamics to statistical mechanics. In the type of reduction of most concern to us here, the notion of a "level" enters in: T_2 is said to be a "higher" level than T_1, that is, it deals with more complex entities of which the entities that are the subject of T_1 are constituents. To reduce would then be to explain the properties of the whole in terms of the properties of the parts and their configuration. An example would be an explanation of Mendelian inheritance in terms of DNA structure and the action of enzymes, in effect reducing Mendelian genetics to molecular biology.[3]

The topic of reduction received a good deal of attention from the logical positivists earlier in the century, and it continues to be a debated issue today. The summary above does not begin to do justice to the subtlety of the distinctions and the detail of the case-studies to be found in the literature.[4] But it may suffice to enable some needed inferences to be drawn. To reduce a higher-level property is not to deny its reality. It is to question the supposed difference of level, since the reducing properties (concepts, mechanisms, etc.) are alleged of themselves to account causally for the reduced property. Historically, however, reduction has often been supposed to warrant a much stronger inference.

The ancient atomists postulated atoms possessing only shape, position, configuration, and motion. Such properties as the taste and color of visible bodies were to be explained by the properties of the constituent atoms. The conclusion drawn was that the sensible qualities, thus explained, were also explained away. They existed, in consequence, only "by convention." With the revival of atomism in the seventeenth century, doubts about the existence of secondary qualities surfaced once again: if they could be explained by the primary qualities of constituent corpuscles, what kind of reality could they have in their own right?

Successful reduction, as already noted, does carry with it an ontological consequence. Since the reduced quality can be fully specified in terms of the reducing qualities and their configuration, it is no

---

3 Philip Kitcher analyzes this particular example in some detail in order to argue that such a reduction does not, in fact, occur; this leads him to a critique of reductionism in biology generally (Kitcher 1984).

4 See, for example, the one-hundred-page chapter devoted to reduction in Schaffner 1993.

longer regarded as an *autonomous* property. It does not have to be separately specified in a description of the complex entity. Yet it is nonetheless real. The color of an object is no less real because it can be explained in terms of the properties and configuration of the constituents of the body's surface layer of atoms. The negative overtone of the terms 'reductionist,' 'reductionism,' derives in part, no doubt, from the widely-shared belief that a reductionist is someone who denies the existence of a strongly evidenced reality.[5] It is true that someone who claims reduction in a particular domain is denying *diversity* of an irreducible sort. But that is a different matter. If explaining self-awareness in neuro-physiological terms were to be equivalent to denying its existence, then reductionism would indeed be a threatening program! The language of "nothing but" on the reductionist side, furthermore, does nothing to dispel this misunderstanding. All of this does not mean that successful neuro-physiological reduction would not have significant ontological and philosophical consequences. But they are not quite what they are often supposed to be.

One further common misapprehension should be noted in regard to reduction itself.[6] When the science of some complex whole is reduced by the science of its constituent parts, in some cases what has happened might be better described by calling it an *enlargement* of the lower-level science. As we have just seen, it was standard belief from the seventeenth century onwards that a secondary quality like color could be reduced in terms of the mechanical qualities and the configuration of the constituent corpuscles of the colored body. John Locke professed himself baffled as to how such an explanation would proceed, but remained confident that it could, in principle, be carried through (Locke 1690, Bk. 4, chap. Three). Since the corpuscles and their motions were causally responsible for the property perceived by us as color, the science of these corpuscles, mechanics, would be sufficient to explain color. But *whose* mechanics? Newton challenged the list of "primary" properties he had inherited from the "mechanical"

---

5 Peacocke (this volume, 352), for example, appears at times to imply this view: "Higher-level concepts. . .must be accorded a *prima facie* status of referring to realities until their respective terms and concepts have been shown unequivocally to be reducible totally to the sciences of the lower levels." "Until"? But surely even after reduction has been achieved, the properties are just as real, just as rooted in the reality of the body, as they were before?

6 I discussed this point in some detail in an earlier treatment of reduction and related topics in McMullin 1972.

philosophers of an earlier generation by adding a factor that appeared distinctly non-mechanical to his critics. Gravity was needed to explain the behavior of such complexes of bodies as planetary systems. It was the behavior of the *complex* that forced the addition of a concept that would revise the notion of "mechanical" action so drastically.

But much worse was to come. It turned out that Newtonian mechanics, fortified by the "unmechanical" notion of gravity, could not explain the most obvious attribute of solid bodies, namely their color. It took a complete revision of mechanics itself, a turning-upside-down once again of its most basic notions of mechanical action, to carry the so-called "reduction" through. Quantum mechanics called upon spectroscopic data, data about the frequencies of the radiation emitted by complex bodies, atoms and molecules. Attention to the properties of the whole ultimately forced an entirely new understanding of the capacities of the parts. What is important to grasp here is that it is not just the properties of the parts in isolation (as it was in Cartesian mechanics) that define the science of the parts, but also their properties when joined with others in complex configurations. And these latter capacities or potentialities can only be discovered by treating the properties of the complex whole as, in some cases at least, epistemologically primary. Quantum mechanics may look like a science of elementary particles, and in a sense, of course, it is. But it must not be forgotten that it was constructed by treating the properties of complexes as clues to capacities that would somehow have to be incorporated in the science of the parts.

The label 'reduction' is thus in this respect equivocal. Two morals may be drawn from this. One is that reduction is not necessarily the simple shifting of epistemological and ontological weight from whole to parts that it is often assumed to be. The other is that the fundamental revisions of the sciences of the parts that attention to the properties of the complex occasionally brings about cannot be anticipated. Thus to speak of reducing biology or psychology to "physics" is not the clear-cut proposal it is usually assumed to be. Endless debates between proponents of such reduction and their critics rarely pay attention to the ambiguity in the fundamental term in the reduction. What is meant by "physics" in such a scenario? The physics of today? The colors of material bodies were not, in fact, reducible to physics or mechanics as these terms were understood in the mid-nineteenth century. How are we to know what physics will look like a hundred years hence? Physicists in 1900 could not have imagined how fundamental a transformation would be worked by theories, then still to come, that today we know as relativity theory and quantum theory. It is risky to set limits in this regard. Consideration of the properties of such complex wholes as the human brain could eventually force a revision in the science of the physical constituents of the brain just as fundamental as the revision that consideration of the color of atomically-constituted bodies forced on the Newtonian science of the atom's constituents. At this point, we simply do not know.

EMERGENCE AND MATERIALITY

The issue of reducibility can also arise in a rather different context, when one asks how complex properties come to be in the beings that display them. One might ask, for example, how the processes of biological growth in the human fetus can bring about the appearance ultimately of such properties as intelligence. Or one might ask, at a further remove, how a new kind, distinguished by various upper-level properties (the first living cell, say) could have come about in the course of evolutionary development. Ordinarily, scientists go about answering requests of this sort for what has been called genetic explanation by looking to the prior constituents of the complex being and the processes in which they engage; they assume that these will causally explain how the upper-level property came to be. But suppose the property is irreducible? Then an explanation of this sort is automatically blocked. It is, no doubt, in large part because of this that claims for the irreducibility of various "levels" in nature have been greeted so often with suspicion from Descartes' time onwards. Nonetheless, belief in such levels persisted even after Darwin's *Origin of Species* posed a strong challenge. It seemed obvious, to some scientists at least, that such levels do exist, and that the transition upwards from one level to the next in the course of development either of the individual or of the kind cannot be the sort of smooth predictable affair that reducibility would, in principle at least, allow the scientist to take for granted in advance.

The issue was much discussed in the later nineteenth century, not only because of the growth of evolutionary forms of explanation but also because of concrete incidents of unexpected reduction in physics, as optical theory for example was subsumed into electromagnetic theory. A number of British philosophers were led to propose a notion that came to be called "emergence."[7] A high-level property is said to be emergent when it is, roughly speaking, irreducible by the sciences governing the constituents from which it derives. To say that it "emerges" implies that in some sense it was already there in potency but that it is in a significant sense new. A favorite example of these early defenders of emergence was water: its properties surely cannot (they reasoned) be reduced by the properties of hydrogen and oxygen alone. Water must be supposed, then, simply to emerge when the material substrate is right, that is, when oxygen and hydrogen are brought together in the proper proportions and in the proper way. With the advent of quantum mechanics, claims for the irreducibility of such chemical complexes as water by the mechanics applying to their constituents have been undermined. But the reducibility of living processes by the concepts of biochemistry is still debated. Authors as diverse as Michael Polanyi, G. G. Simpson, Ernst Mayr, and Marjorie Grene, defend one or other form of irreducibility of level in biology and hence are led to postulate emergence to describe what happens in the regular course of biological development when an upper-level property makes its appearance.

Criticisms of emergence are of two kinds. Some philosophers have argued that the notion itself is incoherent or, at the very least, ill-defined. A certain configuration of the constituents is said to be a sufficient condition for the upper-level property to appear.[8] Why not,

---

7 The term was coined by G. H. Lewes in 1875, but the concept was already implicit in J. S. Mill's *System of Logic* (1843) and Alexander Bain's *Logic* (1870). It was subsequently developed further by Lloyd Morgan and Samuel Alexander, and found its "canonical" form in C.D. Broad's *Mind and its Place in Nature* 1925. See McLaughlin 1992. McLaughlin treats in some detail the variety of ways in which the defenders of emergence tried to clarify this difficult concept in response to incessant criticism from more reductionistically-inclined colleagues. He concludes that emergentism has been "refuted" by the scientific achievements of the twentieth century, notably quantum mechanics (89-91). It is significant, however, that he sets the psychological realm outside the scope of his analysis, restricting himself to the chemical and the biological.

8 Emergence resembles the notion of supervenience introduced by Donald Davidson who suggested that "mental characteristics are in some sense dependent, or supervenient, on physical characteristics," and thus that "there cannot be two events alike in all physical respects but differing in some mental respect..." (Davidson 1970, 88). Elaborating variations on this notion became something of a cottage industry among analytic philosophers in the years following. It would be fairly generally allowed today that emergence and supervenience, despite their apparent

then, say that this causes the property to appear? And if one can say this, is it not to say that there is a law connecting the two, and perhaps even to imply that there must be a theory explaining why the law holds, even if we cannot yet say what that theory might look like? And if this is the case, can one still hold that the upper-level is irreducible? The questions here are complex and have led to intricate debates.[9] Many of the issues arise from the difficulties in defining the notion of reduction to which we have already alluded. We shall have to lay aside both sets of issues, important though they are, for reasons of space. The other sort of criticism of emergence is to say that the concept is not needed, since none of the alleged cases of irreducibility hold up. Committed reductionists, like Paul and Patricia Churchland, have argued full reducibility even of the domain of the mental and the intentional, claiming that it can be reduced by neuro-physiology, in principle at least (Churchland 1984; Churchland and Churchland 1986).

This is, however, a highly controversial position. Closer to the center might be the view of Jaegwon Kim who regards the "current orthodoxy" in the philosophy of mind as being a nonreductive physicalism which, in his view, is equivalently a form of emergentism (Kim 1992, 121). He is, however, dubious about the notion of downward causation to which (he maintains) an emergentist must be committed, arguing that assigning to the mental the capacity to influence that which sustains its very existence "threatens the coherence of this popular approach to the mind-body problem" (ibid., 137).

(contd.)
resemblance, are by no means equivalent, particularly in their relation to irreducibility. Paul Humphreys puts it rather strongly. In dealing with the problems raised, for instance, by "concepts in the mental realm," he asserts: "Reduction has been out and supervenience has been in. This position is only half right. Reduction is still not an option, but supervenience is no good either. It is a notion that is empty of any scientific or metaphysical content, and what anti-reductionists need in its place is emergence" (Humphreys 1997b, 337). See also his Humphreys 1997a; and various essays in Beckermann et al. 1992a, especially Beckerman, 1992b.

9 See, for example, Bechtel and Richardson 1992; Schaffner 1993, esp. 411–516. Bechtel and Richardson focus on the interactive organization of the components of a complex system and conclude that emergence provides a viable middle ground between the extremes of reductionism and holism. Schaffner leans somewhat the other way; he believes that by a careful reworking of the notion of reductionism, one can dispense with "in-principle emergentism," in the domain of the biomedical sciences at least.

To sum up, then, though the mind-body problem is perhaps the main battlefield in contemporary analytic philosophy, there would be fairly strong support for the claim that the realm of the mental (or the intentional) is not reducible by the sciences of the brain's constituents, though there would be wide disagreement about how this should be shown and just what notion of reduction one ought to employ in this context. The further question as to how mental processes or properties make their appearance leads to the admission of the notion of emergence, allowing once again for a degree of variation, this time in how this latter concept is defined by its proponents.

One further question needs at least brief treatment. Recent defenders of emergence in the philosophy of mind tend to characterize their view as "non-reductive materialism." Is it, in fact, materialist? Not if materialism be supposed to exclude the existence of spirit, of a non-material form of existence. Peacocke, a theist as well as an enthusiastic proponent of emergence, would certainly not qualify as a materialist in this sweeping sense. Emergentism is, however, consistent with a broadly materialist account of the mind-body relationship. An emergent property is said to emerge from the "matter" that prepares its way. And the material configuration of the substrate furnishes a sufficient condition for the property's appearance. The term, 'material,' is an exceedingly elastic one (McMullin 1995; 1999). But if one harks back to its original sense in Aristotle, it connotes the potentiality of a thing to take on a different property or even to become a different kind of thing entirely, while maintaining the continuity conveyed by the notion of *change* (rather than replacement). If, in the process of change that a complex entity undergoes, an emergent property makes its appearance, the potentiality for that property must in some sense have been there in advance. But in what sense?

This is one of the places where the debate about the credentials of emergence as an explanatory notion has been joined. Emergence is not an ordinary sort of change where the outcome can be anticipated and causally explained. There is a degree of discontinuity, a shift from a lower to a higher level. The traditional notion of potentiality carried with it the implication of a smooth process ready to be set loose by the appropriate external agency. Emergence does not seem to be like that. Perhaps a certain expansion of the notion of materiality is needed if emergence be allowed.

Throughout much of the history of the concept of matter, it seems to have been presumed that one somehow knows in advance what the limits set by "materiality" are. Matter is corruptible; matter is extended; matter has location; matter cannot act at a distance. . . . In discussions in medieval philosophy of the "immateriality" of intellect, it was assumed that materiality could be defined in advance as a contrast term.[10] Closer to the modern usage of the term, the natural philosophers of the seventeenth century equated "matter" with whatever obeys the basic laws of mechanics (McMullin 1978, 52–55). In the heyday of Newtonian mechanics, this set reasonably tight bounds on what could count as "material" action, though even then gravity posed something of a challenge. But the upheavals in mechanics of the past century should warn one against such easy assurance today. In particular, the unresolved debate about how best to interpret the quantum formalism in ontological terms makes it difficult, if not impossible, to specify the limits of "material" action in terms of today's science. And we have little idea of what the mechanics of the future will look like at the deepest level when gravity and the remaining fundamental forces are brought under a single formalism. It is well to keep in mind that it is precisely at the deepest level of what constitutes "mechanical," effectively "material," action that the greatest surprises have come in the past, witness the paradigm-shifting contributions of Newton, Einstein, and Bohr.

If emergence be admitted, the notion of the "material" might have to be loosened up even further. An emergent property is one that could not have been anticipated on the basis of the prior science of the

---

10 And perhaps it could, because of the special features of the Aristotelian doctrine of matter and form. See, for example, how Thomas Aquinas relates "matter" and mind: "But we must observe that the nobler a form is, the more it rises above corporeal matter, the less it is subject to matter, and the more it excels matter by its power and its operation. Hence we find that the form of a mixed body has an operation not caused by its elemental qualities [is emergent?]. And the higher we advance in the nobility of forms, the more we find that the power of the form excels the elementary matter; as the vegetative soul excels the form of the metal, and the sensitive soul excels the vegetative soul. Now the human soul is the highest and noblest of forms. Therefore in its power it excels corporeal matter by the fact that it has an operation and a power in which corporeal matter has no share whatever. This power is called the intellect." *Summa Theologica*, I, q. 76, a. 1, c. For Aquinas, normal growth, of the human fetus, say, involves progression from the lower form to the higher form, but this might not qualify as an analogue of emergence because of the role played by teleological causation in the growth pattern overall.

constituents only.[11] A form of "materialism" that allows emergence is, therefore, a pretty capacious position. It would admit properties like intentionality and self-awareness that, on the presumption that they are emergent, would lie outside the scope of the basic science of the world's constituents. On this reading, it would seem, on the face of it, to be difficult to say of any property in advance that it lies forever outside the limits of materiality. The "material" is simply the realm of becoming, whose capacities for new sorts of otherness only time can reveal. The traditional assumption that "materialism" confines one to categories that are immovably fixed in advance would thus have to be rejected. But, as we have seen, there is another reading that would begin not from matter but from spirit and would make the categories of the "material" and the "immaterial" independent of developments in the natural sciences. The theologian, Karl Rahner, expresses this view succinctly:

> What spiritual means is an immediate non-empirical datum of human knowledge. . . . It is only on the basis of that knowledge that it is possible to determine the actual metaphysical meaning of 'material.' It is an unmetaphysical and ultimately materialistic prejudice common among scientists to suppose that. . .[they] know precisely what matter is, and then subsequently and laboriously and very problematically have to discover spirit in addition, and can never properly know whether what it signifies cannot after all be reduced to matter in the end. (Rahner 1965, 47)[12]

---

11 Assuming that the correlation between the material configuration and the property which is claimed to be emergent has not already been built into the science of the constituents. To avoid vacuity in discussions of emergence, one has to block from inclusion in advance in the reducing science a simple addition of the capacity to set the stage for the emergent property. If, of course, this capacity can eventually be incorporated in a coherent way into the reducing science (recall what was said above about reduction's being in some contexts closer to enlargement), then the property is not in fact, emergent. But prior to achieving such a reduction, a capacity of this sort could be no more than an *ad hoc* addition. This point is already implicit in Broad's attempts to give a more precise definition of emergence, and is made explicitly by Hempel and Oppenheim in their influential discussion of emergence (Hempel and Oppenheim 1965, 260). See also Beckerman 1992b, 104.

12 Elsewhere, he says: "What matter is in general. . . is not a question for natural science as such but a question for ontology, on the basis of an existential metaphysics; such an ontology can answer this question because it already knows what spirit is, and thus. . .what the material is as such, viz. that which is closed in its individuality to the experience of the transcendence of being as such" (Rahner 1974, 162). See also Wright 1996, 127.

Though this would seem to make the border between matter and spirit impassable, Rahner will also say "without scruple, that matter develops out of its inner being in the direction of the spirit" (Rahner 1966, 164). In his view, the Creator has endowed matter with the capacity to "transcend itself"; the "power of self-transcendence" may allow a "leap to a higher nature" (ibid., 169, 165).[13] It sounds like emergence, in a different, more metaphysical, idiom.[14]

DUALISM

We are now in a better position to assess, in a very brief way, the strengths and weaknesses of two very different accounts of the human soul, dualist and emergentist. What makes such an assessment more than ordinarily problematic, however, is that it has to take into account three quite diverse sorts of knowledge-claim, scientific, philosophical, and theological. Each of these possesses its own characteristic notions of evidence, its own modes of procedure. Even this is to oversimplify, since the sciences and more especially philosophy and theology are themselves very far from agreement internally on matters of evidence and procedure. And in the event of disagreement, how is one to assign relative weights to the three different sorts of consideration?

What about the dualist claim that the sciences need not be taken into consideration in this context, on the grounds that the activities on which the existence and nature of the soul are predicated are not accessible to the methods of the biologist or the psychologist? On the face of it, such a claim seems problematic. Why should the psychologist be barred from investigating self-consciousness, free choice, and the like? Why should the neuro-physiologist or the molecular biologist be deemed incapable of investigating the material basis for such activities in the brain? Perhaps more tellingly, might not the inaccessibility argument be reconstrued to point to the *non-reductive* charac

---

13 These quotations do not do justice to the nuanced metaphysical argument he offers here, and in his *Hominisation*, chap. Three (Rahner 1965a).

14 He gives priority to metaphysics on the grounds that "spirit is a reality that can only be understood by direct acquaintance, having its own proper identity derived from no other" (ibid., 53). This makes it, equivalently, irreducible to the language of the physical sciences, a claim echoed, of course, by the proponents of emergence.

ter of certain aspects of mind, and hence to be satisfied by emergentism, say?

The emergentist agrees with the dualist that the resources of physics and biology are of themselves inadequate to deal with the intentional character of human consciousness. But emergentists do not take the further step of supposing that conscious activities are not rooted in the physical structures of the human organism, nor do they bar an appropriately scientific treatment of the intentional aspects of mind. The science here would not, of course, reduce to physics or biology. But it would not impose the absolute ban on "scientific" investigation that renders dualism suspect in the eyes of most scientists and many philosophers. Let us assume, then, that the sciences are not to be ignored in an investigation of the soul.

How should such an investigation proceed? Let us begin with dualism, in both its stronger (Platonic or Cartesian) and its moderate (Thomistic) forms. The original arguments in favor of dualism were philosophical in character. This is not the place to analyze these arguments in any detail, though of course this would be necessary were our investigation to be anything more than schematic. The capacity of the mind to reach unchanging truth (Plato), the reception of the forms of material things into the mind as the means by which these things are to be grasped intellectually (Aristotle), convinced the most influential among Greek philosophers that the power of thought must transcend the "material" order of the changeable and the singular. Since the senses are clearly corporeal in their mode of operation, however, there was an enduring difficulty about how to conceive the relationship between sense and intellect, a difficulty that Plato and Aristotle addressed in very different ways. Aquinas for his part allowed that the body is necessary for the normal operation of the human intellect because of the dependence of the intellect on the senses for the starting-point of knowing (the "phantasm"). But he also insisted that the intellect is "a power in which corporeal matter has no share whatever" (Aquinas *ST* I, q. 76, a.1.c). Here is an issue where a great majority of those engaged in the mind-related sciences today would disagree.[15]

Another challenge would come from the continuities of structure

---

15 The late John Eccles, a rare dualist among neurophysiologists, insisted on the interactionist character of his dualism and believed that recent developments in quantum mechanics could explain the nature of the interaction (Eccles 1993).

and behavior proclaimed by evolutionary science. Soul-body dualism affirms a sharp ontological discontinuity between the pre-human and the human. It is easy enough to claim the synchronic uniqueness of human beings, that is, the distinction that can be drawn between them and the other living species of today, on such grounds as self-consciousness. How sharp the distinction is and especially whether it testifies to the sort of discontinuity that dualism requires would be a matter of debate. But the case for *diachronic* uniqueness, that is for the claim that there was a sharp discontinuity of a non-material sort marking the advent of the first humans would presumably be impossible either to make or to refute on scientific grounds.[16] Since dualists do not believe that mind requires a physical organ for its operation, they would not expect a macromutation or anything of the sort to mark the transition from the pre-human to the human, from non-mind to mind. Should there not, however, have been abrupt changes at the behavioral level? Ought not the sudden appearance of intellect have been marked by a rapid appearance of culture? Anthropologists do not seem to have found any evidence of this so far. The story they tell is, rather, of gradual changes spanning aeons as the hominid line developed.

How does dualism stand in contemporary philosophy? Though it has notable defenders (Foster 1991; Swinburne 1986), it seems fair to say that in Anglo-American philosophy, at least, defenders are far outnumbered by critics. Though many of the latter would defend a version of reductive materialism, it is well to underline once again that the rejection of dualism in no way entails the acceptance of a reductionist account of mind. Some of the philosophers' objections to a strict dualism are prompted by the seeming paradoxes involved in the claim that the human being is a non-physical thing, possessing mental properties only (van Inwagen 1993, chaps. Nine,Ten). Another set of criticisms focuses on the issue of how the two substances are supposed to interact, though these criticisms have lost some of their force in the light of recent advances in quantum mechanics where the interactions

16 Scientists who discuss the historical development of mind tend to take the continuity of the process more or less for granted. See, for example Deacon 1997. Non-dualist philosophers make the same assumption. Suzanne Cunningham argues for an expanded notion of intentionality that would allow the construction of a plausible evolutionary account of how intentionality might have gradually developed (Cunningham 1997).

between elementary particles appear to be very odd indeed. Some objections urge a substantial continuity, both at the physical and the mental levels, between humans and the rest of the animal world (Hasker 1974). In the face of this last challenge, the dualist options appear to reduce to three: either to attribute immortal souls to higher animals,[17] to restrict mind to humans only, as Descartes did, or to base the argument for dualism not on the abilities of mind in general but only on a small subset of those abilities purported to be unique to humans alone.

The qualified dualism of Thomas Aquinas escapes many of these objections but faces others on its own account. What exactly is the ontological status of the human soul in his view? Is it or is it not a substance? At first sight, Aquinas appears to be on both sides of this issue. But he is quite explicit in concluding to the incoherence of the view that the human soul is a substance in its own right,[18] a view that was in fact held by many of his contemporaries and that has been defended in more recent times by many Thomists. Such a view, he urges, would make the union of soul and body accidental, instead of substantial, as the unity of the human demands.[19]

Though the soul is not strictly speaking a substance, it can however be called "subsistent," Aquinas argues, because its intellectual operation is independent of matter. He needs this in order to secure the theologically significant corollary that the soul survives the death of the person. But if the soul is defined as the form of the human body, how can its operation be independent of body? Forms are co-principles with matter, but cannot on that account be called "immaterial" without risking serious ambiguity.[20] The problem lies with the at-

---

17 Swinburne (1986, 198–99) feels compelled to take this rather daring step.

18 For references and a helpful overview of this issue, see Farmer 1997.

19 Aquinas, *Quaestiones de Anima*, a. 1, response (Aquinas 1984). See also *ST*, 1, q. 29, a.1, ad 5: "The soul is a part of the human species; and so, although it may exist in a separate state, yet since it retains its nature of unibility, it cannot be called an individual substance, which is the hypostasis or first substance, as neither can the hand, nor any other part of man. Hence neither the definition nor the name of person belongs to it." He does allow that the soul may be called a substance in an extended sense of that term, because it is "that by which a primary substance is what it is." In this sense, of course, *all* substantial forms could be called substances.

20 Eleanore Stump concludes that "the immateriality of the soul is thus for him [Aquinas] a direct consequence of his view of the soul as a form" (Stump 1995, 511), referring in particular to *ST*, I, q. 75, a.5. But would not this apply to *all* forms, and hence undermine the argument from immateriality to immortality?

tempt to extend the matter-form framework to allow for the possibility that the intellect might properly be called a form and even "the form of the human body" (Aquinas *ST* 1, q. 76, a.1.c.), though it does not serve to determine any matter.

Such objections to dualism as these are by no means conclusive. But they lead us finally to ask whether the *theological* warrant might not often be, for Christians at least, the primary motivation for dualistic belief.[21] It is true that there is little trace of a specific soul-body dichotomy in the Old Testament until one comes to the late Book of Wisdom, written less than a century, perhaps, before Christ's birth. The Greek provenance of the dichotomy in this text is unmistakable; not only is the soul, rendered by the Greek term '*psuche*,' regarded as an immortal part of the human being, but it is also described as pre-existing the body (8:19) and as burdened by the body (9:15). In the New Testament, '*psuche*' often corresponds roughly with '*nepes*', the traditional Hebrew term for the living being, the principle of life, the self (Lynch 1967, 450). But it also sometimes carries the Greek sense of soul: opposed to body and immortal. There are, however, relatively few texts where the dichotomy is clearly conveyed.[22]

The early Church Fathers found the prevailing Greek views on the natural immortality of the soul congenial to the development of Christian doctrine generally. It allowed one to picture, for example, the souls of the just and unjust remaining in existence awaiting the Last Judgement and reunion with their bodies in resurrection. True, if the soul were already naturally immortal, it was not altogether clear, in the Greek view of the human, at least, what the advantage would be in rejoining it with its one-time attendant body. But as time went on, theologians took the Greek dichotomy for granted;

---

21 Peter van Inwagen finds this warrant to be much less strong than Christians usually assume it to be. What he wants to establish is that materialism is "a possible point of view for a Christian to adopt." His own conviction, based on both philosophical and theological considerations, is that "dualism represents a false picture of human nature," though not (he cautiously adds) a "perniciously" false one (Van Inwagen 1995, 486–87).

22 The clearest one might be: "Do not be afraid of those who kill the body but cannot kill the soul, but rather be afraid of him who is able to destroy both soul and body in hell" Matt. 10:28. Significantly, however, the corresponding passage in Luke is: "Do not be afraid of those who can kill the body and after that can do no more. I will tell you whom to fear: fear him who has power not only to kill but, after he has killed, to cast into hell." Luke 12:4–5. See Van Inwagen 1995, 482.

Augustine, in particular, made it a key element of his theology. The supposition was, of course, that the dichotomy could be validated philosophically, though in the later medieval period, some distinguished theologians (Scotus and Cajetan, for example) were doubtful of this, though not in the least questioning its theological warrant.

To review that warrant as it stands today would be a massive task. And one would have to separate the Catholic and Protestant traditions. The warrant seems much stronger in the former. Not only is a very heavy weight attached to Patristic tradition there, but there have been numerous declarations by Church Councils and by explicit Papal statements of varying doctrinal weights, like the one quoted above from Pope John Paul II. True, the language in which dualism is expressed in each case followed the prevailing philosophical idiom of the day.[23] What remained invariant was the evident conviction that a dualistic distinction of one sort or another should be maintained. At a time when philosophical arguments for the natural immortality of the soul had come to be widely challenged (1844), Pope Gregory XVI, condemned the view that this doctrine could not be demonstrated "by reason alone," though he did not indicate how the demonstration should proceed (Denzinger 1962, n. 2766). Taken together, these declarations and others that could be added constitute strong warrant indeed, as theological warrant is understood in the Catholic tradition.[24] In other Christian traditions, the matter might be less clear, but there too the handful of New Testament references and the testimony of the early Church Fathers could carry significant weight.

---

23 And so the Council of Toledo (688 A.D.) speaks of "two substances, soul and body"; the Fourth Lateran Council (1215 A.D.) declares that man is "constituted from spirit and body"; the Council of Vienne (1312 A.D.) asserts that "the soul is the form of the human body"; the First Vatican Council (1870) returns to the older formula: "by spirit and body constituted"; the Holy Office under Leo XIII, a strong advocate of the Thomistic revival, chose traditional Thomistic language: the soul is "the substantial form of the body" (1887); Pius XI in his encyclical *Divini Redemptoris* (1937) declares that in man there is a "spiritual and immortal soul," a strongly dualistic formulation. These declarations will be found in Denzinger and Schönmetzer 1962, nn. 567, 800, 902, 3002, 3224, 3771. For a more detailed account, see Rahner 1965b.

24 Catholic theologians warn, however, that one must be very careful in differentiating between different sorts of Papal and conciliar statement, between, for example, Papal addresses (like the one from John Paul II quoted above) and dogmatic definitions. See Sullivan 1996, chap. Two, "Evaluating the level of authority exercised in documents of the magisterium."

In short, then, the primary argument in support of the dualistic position today, at a time when this position is challenged both from the scientific and the philosophical sides, might well (for Christians) seem to be the theological one, raising an acute, and not entirely novel, question as to how such a situation is to be handled. The simplest approach, of course, would be to challenge the relevance of the sciences in matters concerning the soul, and to make as convincing a case as possible for some form of dualism while responding to the objections to dualism generally. This is what Christian thinkers have most often chosen to do.

In his essay in this volume, Peacocke sketches a non-reductive alternative to dualism and lays aside the matter of theological warrant. This latter strategy is perhaps understandable, given the constraints of space. Still it does leave the reader with a question: how would one deal with the theological arguments for the dualist position he dismisses?[25]

EMERGENTISM

A comment first on the scientific and philosophical considerations for and against Peacocke's chosen alternative, emergentism. The arguments in its favor have been no more than hinted at above. The difficulties it faces lie, first, in the notion of emergence itself which some critics have called incoherent. The debate is not so much between dualists and emergentists as between emergentists and reductive materialists. The issues are extremely complex, too complex for adequate review here. My purpose in drawing them to attention is only to emphasize that the rejection of dualism by no means entails emergentism. There is a lot of work to be done first if this is to be the chosen alternative from the philosophic and scientific standpoints.

But what about the theological objections? The stakes are high. Are human beings naturally immortal or not? Is it the case that, independently of Christ's coming, the individual human mind is such that

---

25 Peacocke in one of his earlier works emphasizes the Greek origins of dualistic belief and notes that dualism is alien to the Hebraic traditions of the Bible. He relies on a number of Biblical scholars, notably W. Eichrodt (Eichrodt 1961), to claim "an affinity between the view of man as a psychosomatic unity in the Biblical tradition and that stemming from science, an affinity which has been obscured by the strong influence of some elements in Greek thought on the development of Christian ideas" (Peacocke 1971, 152). See also Peacocke 1985, 154–56.

once it exists, it necessarily confers immortality on that individual? Or does survival after death depend on the Christ's promise of resurrection? Does immortality apply to the whole person, rather than in the first instance to the soul? Is immortality gratuitous or ontologically necessary? If the philosophical arguments in favor of dualism and its corollary immortality fail, does the theological warrant still stand? Can it survive the philosophical critiques of dualism? How, in short, might an emergentist who is also a Christian defend himself or herself against obvious theological objection?

Several possible lines of argument suggest themselves. The first is implicit in what has already been said about Greek influence on early Christian dualistic formulations. How does this affect the strictly *theological* credence to be given to these formulations? Does their adoption by early Christian writers as well as by later Christian theologians afford them an *independent* theological warrant, so that if their philosophical credentials come to be challenged, theologians ought rally in their support? There has been a tendency, more marked in the Catholic tradition, to extend the authority of theology to a philosophy that successfully explicates Christian belief in philosophic terms. A whole series of Papal pronouncements around the turn of the last century seemed to accord Thomistic philosophy, then enjoying an active revival, a status little short of definitive truth. But might one call upon the same principle of accommodation here that theologians recognize in other contexts, and argue that the writers of Scripture were simply accommodating their mode of expression to the idiom and popular belief of the day? The issue is clearly a delicate one, and is in fact so described in a recent message from Pope John Paul II to the participants in a conference on the relations of the physical sciences and theology:

> Theology is not to incorporate indifferently each new philosophical or scientific theory. As these findings become part of the intellectual culture of the time, however, theologians must understand them and test their value in bringing out from Christian belief some of the possibilities which have not yet been realized. The hylomorphism of Aristotelian natural philosophy, for example, was adopted by the medieval theologians to help them explore the nature of the sacraments and of the hypostatic union. This did not mean that the Church adjudicated the truth or falsity of the Aristotelian insight, since that is not her concern. It did mean that this was one of the rich insights offered by Greek culture,

that it needed to be understood and taken seriously and tested for its value in illuminating various areas of theology. (John Paul II 1990, M10–M11)[26]

Soul-body dualism was assuredly another one of the "rich insights offered by Greek culture." Ought it, however, be ascribed a different role than hylomorphism, one touching more closely on the substance of Christian faith? There may be room for disagreement here. Did the dualist beliefs of the Greek world help Christian theologians to a truth about the soul that could not be otherwise expressed? Or might the nature of the human be conveyed as well or better, from the theological standpoint, by a non-dualist doctrine like emergentism, one that retains the distinctiveness of the human without committing to the natural immortality of every possessor of intellect?

One advantage that emergentists can claim in this regard is that their view respects the gratuity of the gift of Resurrection, which is, according to Christian faith, the outcome of Christ's mission among humankind. Immortality would pertain to the order of grace alone, a gift of the Creator supernaturally conferred through the mediation of Christ. According to dualist doctrine, on the other hand, we would have been immortal anyway, independently of Christ's saving act. Resurrection for the strict dualist amounts only to the enlargement of an immortality already a necessary commitment of mind, a problematic enlargement, it might seem, for those who hold the soul to be already a substance in its own right. For those who defend the qualified dualism of Aquinas, however, resurrection is apparently called for by the very nature of things: "To be separated from the body is not in accordance with [the soul's] nature" (Aquinas 1945, *ST* I, q. 89, a.1.c). Aquinas argues that since "it is contrary to the nature of the soul to be without the body," and "nothing which is contrary to nature can be perpetual," it was *necessary* that the soul be united again with the body: "The immortality of soul seems to *demand* a future resurrection of bodies" (Aquinas 1957, Bk. 4, chap. 79, par. 10, 299 emphasis added). Might not this, however, call into question the gratuity of Christ's saving action, and compromise the distinction between the

---

26 Russell et al. 1990 have assembled a lively set of responses to the message from a variety of sources, Christian and non-Christian. The text of the message is also found as the preface to the proceedings of the conference to whose participants it was addressed (Russell et al. 1988).

order of grace and the order of nature, so fundamental a feature of Aquinas's own theology? The strain here between theological and philosophical commitments seems evident. In the emergentist perspective, on the other hand, it is the whole person that is resurrected; resurrection of the person belongs entirely to the order of grace.

Emergentists do not have to suppose, as dualists do, that a "special" causal action on God's part is required at the moment of origin of each and every human being. God as Creator already sustains in being the entire natural order of interconnected causes and effects, each effect being directly traceable to causal agencies in the world around us. Much has been written in recent years about the integrity of the natural order, about the wholeness one would anticipate in the work of an omnipotent Creator (Baker 1995, 501; Van Till 1996; McMullin 1993). Dualists are forced to postulate an insufficiency in the order of secondary causes: the Creator apparently has to supplement this order at the moment of origin of each and every human being.[27] Dualism, in fact, seems to require a Creator who intervenes to bring each human soul to be, not from the potencies of the created world, but directly *ex nihilo*.[28] Such intervention is not miraculous in the ordinary sense. It is regular; it can be relied on when the material antecedent conditions are satisfied. But it is not natural either; it transcends the causal powers implanted by the Creator in the natural order. Nor does it pertain to the order of grace, to salvation history, to the extraordinary offering of a covenant between Creator and creature that began with Abraham and came to its climax in the death of Christ.

---

27 W. Norris Clarke, in an eloquent defence of the moderate dualist position of Aquinas, argues, to the contrary, that it offers a "far richer vision of the peculiar dignity of the human person" to suppose that, at the origin of each human being, "two great lines of causality" converge, one "the ascending line of the intrinsic forces of the earth," and the other "the descending one of God's own special causal influx" (Clarke 1998, 15).

28 We already saw that theology offers dualism a warrant of sorts. It seems that dualism returns the favor. One of the reasons for the decline of support for dualism among philosophers is surely the decline of theistic belief in their ranks. A dualist who is *not* a theist has to extemporize when explaining how souls came into being. Swinburne remarks: "The ability of God's actions to explain the otherwise mysterious mind-body connection is just one more reason for postulating his existence" (Swinburne 1986, 198).

Rahner asks:

> Is not the essential difference blurred between natural and secular
> history on the one hand and the really personal, sacred history of
> redemption on the other, if God's action even outside the history of
> redemption be attributed to a definite location in space and time, because
> a particular individual reality, in distinction to others, receives a
> privileged direct relation to God? Must science not perpetually try to
> remove this stumbling-block, by reason of the very principles of its
> method?(Rahner 1965a, 66)[29]

God is still "of course, the cause of the soul as God is the cause of
everything." And God's creative power is especially evident in the
manifestation of self-transcendence on the part of finite being at the
origin of each human. What Rahner questions is the explanation of
human origins as "an exceptional, extraordinary occurrence whose
special ontological features contradict everything that is otherwise
understood regarding the relation of the first cause to second causes"
(ibid., 67–68).

In the emergentist perspective, this peculiar sort of supplementation on God's part of the orders of nature and grace would not be
needed. In this perspective too, the parents of a human child would
not simply beget a body; they would truly be parents of the whole
person who is their child. Resurrection would pertain exclusively to
the order of grace; it would serve as the culmination for each individual of salvation history. It would also follow that the resurrected
person would no longer be governed by the measured temporality of
earth, by the limitations of a past that is gone and a future that is not
yet. What would take its place we have no idea; analogical inference
from the order of nature to the order of grace is risky at best. Thus
queries about a time-lapse between death and judgement would simply
be out of place.

Challenges to human uniqueness have come from many quarters in
recent decades, from molecular biology, from physical anthropology,
from cognitive psychology. . . . Yet uniqueness is crucial from the
theological standpoint. The promise of resurrection extends only to
the human; there can be no half-way houses, no gradual shadings into
immortality. For the dualist, this is simple: God supplements the or-

---

29 Translation has been slightly simplified.

der of created causes to infuse a soul into each body to make of the composite a human being. There is an unambiguous ontological distinction between the human and the non-human. No matter how similar their bodies may be, it is the infused soul that makes the human difference. At what point in the hominid line would God have infused the first souls? Were there necessary conditions to be fulfilled on the evolutionary side first, as matter gradually became more organized? Was there an ontological break between the pre-human and the human on the material side also? Did there have to be?

In answering questions such as these, it would make a difference which form of dualism is being proposed. If the soul is the form of the human body, it would seem that there should have been a physical difference accompanying the transition from the pre-human to the human. But, of course, if intellect is the differentia, and if it acts independently of any physical organ, the infusion on God's part that makes the human form something more than a substantial form in the ordinary sense marks the transition that the salvation story requires, even though no physical change, in the brain for example, may accompany this transition.

The emergentist cannot point to so definite a marker. Yet since emergentists accept that the distinctively human features are irreducible, there still remains an ontological difference between the pre-human and the human, though obviously not so neat a one. Might the evolutionary line shade from reducibility into irreducibility? Could there have been a substantial continuity in the hominid line as the irreducible features that today distinguish the human quite sharply from other primates gradually made their appearance over a lengthy period? A more sustained analysis of the emergentist option would be required in order to give a plausible response to these questions.

One might, of course, suppose that there could have been a genetic mutation that brought about sudden large-scale phenotypic changes in brain-structure, say. Macromutations are not much favored among geneticists nowadays, but the ability of gene defects to occasion such fateful phenotypic alterations as Down's syndrome is all too well documented. An even more speculative possibility might be envisaged on the theological side. In the non-dualist perspective, continuance of the person in existence after bodily death is a gift of God's grace, not a natural concomitant of intellect. Is it not conceivable, then, that God might choose to begin conferring that gift at a point in the evolutionary line that is not marked off by some intrinsic ontological difference between the parent organisms and their descendants? In such an event, God would be directly involved in the origin of each human creature but not by a supplementation of the natural order. Rather, it would be by an act of election. God would breathe life into the "dust" of earth, as the *Genesis* metaphor puts it, by freely raising up to eternal life the organism that the evolutionary process had over the long years prepared for that moment. There would still be something quite special about the origin of each properly human life, but it would not entail the *ontological* intervention on God's part that Greek dualism requires.

This is clearly quite speculative from the theological standpoint. But it might be sufficient to indicate that there may, in fact, be theologically viable alternatives to the dualisms that the early Christian Church inherited from the philosophers of Greece. These alternatives would have to be worked out in detail before any kind of judgement could be passed on them. And judgement, as we have already seen, would even in that event be difficult to render, not least because of the profound differences between Christian theologians regarding the weight to be given to the different sources of theological warrant.

If it accomplishes nothing else, this discussion may at least have convinced the reader that easy assumptions about the essential separability of scientific from theological concerns must be challenged in one crucial domain of contemporary inquiry, at least. Many of the second-order epistemic issues about scope, warrant, and authority come into question once again that the Copernican issue raised long ago. We have learnt much since Galileo's day about the constitution of the world around us and about ourselves as part of that natural world. But the second-order issues remain and are still far from resolution.

14

Philosophical Anthropologies and the Human Genome Project

Kevin T. FitzGerald. S. J.

In considering the Human Genome Project [HGP] and the potential applications of its findings, the connection between concepts of human nature and morality is especially significant.

Ethicist James Gustafson has stated, "the question of what constitutes the normatively human is the most important issue that lurks in all the more specific and concrete problems we face when ethical issues are raised about developments in the field of genetics," (Gustafson 1974, 274). In fact, he remarks, the importance of this issue is not confined to the analyses done by ethicists or moral theologians, it is also a concern for scientists, for "a biologist's beliefs about the essence of humanity, and the values derived from such an essence, will affect the kind of experimentation and therapy he or she supports" (Gustafson 1993, 4).

When questions and beliefs concerning fundamental human characteristics are studied systematically and comprehensively, such a study can result in a theory of human nature or philosophical anthropology.[1] The present state of philosophy makes any comprehensive definition of philosophical anthropology difficult, if not

---

1 For more on philosophical anthropologies and theories of human nature see Pappé 1967; Jaggar and Struhl 1995.

impossible (Cockburn 1991, 1–2). Traditionally, philosophical anthropologies have been seen as attempts to provide a complete framework for pulling together information about human nature from the various disciplines which investigate it. Using this framework, conclusions have then been derived as to which human characteristics are most significant or unique, which methodologies are appropriate for discovering these characteristics, and which actions or ways of living are best (Jaggar and Struhl 1995).

Today there are serious questions and challenges about the role and function of philosophical anthropologies. One approach to addressing these challenges is to lessen one's claim as to the ability of philosophical anthropologies to mediate or adjudicate among the various disciplines investigating human nature. Instead of claiming that philosophical anthropologies can provide an architectonic structure capable of accurately evaluating every insight from any discipline with respect to the contributions of the other disciplines, one could propose that a philosophical anthropology might serve as a foundation for genuine conversation among the various disciplines.

Regardless of the success at delineating and justifying the role and function of philosophical anthropologies in the present academic climate, ethical analyses of new genetic technologies, such as the HGP, still often hinge upon implicit, if not explicit, theories of human nature operative within the ethical frameworks used to address the issues raised by these technologies. Thus, the more theoretical debate over the exact role and function of philosophical anthropologies can be left unresolved for the moment in order that a closer investigation of the actual impact of philosophical anthropologies in the ethical analysis of the HGP can be pursued.

Genetic information and technology have begun to offer humankind the opportunity to categorize and manipulate directly the biological basis of human nature. This categorization and manipulation could alter what people believe is normative for human beings by providing new scientific information which may be used in constructing the philosophical anthropologies which delineate these norms.

For example, gender issues have undergone a significant evolution within the last thirty years in Western culture, and worldwide more recently. Genetic research is presently underway to delineate further what is the same, and what is different, between male and female.

How this research will impact norms concerning gender roles is difficult to predict, but the potential for altering or modifying traditional norms is equally difficult to deny. Issues of race, sexual orientation, intelligence, and even physical stature also are potentially explosive, depending on how the genetic information is interpreted and integrated into people's concepts of human nature (see Schaffner essay this volume). This interpretation and integration are, in turn, dependent upon current operative philosophical anthropologies. Therefore, a truly significant interaction between science and philosophy is taking place within philosophical anthropology.

Historically, there has been a full spectrum of philosophical anthropologies, ranging from the denial of any real human nature as found in the Upanishads, to the superhuman or self-deifying natures proclaimed by Nietzsche and Ayn Rand. Theories from across this spectrum have been and are being used, implicitly or explicitly, as the basis for ethical evaluations of the efforts of human genetic research and its potential applications. It is not within the scope of this essay to survey each of these philosophical anthropologies.

The specific focus of this paper will be to show some of the inadequacies of theories of human nature which do not integrate contemporary scientific information with knowledge from other disciplines in order to form a basis for the ethical analysis of genetic research or intervention. These inadequacies will be contrasted with the advantages of a philosophical anthropology which is open to integrating the most up-to-date scientific data about human nature. This criterion of integration has been selected because of the undeniable benefit scientific data has given and can give to the understanding, expression, and development of human nature.

FOUR PHILOSOPHICAL ANTHROPOLOGIES

From contemporary Western philosophy it is possible to identify four types of philosophical anthropologies as representative of major strains of thought regarding human nature: (1) static, (2) scientistic, (3) dichotomized, and (4) dynamic. Examples from each of these four types will be used to evaluate their ability to serve as a basis for the ethical analysis of the HGP and its applications.

First, *static* philosophical anthropologies. These are based on primarily philosophical or theological beliefs about characteristics,

398   Kevin T. Fitzgerald, S.J.

including physical characteristics, of human nature which are fundamentally unchanging. Such beliefs result in the conclusion that changing these characteristics would lead to the creation of non-human or deficient human beings. Hence, these philosophical anthropologies proscribe any such alterations. Though obviously informed at some point in the past, even the very distant past, by biological information, these philosophical anthropologies resist integrating the findings of more recent biological research because such data may undermine the most fundamental claims about human nature.

Ethicist Leon Kass, M.D., presents such a view in his book, *Toward a More Natural Science*. He sees science as primarily in conflict with morality and ethics. "The knowledge of nature that science finds has been, for ethics, at best only negatively helpful, or if you prefer, subversive insofar as it embarrasses the ethical claims of traditional philosophical or religious teachings, by contradicting what they say about nature, man, and the whole" (Kass 1985, 320). Rather than changing his purported fundamental human characteristics, Kass's solution to this conflict is to change the way biological science is done by making it more "natural" (ibid., 346–47). This more natural science would be guided in its investigations and conclusions by the philosophical anthropology Kass defends.

In his philosophical anthropology, Kass sees the changes offered by genetic technologies as unlikely to produce a better life; and even if they did, Kass asserts that people have a proprietary interest in surviving as they presently are (ibid., 78). The question arises: is the image of human nature put forward by Kass accurate?

Such a static philosophical framework, as the one Kass presents, operates on an image of human nature drawn largely from philosophical or theological tenets supported by selective (and often dated) biological information. This approach favors philosophical and theological knowledge over scientific knowledge. Some might be inclined to regard static philosophical anthropologies as advantageous since these concepts of human nature are not affected by the ever-changing results of scientific research and would therefore provide a more stable foundation for ethical analysis. The validity of such an interpretation, however, relies on a justification for giving such importance to stability and the status quo that contemporary scientific information about human nature becomes secondary, or is considered subservient to the philosophical or theological concepts of humanness which undergird that moral framework.

A different perspective of human nature comes from an ethical framework based solely or primarily on scientific information. Such is the second type of philosophical anthropology under evaluation here: *scientistic*. Scientistic philosophical anthropologies employ a kind of epistemic chauvinism similar in many respects to that employed by a static philosophical anthropology. The principal difference between the two approaches is that each chooses a different type of knowledge as preeminent and privileged. However, as with the static type, scientistic philosophical anthropologies also require a justification for making other types of knowledge, such as moral or religious, subservient in structuring its underlying framework.

E. O. Wilson, founder of Sociobiology, is a well-known proponent of scientistic anthropology. "The organism is only DNA's way of making more DNA," Wilson states in his oft-quoted volume, *Sociobiology, The New Synthesis* (Wilson 1975, 3). In contrast to the static view, Wilson sees biology as the very foundation of ethics:

> Human behavior like the deepest capacities for emotional response which drive and guide it is the circuitous technique by which human genetic material has been and will be kept intact. Morality has no other demonstrable ultimate function. (Wilson 1978, 167)

Certainly, one cannot accuse Wilson and other proponents of scientistic philosophical anthropologies of not attempting to incorporate the latest scientific data into their moral frameworks. The problem is that they include little else. Their scientific reductionism leaves no room for data from other branches of knowledge. Hence, no integration of these various types of data occurs and one is left with a concept of human nature so thin as to cause philosophers such as Mary Midgley to lament that the discussion is now handed over entirely to the biological experts who emphasize genetic determinism. "What little authority we once thought we had for speaking about our own lives has vanished" (Midgley 1994, 79).

A third type of philosophical anthropology has been proposed in various forms. It presents a dichotomized interpretation of the human nature by separating human biology from moral characteristics or personhood in order to address ethical issues. This approach minimizes or rejects humanness as an important aspect in the ethical evaluation of genetic interventions. Personhood, as separate from humanness, is seen as the guiding concept toward which genetic research and intervention should be oriented.

Ethicist H. Tristam Engelhardt makes such a distinction between personhood and humanness:

> The difference and distance between us as persons, as manipulators of our nature, and us as human, as objects to be manipulated gives us our destiny as self-refashioners, self-manipulators. Being self-conscious and rational, we can always objectify our bodies and in so objectifying them bring their shortcomings into question. In seeing ourselves as objects, we then raise for ourselves the moral problem of all creators, namely, to create prudently and responsibly. (Engelhardt 1984, 293)

Others go even beyond Engelhardt in emphasizing this dichotomy. Some argue that to posit the notion of humanness as morally significant is to be guilty of a bias akin to racial prejudice. Others find no difficulty in supposing that an artificially created being, i.e., a machine, could be a person. Again, it has been suggested that one's thinking is rational only in so far as it is unconditioned by the fact that one lives the life of a human being (Cockburn 1991, 1).

One immediate difficulty with the dichotomized approach is that the distinction between humanness and person is not philosophically well developed. For instance, if self-conscious rationality is used as the distinguishing characteristic of personhood, as Engelhardt suggests, then the separation of human rationality from human animality results in the estrangement of the human good from human biological flourishing. This estrangement obstructs the legitimate contributions to understanding human nature that can be made by areas of investigation such as sociobiology or the HGP.

Additionally, however one draws the line between humanness and personhood, arguments for a dichotomized approach lead to the conclusions that humanness is a completely malleable aspect of an individual's nature while personhood remains unchanging in the application of genetic technology. Such conclusions fit well with an ethical reductionism which emphasizes personal autonomy and market forces as the premier principles in moral decision-making. Hence, the discoveries of research projects such as the HGP merely become the basis for providing more genetic options to be purchased by those individuals or societies who can afford them. Questions of justice and beneficence are then applied secondarily in order to assist in the adjudication of conflicts among autonomous persons. Though appearing simple, such ethical reductionism runs counter to common moral sense when individuals or groups are allowed to select genetic technologies in order to gain an advantage over others.

The last of the typologies considered here is the *dynamic* philosophical anthropology. A dynamic theory of human nature attempts to integrate socio-historically conditioned concepts of human nature with the findings of contemporary scientific research.

Theodosius Dobzhansky, one of the founders of evolutionary genetics, recognized the possibility of uniting human biology and human culture in a philosophical anthropology which emphasized the dynamism found in both.

> The most important point in Darwin's teachings was, strangely enough, overlooked. Man has not only evolved, he is evolving. This is a source of hope in the abyss of despair. In a way Darwin has healed the wound inflicted by Copernicus and Galileo. Man is not the center of the universe physically, but he may be the spiritual center. Man and man alone knows that the world evolves and that he evolves with it. By changing what he knows about the world man changes the world that he knows; and by changing the world in which he lives man changes himself. Changes may be deteriorations or improvements; the hope lies in the possibility that changes resulting from knowledge may also be directed by knowledge. Evolution need no longer be a destiny imposed from without; it may conceivably be controlled by man, in accordance with his wisdom and his values. (Dobzhansky 1962, 346–47)

Control of human evolution was of great importance to Dobzhansky because the ability to change had been the very key to the success of human development. "The human species is biologically an extraordinary success, precisely because its culture can change ever so much faster than its gene pool" (Dobzhansky 1962, 319).

Though he showed great interest in fostering societal values such as freedom, health and creativity, Dobzhansky never really developed a philosophical system capable of grounding these values and demonstrating their interconnection. In other words, he does not offer an heuristic framework which can be applied to the ethical analysis of genetic research and technology.

One example of philosophical anthropologies which do provide a heuristic framework for ethical analysis, and strongly insist upon the dynamism of human nature, is the philosophical anthropology developed by Karl Rahner, the German systematic theologian. Rahner presents this dynamism as more than just external forces shaping human nature, or humans shaping their external environment. Humans directly and fundamentally shape their own nature, and, therefore, in this sense must manipulate themselves (Rahner 1972a, 227). Human beings achieve this self-manipulation because human nature is both radically incomplete and open to change. Change can occur because human nature has the freedom to determine what it is to become.

Rahner understands freedom and self-determination to be at the basis of human nature (Rahner 1972b, 213). Freedom, for Rahner, does not mean choosing this or that, or doing this or that. Instead, it is the fundamental capacity for self-determination. Self-determination, likewise, is not simply a single free act. It is a lifelong process whereby an individual fashions himself or herself through choices and actions made in the light of that which one wants to become. Hence, freedom and self-determination involve the shaping of one's self as a whole, not merely a molding of one's body as an external vessel to meet the needs of one's rationality or genetic fitness.

This capacity for self-determination, however, is limited. It operates within parameters, constrained by social and biological limits, which act like guardrails to keep individuals and societies from taking self-destructive turns contrary to the exercise of human freedom (ibid., 217). One such parameter is human embodiment. Being embodied limits the manipulations proper to self-determination. For example, any manipulation which resulted in a significant reduction in genuine human intercommunication, even if it gave an individual a longer and less disease-ridden life, would be contrary to human freedom because it would reduce that individual's fundamental capacity for self-determination (Rahner 1972a, 233).

Ultimately, though individuals exercise their freedom and shape themselves as a whole, they can never exercise that freedom or shape themselves totally. Human nature remains embodied and historical, and, therefore, incomplete and contingent. No individual can fully construct himself or herself all at once at any one time, nor can human society reconstruct itself or nature to the extent that either becomes entirely a product of human manipulation. Since such complete change is not within human capabilities, it is not the direction nor goal of human self-determination.

These insights led Rahner to the conclusion that human self-determination is possible only within the context of interpersonal community, since others are necessary for human development and the exercise of freedom (Rahner 1969). Personhood, therefore, is also defined in terms of relation and freedom, rather than the often used concepts of substance and reason. In contrast, then, to dichotomized philosophical anthropologies which emphasize reason in defining personhood, Rahner's approach does not allow for the removal of humanness from the moral equation because human embodiment shapes any individual's relationships and self-determination. Hence, the litmus test for genetic research and its applications is: do they serve to promote human freedom and self-determination? This standard is the basis of the heuristic framework in Rahner's philosophical anthropology.

Reacting against static philosophical anthropologies, Rahner contends that these anthropologies

> . . .often used a concept of nature (natural, according to nature) which ignores the fact that, although man has an essential nature which he must respect in all his dealings, man himself is a being who forms and moulds his own nature through culture, i.e. in this case through self-manipulation, and he may not simply presuppose his nature as a categorical, fixed quantity. (Rahner 1972b, 215–16)

Instead, Rahner concludes that if an essence of human nature is to be discussed, then one should consider this essence as not an intangible something, essentially permanent and complete, but the commission and power which enable one to be free to determine oneself to one's ultimate final state (ibid., 212).

Knowledge about human nature is significantly informed by contemporary scientific data. Acknowledging the need for such information, Rahner welcomed an open dialogue between science and other areas of human knowledge and research, including philosophy and theology (Rahner 1985, 8).

As noted earlier, the scientistic philosophical anthropologies also make use of contemporary scientific information in formulating their theories of human nature, but they fall short of developing heuristic frameworks which integrate that information with other types of knowledge. As a result, they do not delineate standards for the use of genetic research based on a broader scope of human thought and experience. This broader scope is particularly important because the application of genetic research will affect not just biological human nature but the whole human being.

From this sampling of four types of philosophical anthropologies, one sees different approaches to the ethical analysis of genetic research and its applications. The question, then, is this: Are all four approaches equal to the task? As was stated earlier, the answer is: no.

The static philosophical anthropologies, concerned with preserving the value of the human being, are hindered by their selection of crucial human characteristics based on philosophical or theological tenets supported by selective, and even dated, scientific information. Consequently the heuristic frameworks employed cannot adequately assess contemporary scientific information without running the risk of this information contradicting what static philosophical anthropologies consider to be essential to an understanding of human nature.

Scientistic philosophical anthropologies appear to overreact in responding to the flaws in static philosophical anthropologies. In their eagerness to embrace contemporary scientific knowledge, they exclude other types of knowledge (as do the static types). This exclusion leaves these anthropologies with an impoverished picture of humanity. Reducing the human to the merely biological prevents their heuristic frameworks from taking into account the rich tapestry of human experience and interpretation. How, then, can these frameworks assess the total impact genetic research and its applications will have on human nature and society?

In order to reduce the complexity of the issues or to simplify the lines of argument from other areas of philosophy, dichotomized philosophical anthropologies seek to remove totally the biological component from the moral equation. Human biological characteristics are not considered crucial in the formulation of a philosophical anthropology. Consequently, the dichotomized frameworks cannot properly evaluate the effect changes in human bodies will have on how humans behave toward themselves and others. In the face of human experience, this denial of the importance of the human body in the moral equation requires a greater justification than its usefulness in solving certain philosophical problems.

What about dynamic philosophical anthropologies? As has been seen above, dynamic philosophical anthropologies do not deny the fundamental role of scientific information in ascertaining the significant characteristics of human nature. They do not deny the importance of data and information derived from all areas of human knowledge and experience. And, finally, they do not deny the impact our bodies have on who it is human beings consider themselves to be. This openness makes it possible for the heuristic frameworks derived from dynamic philosophical anthropologies to take into account and to evaluate the importance of all areas of human experience and, consequently, to undergird a more thorough and complete ethical analysis of scientific and technological advances.

THE GENOME PROJECT

With the advantages of the dynamic philosophical anthropologies now made evident, the question becomes: how successfully can they be applied to an ethical analysis of genetic research such as the HGP?

The HGP is a massive research program with the goal of sequencing and mapping the human genome. Even more challenging than its research agenda are the ethical issues raised by this research and its potential applications. Questions about public access to genetic information, genetic screening, access to genetic technology, and genetic engineering are just some of the concerns people have (see Kitcher, this volume). The limits of time and space here obviously preclude a thorough analysis of each of these issues. However, the application of the four types of philosophical anthropologies that have been reviewed above to an ethical analysis of one of these issues may suffice to show the strengths of dynamic philosophical anthropologies over against the weaknesses of the others in addressing these problems.

The prospect of human genetic engineering arouses both great hopes and fears. Crippling diseases such as muscular dystrophy and cystic fibrosis may finally be cured. Lethal conditions like Tay-Sachs and Lesch-Nyhan could be prevented, and even the widespread afflictions of cancer and HIV are targets for possible genetic intervention. Conversely, the specter of societal eugenics programs and genetic experimentation gone horribly awry also appears whenever human genetic interventions are considered (see Pernick,Caplan, this volume).

Before proceeding to the application of the four types of philosophical anthropologies to an ethical evaluation of genetic engineering, the limits of time and space require a finer focus on the most relevant issues. One distinction that is often made in discussions about genetic engineering is the difference between gene therapy and genetic enhancement. This distinction suggests a significant moral difference between the acceptability of genetic therapy to cure or prevent disease and that of enhancing human capabilities. This distinction is similar to distinctions made earlier this century between negative and positive eugenics, except that present genetic interventions also include decisions concerning individuals without necessary reference to societal programs or goals (Wachbroit 1995).

Though the distinction between therapy and enhancement has been employed by many writers attempting an ethical analysis of human genetic interventions, no consensus has been achieved as to where a boundary might be drawn between the two, or what basis might justify drawing that boundary. This problem results, in part, from the similar ambiguity encountered when one attempts to delineate health and disease the concepts upon which any distinction between therapy and enhancement must rest, since therapy refers to the treatment of disease. It is within this struggle to distinguish therapy and enhancement, and, therefore, health and disease, that differences in conclusions derived from the four types of philosophical anthropologies may be elucidated.

It is important to note that issues concerning the safety of genetic interventions will be set aside here. Some authors tend to argue for a restriction of human genetic interventions based on the lack of evidence available at present regarding the use of genetic technologies within acceptable limits of risk. These arguments are primarily technical in nature; once the safety problems are solved, the ethical issues about which genetic interventions should be done still need to be addressed. It is precisely these ethical issues which are the focus of the following comparison among the four types of philosophical anthropologies.

Before drawing a line between therapy and enhancement, or health and disease, one has to establish some kind of norm for human health so that it can be contrasted to conditions other than health. This norm could be based solely on medical statistics or scientific information, such as some authors have concluded the HGP is intended to provide rightly or wrongly. This approach might be seen as consistent with a scientistic philosophical anthropology. One advantage of such a norm, some have argued, would be its relatively value-free position. The data is gathered according to the best scientific methodology and a picture of normal human nature is formulated which can then be employed in correcting abnormal, presumably unhealthy, conditions.

Many authors view this approach to defining health as riddled with difficulties (Peterson 1992, 75–84). Some see methodological difficulties in establishing a scientific norm for health because health, even as defined by science—e.g., as survival and reproductive rates—seems to vary significantly according to the people's situation. Hence, under different conditions, the same genetic trait combinations result in varying rates of survival and reproduction.

Which particular combination of which traits, such as body type or metabolism, will the scientistic approach choose as normal? Will these traits be varied according to geographic and climatic conditions? If so, how many types of normality will there be? What about cultural and historical conditions, which also affect survival and reproduction? If these latter conditions are included, then the scientistic health norm is no longer so value-free. If cultural values are not included, the survival and reproduction rates of the scientistically healthy may not be as good as they could be, leaving the goals of this approach to health less attainable.

In addition to these methodological problems, others argue that the scientistic approach is not capable of delineating distinctions between health and disease because these concepts are extensively shaped by human goals. Hence, science can provide information about biological functioning, but health also involves the choices people make in deciding how to evaluate and respond to their biological functioning according to these goals, both individual and societal (ibid., 81–84).

With socio-historical aspects and human goals significantly shaping the concepts of health and disease as they are presently used in human society, scientistic philosophical anthropologies appear to be severely inadequate as a basis for distinguishing health from disease, and, consequently, justifying and defining a distinction between genetic therapy and enhancement. Furthermore, any arguments for the use of scientistic philosophical anthropologies as such a basis in some ideal future would have to answer the pivotal questions raised about the exclusion by these philosophical anthropologies of all non-scientific information, much of which presently informs the concepts of health and disease as well as human nature itself.

Static philosophical anthropologies, as mentioned above, do include other types of knowledge, principally philosophical and theological.

They use philosophical and theological tenets to provide a basis for defining health and disease, and distinguishing between the two. Subsequently, a distinction between genetic therapy and enhancement is derived using a fixed application of this notion of normal human health beyond which any intervention would always be considered enhancement. Drawing this line between therapy and enhancement can then be justified as part of the overall concept of health which includes restoring one's capacities to a level designated as appropriate for human nature.

Two questions arise concerning this approach: (1) how are the appropriate levels of human capacity determined and justified?; and (2) if it is justified to increase someone's capacity to some designated level, why stop there? (Peterson 1992, 102). These questions call for information about biological functioning, as well as about people's goals and values. Use of the best scientific knowledge, which changes and progresses with time, provides a better foundation for decisions concerning biological functioning and the nature of disease, as well as where and why to draw a line between genetic therapy and enhancement. New scientific information will eventually challenge any distinctions drawn between health and disease, or genetic therapy and enhancement if these distinctions are based on dated or incomplete scientific data. Since static philosophical anthropologies use scientific information selectively to support their philosophical and theological positions of maintaining a steady state of biological function, their designations of health and disease (as well as what counts as genetic therapy or enhancement) will run counter to the scientific information gleaned from such projects as the HGP. This conflict will result in either the rejection of the static philosophical anthropology or the rejection of the scientific data and methodology as seen in the position held by Kass.

Dichotomist philosophical anthropologies seek to protect and enhance the autonomy of persons to choose how to respond to their biological functioning according to their goals. This defense of personal autonomy requires bracketing the ever increasing body of scientific knowledge away from the moral discussion about what is normal biological functioning since this information might limit one's autonomy. The dichotomist approach accomplishes this avoidance by defining health and disease according to an individual's (or a society's) goals (ibid., 82). Therefore, distinctions between genetic therapy and enhancement depend entirely on what the individual believes helps or hinders the achievement of these goals.

Various difficulties are encountered in this dichotomist approach to drawing a line between genetic therapy and enhancement. One major problem is the possibility of an endless list of different diseases to be treated, since individuals get to choose what they consider to be a disease for them at any given time (ibid.). Secondly, decisions concerning the allocation of limited medical genetic resources can no longer be made using biological severity as a criterion since the seriousness of an individual's disease will depend on that individual's evaluation of how critical treatment is to attaining one's goals. Such problems can easily lead to scenarios which trouble the moral sensibilities of most people, e.g., wealthy parents paying for expensive genetic interventions on their children in an attempt to provide them with the athletic abilities or physical features guaranteed to insure their success in a given culture (see Caplan this volume).

Dynamic philosophical anthropologies recognize the importance of scientific information about human biological functioning for delineating health and disease. Moreover, the dynamic approach acknowledges that this information continues to increase and develop, and so must continually be integrated into any norms concerning human nature and health. Finally, dynamic philosophical anthropologies incorporate the findings of scientific knowledge which reveal the evolving and developing properties of human nature itself.

Using the dynamic perspective, norms for human nature and health can be derived from various types of knowledge about human beings. These norms would include current human physical characteristics, but would not set an upper limit on those characteristics. Hence, a clear distinction could be drawn between genetic therapy and enhancement at any given time, yet the line between the two would shift as better information became available. Furthermore, though the distinction between therapy and enhancement could remain clear, since therapy refers only to disease treatment, genetic enhancement itself would not necessarily be unjustifiable.

Though not justified by the traditional medical reasons for combating disease, enhancement could be justified if it served the overall needs of humanity in the self-determination of its own nature. One possibility for such a situation would be the development of a technique for genetically improving the human immune system. This improvement could have the general benefit of making the immune system more effective in preventing and fighting cancers, or be specifically tailored to prevent a certain auto-immune disease or enhance resistance to deadly viruses. Such interventions would enhance people's lives by reducing the burden of disease on both individuals and society, and by increasing the scope of freedom and self-determination experienced by those who would have otherwise been detrimentally affected by one of the targeted diseases.

One key advantage of the dynamic philosophical anthropologies that becomes apparent from this analysis of genetic engineering is the balance these philosophical frameworks bring to the evaluation of how to apply this new technology. The progress and opportunity which science offers is not denied, but it is also not accepted uncritically. Different areas of knowledge are brought together in order to generate a more comprehensive picture of human existence and the choices humanity has in determining its own future, genetically and otherwise. The dynamic philosophical anthropologies provide both clarity and stability to an ethical analysis of human genetic interventions at any given time, while remaining open to future change and progress.

From the brief analysis and application of the four types of philosophical anthropologies presented in this paper, it can be seen that these four types differ significantly in their openness to integrating contemporary scientific knowledge along with other sources of knowledge about human nature. If the tremendous potential benefits of genetic research such as the HGP and its applications are to be realized, only dynamic philosophical anthropologies, which are open to integrating information from this research with insights drawn from a broad range of human thought and experience, will provide an adequate basis for the ethical analysis so necessary.

Commentary on "Philosophical Anthropologies and the HGP"

John M. Staudenmaier, S. J.

What most caught my eye in Dr. Fitzgerald's paper is an intersection between mainstream rhetoric about "Science" (and *a fortiori* about the rhetoric of the Human Genome Project) on the one hand, and a key element of Fitzgerald's preferred ethical theory, on the other. In his very helpful schema of four potential ethical frames of reference for assessing the HGP, Fitzgerald comes down in favor of what he calls "dynamic philosophical anthropology." What he most likes in the thinking of Dobzhansky and Rahner, if I have read him rightly, can be sorted into two dimensions. On the one hand, he quotes Dobzhansky's observation that in human evolution the definition of the human species changes as humans create and recreate their cultures. "The human species is biologically an extraordinary success, precisely because its culture can change over so much faster than its gene pool" (Dobzhansky 1962, 319). Thus to reflect on the moral significance of any human endeavor would require an understanding of "the human" that recognizes a dynamic of core human qualities as they get reshaped by culture. More specifically for Fitzgerald, human culture making includes human science making so that an ethical assessment of an endeavor embedded in scientific praxis like the Genome Project must take into account the changing understanding of humanity that flows from the very endeavor to be ethically assessed.

We have a genuine problem here, of course, and Fitzgerald is aware of it. Including the object of ethical inquiry within the definition of terms that governs that inquiry risks losing purchase on what you mean by ethical inquiry at all. He rightly critiques such an approach as scientistic and aptly cites E. O. Wilson as representing the pure position. Still, the problem does not easily go away. Like most of us, Fitzgerald adopts on a commonplace rhetoric when talking about Science. Thus on page 409 he argues that a valid ethic must "integrate contemporary scientific information to serve as a basis for ethical analysis" and integrate "the most up-to-date scientific data about human nature." The use of "scientific" here, linked with the chronological direction terms "contemporary" and "most up-to-date" calls to mind the longstanding Western presentation of Science—used in the singular with a capital S—as a meta-historical, trans-cultural process that produces findings and data and information from within itself, independent of any larger context of meaning. In particular, it is represented as independent of any larger context of meaning to ground an ethical inquiry. It is hardly surprising to encounter the rhetoric of transcendent Science in discussions like these. That rhetoric is so deeply embedded in the Western imagination and in public discourse over the past several centuries that it is difficult to avoid. That Fitzgerald is aware of the underlying problem to which I am calling attention will become evident when I return to the second key element in his preferred ethical approach. First, however, let me sharpen the issue by a few observations on the rhetoric of the Genome Project itself. To summarize in advance, I take the ordinary rhetoric of HGP proponents to be much closer to Fitzgerald's scientism than to his dynamic philosophical anthropology.

Note first the very use of the term "Project." Why the singular rather than the plural? What would happen if we conceptualized HGP as a federation of laboratories, with their differing agendas and funding sources, moving at varying rates depending on the parameters of their research? Such a shift would emphasize the local and specific character of laboratories and research teams, would highlight the distinctive flavors brought to such work that emerge in different environments. It would, in short, call attention to the way science gets done in ordinary practice, by groups and individuals who serve many goals while defining, raising money for, carrying out, and publishing their research. By calling HGP "The Project," in the singular, proponents tend to blur outsider's vision of what goes on among their several worksites.

The term "Project" also carries remarkable harmonic resonances with the first great scientific/engineering project of the contemporary Western era. Unlike the HGP, that first project was accurately named in the singular because it was orchestrated from the top with the strict military procedures adopted in wartime. The thousands of scientists and engineers who participated observed an extraordinary docility to higher command and maintained the security of the entire multi-site endeavor with a discipline so near perfect as to be hard to credit today. I refer, of course, to The Manhattan Project, that profound symbolic event which has so deeply influenced the self understanding of the larger public about what Science and Technology can do. I have been impressed at the way HGP rhetoric echoes Manhattan Project rhetoric, an issue explored in detail in this volume in the contributions of John Beatty and Timothy Lenoir. We are told that the HGP has a target date for mapping the Genome. Why is that important? What pressing need requires the completions of these multiple lines of research by a stated date? The reason for this ticking clock in the Manhattan Project was straightforward, at least at the outset. Because German physicists, Hann and Strassman, had achieved fission in 1938, U.S. physicists feared that the Nazi war machine might move from that breakthrough to a working bomb during the war. The U.S.-British effort was correctly perceived to be mounted against a pressing time constraint. It was that same pressure, fear of Nazi technical and scientific capabilities combined with virulent Nazi ideology, that motivated the entire U.S. war effort, including the efforts of those who engaged in the Manhattan Project.

The HGP does not face such a single-minded enemy. What it does share with the Manhattan Project, of course, is a full plate of moral ambiguities that require serious thinking. Will the lines of research currently in play blend their improvements of certain disease treatments with a notable increase of police-state style eugenics? As Fitzgerald so aptly observes, the world of biological practice does not lack rhetoricians who would invoke an inevitable and trans-historical force to legitimate the silencing of critical debate that is not completely contained within the insider community of elite practitioners. "Their scientific reductionism leaves no room for data from other branches of knowledge" (Fitzgerald this volume, 413). He sees a key difference between the resignation expressed by Mary Midgley ("What little authority we once thought we had for speaking about our own lives has vanished." [ibid.]) and Dobzhansky's hope "that changes resulting from knowledge may also *be directed by knowledge*" (ibid., 415, my emphasis).

How does a society "direct. . . changes resulting from knowledge?" How does a society find the resources needed for extremely complex and contentious conversations about matters as far reaching as the potential redefinitions of human life that are gradually becoming available through Genome mapping work? Such a conversation requires, at a minimum, a cultural capacity to regularly renew in its members the qualities needed for substantive debate: patience with ambiguity, receptivity to perspectives other than one's own, a sense of humour, the willingness to take stands and risk controversy, hope. How, in short, might we best imagine the relationship between the societies in which we are citizens and this fast moving and very complex body of biological practice called HGP? Fitzgerald wisely cites Rahner here: "human self-determination is possible only *within the context of interpersonal community*" (ibid., 416, my emphasis).

Indeed. Imagine a rhetorical and symbolic environment in which all of the active players in Genome mapping projects were presented—to themselves as well as to the larger public—as human practitioners whose humanity flavored and shaped and leavened their research. Imagine, in short, Genome projects whose inherent humanity constituted a key element of their public symbolic presence. Imagine researchers whose funding proposals included the language of fatigue and confusion along with milestones and benchmarks and breakthroughs. Imagine a rhetoric of genetic research that invited non-geneticist voices into the conversation about the pace of the work, about the often subtle choices made in defining the next steps and about how to talk about the meaning of the researches already extant. Such a rhetorical climate would meet Fitzgerald's requirement for an ethical inquiry that takes scientific practice seriously and listens to it closely all the while situating the inquiry "within the context of interpersonal community."

A conference such as this one encourages me. I am sometimes discouraged by the absolutizing claims of Genome research advocates who seek to insulate genetic research from serious debate by clothing it in the symbols of inevitable progress and that most untypically popular of America's wars, congering an amorphous Hitlerian enemy to be fought and defeated. Conversations among scholars, from various disciplines, who pay serious attention to genetic research may be one of the most helpful processes available in our society to coax the rhetoric and symbols of Genome research further into the messy tangles of serious public discourse.

15

Moral Theology and the Genome Project

Richard A. McCormick, S. J.

The explicit purpose of the Human Genome Project (HGP) is to sequence all the DNA in the human genome. When this has been completed, it is clear that we will possess an enormously powerful information base. The very first ethical question such an information base generates is: how will we and should we use it? These are two remarkably different questions. The "will use" question forces us to analyze rigorously our context and culture, its proclivities and biases, its priorities, its strengths and weaknesses and those of the human agents who are *shaped* by these cultural proclivities and biases, and live them out and reinforce them in their actions and policies. If these factors are faced and analyzed honestly, we will be positioned to develop policies that can act as controls of and barriers against our own corporate weaknesses.

The "should use" question will force us to look deeply into the structure of the human person to identify those goods that define our flourishing. Once these goods (really facets of the human person) are identified, it will be possible and necessary to determine what use of the information provided by the Genome Project will support and promote these goods, what use will undermine them. In other words, we will have the basic criteria for determining the morally right ("should use") and morally wrong use of Genome Project's information.

Another way of putting this is drawn from Vatican II. *Gaudium et spes* (n. 51) asserted that the "moral aspect of any procedure. . .must be determined by objective standards which are based on the nature of the person and the person's acts" (Vatican 1965, 37).[1] The official commentary on this wording noted two things: (1) In the expression there is formulated a general principle that applies to all human actions, not just to marriage and sexuality. (2) The choice of this expression means that human activity must be judged insofar as it refers to the "human person integrally and adequately considered" (ibid.).[2]

The "human person integrally and adequately considered" is, then, the criterion to be brought to bear as we face into the Genome Project and the problems it generates.

But what does it mean to use as a criterion "the human person integrally and adequately considered"? There are any number of ways of fleshing this out. For instance, Louis Janssens states that this phrase refers to the human person in all her/his essential aspects. He lists eight such aspects. The human person is (1) a subject (normally called to consciousness, to act according to conscience, in freedom and in a responsible way). (2) A subject in corporeality. (3) A corporeal subject that is part of the material world. (4) Persons are essentially directed toward one another (only in relation to a Thou do we become I). (5) Persons need to live in social groups, with structures and institutions worthy of persons. (6) The human person is called to know and worship God. (7) The human person is a historical being, with successive life stages and continuing new possibilities. (8) All persons are utterly original but fundamentally equal (Janssens 1980).

Another way of explicating the notion of "the person integrally and adequately considered" is that proposed by the Grisez-Finnis school (see McCormick 1980, 156–73). They proceed by asking what are the goods or values we can seek, the values that define our human opportunity, our flourishing. We can answer this by examining our basic tendencies. For it is impossible to act without having an interest in the object, and it is impossible to be attracted by, to have interest in something without some inclination already present. What then are the basic inclinations?

With no pretense at being exhaustive, we could list some of the following as basic inclinations present prior to acculturation: the ten-

---

1 *"objectivis criteriis ex personae ejusdemque actuum natura desumptis."*
2 *"personam humanam integre et adequate considerandam."*

dency to preserve life; the tendency to mate and raise children; the tendency to explore and question; the tendency to seek out other men and obtain their approval—friendship; the tendency to establish good relations with unknown higher powers; the tendency to use intelligence in guiding action; the tendency to develop skills and exercise them in play and the fine arts. In these inclinations our intelligence spontaneously and without reflection grasps the possibilities to which they point, and prescribes them. Thus we form naturally and without reflection the basic principles of practical or moral reasoning.

The morality of our conduct is determined by the adequacy of our openness to these values. For each of these values has its self-evident appeal as a participation in the unconditioned Good we call God. The realization of these values in intersubjective life is the only adequate way to love and attain God.

What does it mean to have an adequately open and supportive posture vis-a-vis these values?

First, we must take them into account in our conduct. Simple disregard of one or other shows we have set our mind against this good. Second, when we can do so as easily as not, we should avoid acting in ways that inhibit these values, and prefer ways that realize them. Third, we must make an effort on their behalf when their realization in another is in extreme peril. If we fail to do so, we show that the value in question is not the object of our efficacious love and concern. Finally, we must never choose against a basic good in the sense of spurning it. What is to count as "turning against a basic good" is, of course, the crucial moral question and it is one that separates Grisez and Finnis from many philosophers and theologians who otherwise accept their basic structure. Certainly it does not mean that there are never situations of conflicted values where it is necessary to cause harm as we go about doing good. Thus there are times when it is necessary to take life in the very defense of life, in our very adhering to this basic value. That means that taking life need not always involve one in "turning against a basic good." Somewhat similarly, one does not necessarily turn against the basic good of procreation (what Pius XII called a "sin against the very meaning of conjugal life") by avoiding child-bearing. Such avoidance is only reproachable when unjustified. And the many conflicts (medical, economic, social, eugenic) that justify such avoidance were acknowledged by Pius XII. Suppressing a value, or preferring one to another in one's choice cannot be simply identified with turning against a basic good. My only point here is that particular moral judgments are incarnations of these more basic normative positions, which have their roots in spontaneous, prereflective inclinations.

The two approaches I have outlined to the idea of "the person integrally and adequately considered" are not that dissimilar. Indeed, there are large areas of overlap.

But now we must ask: what does all this have to do with *moral theology*? And with the Genome Project? Undoubtedly, there would be a pluralism of responses to this question. Drawing on some of my past writings, I would outline the matter as follows. I begin with a citation from James Gustafson, surely one of the most distinguished moral theologians of our time.

> For theological ethics. . .the first task in order of importance is to establish convictions about God and God's relations to the world. To make a case for how some things really and ultimately are is the first task of theological ethics. (Gustafson 1984, 98)

I agree with Gustafson's description of the first task of theological ethics, though it is clear we would disagree significantly about "how some things really and ultimately are." How, then, are they, "really and ultimately"? In its *Declaration on Euthanasia* the Congregation for the Doctrine of the Faith made reference to "Christ, who through his life, death, and resurrection, has given a *new meaning to existence*" (Vatican 1980, 4). If that is true (and Christians believe it is), then to neglect that meaning is to neglect the most important thing about ourselves, to cut ourselves off from the fullness of our own reality.

"A new meaning to existence." Those are powerful words. If Christ has given "a new meaning to existence," then presumably that new meaning will have some relevance for key notions and decisions in the field of bioethics. At this point it is a fair question to ask: what is this new meaning?

Theological work in the past decade or so has rejected the notion that the sources of faith are a thesaurus of answers. Rather they should be viewed above all as narratives, as a story. From a story come perspectives, themes, insights, not always or chiefly direct action guides. The story is the source from which the Christian construes the world theologically. In other words, it is the vehicle for discovering and communicating this new meaning.

At this point let me attempt to disengage some key elements of the Christian story, and from a Catholic reading and living of it. One might not be too far off with the following summary:

- God is the author and preserver of life. We are "made in His image."

- Thus life is a gift, a trust. It has great worth because of the value He is placing in it (Thielicke's "alien dignity").

- God places great value in it because He is also (besides being author) the end, purpose of life.

- We are on a pilgrimage, having here no lasting home.

- God has dealt with us in many ways. But His supreme epiphany of Himself (and our potential selves) is His Son, Jesus Christ.

- In Jesus' life, death, and resurrection we have been totally transformed into "new creatures," into a community of the transformed. Sin and death have met their victor.

- The ultimate significance of our lives consists in developing this new life.

- The Spirit is given to us to guide and inspire us on this journey.

- The ultimate destiny of our combined journeys is the "coming of the Kingdom," the return of the glorified Christ to claim the redeemed world.

- Thus we are offered in and through Jesus Christ eternal life. Just as Jesus has overcome death (and now lives), so will we who cling to Him, place our faith and hope in Him, and take Him as our law and model.

- This good news, this covenant with us has been entrusted to a people, a people to be nourished and instructed by shepherds.

- This people should continuously remember and thereby make present Christ in His death and resurrection at the Eucharistic meal.

- The chief and central manifestation of this new life in Christ is love for each other (not a flaccid "niceness," but a love that shapes itself in the concrete forms of justice, gratitude, forbearance, chastity, etc.). Especially for poor, marginal, sinners. These were Jesus' constant companions.

For the Catholic Christian, this is "how some things really and ultimately are." In Jesus we have been totally transformed. This new life or empowerment is a hidden but nonetheless real dimension of our persons, indeed the most profoundly real thing about us (Barth 1938, 95).

Now, what does all of this have to do with bioethics? I want to reject two possible extremes from the outset. The first extreme is that faith gives us concrete answers to the problems of concrete normative ethics. Josef Fuchs, S. J., is surely correct when he writes:

Medical ethics is theological, and hence Catholic-theological, ethics if it proceeds from faith, i.e., from the Catholic faith. This faith is ultimately not the assertion of the truth of certain faith propositions, but an act, in the depth of the person, of giving and entrusting oneself to the God who reveals and imparts himself to us. Naturally, no concrete ethics—and therefore no medical ethics—can be developed out of faith understood in this way (Fuchs 1987, 304).

The second extreme is that faith has no influence whatsoever on bioethics. It would seem passing strange indeed if what Sittler calls "the invasion of the total personality by the Christ-life" had no repercussions on one's dispositions, imagination and values (Sittler 1958, 45).

How, then, does faith exercise its influence? I will take my lead from Vatican II. In an interesting sentence, "The Constitution on the Church in the Modern World" states: "Faith throws a new light on everything, manifests God's design for man's total vocation, and thus directs the mind to solutions which are fully human"(Flannery 1996, 173n. 11).

Here we have reference to a "new light." The nature of this "new light" is that it reveals human existence in its fullest and most profound dimensions ("God's design. . .total vocation"). The effect of this new light is to "direct the mind." To what? "Solutions which are fully human." The usage "fully human" I take to mean a rejection of any understanding of "a new meaning to existence" that sees it as foreign to the human, and radically discontinuous with it.

The Catholic tradition has encapsulated the way faith "directs the mind to solutions" in the phrase "reason informed by faith." Reason informed by faith is neither reason replaced by faith, nor reason without faith. It is reason shaped by faith. Vincent MacNamara renders this as follows: "Faith and reason compenetrate one another and form a unity of consciousness which affects the whole of the Christian's thought and action" (MacNamara 1985, 116). What is the effect of such a Christian consciousness? MacNamara uses several phrases: "people who see things in a particular way, because they are particular sorts of people." "It [faith] determines, to some extent, one's meanings, what one sees in the world, what are the facts of life, and what among them are the most prominent and relevant facts" (ibid.). He refers to the "total background out of which judgments and choices are made" (ibid., 119) Or again, there is reference to a Christian's "interpretative self-awareness" (ibid. 121).

In summary, "reason informed by faith" is shorthand for saying that the reasoner (the human person) has been transformed and that this transformation will have a cognitive dimension through its invasion of consciousness. I think it true to say that the more profound the faith, the greater and more explicit will be the Christian consciousness—which is a way of saying that how faith (and theology) affects ethics can be seen best of all in the saints. But even we non-saints ought to be able to give an intelligible account of theology's influence. But that account is destined to be more or less incomplete because the transformation worked by faith is at a very profound level not totally recoverable in formulating consciousness.

"People who see things in a particular way." "What one sees in the world." What *does* one see as one contemplates the Genome Project? What is this *particular* way? From a Catholic point of view, I believe that Cardinal Joseph Bernardin's notion of a "consistent ethic of life" should play a dominant role in shaping Catholic consciousness. Bernardin describes this idea as "primarily a theological concept, derived from biblical and ecclesial tradition about the sacredness of human life, about our responsibilities to protect, defend, nurture, and enhance God's gift of life" (Bernardin 1988, 58). The key word is "consistent." What this means is that we bring to a whole spectrum of very different life issues (war, capital punishment, abortion, care of the dying, genetics, sexuality) the same basic attitude: respect. To the extent that this respect is weakened or absent in one area, the entire life-ethic is weakened, and other areas of the protection and enhancement of life are threatened. Bernardin makes it very clear that the consistent ethic of life applies to life-*enhancing* issues as well as life-*protecting* ones, to the quality of life as well as to the *right to life*.

I believe that John Paul II's *Evangelium Vitae* (25 March 1995) lends powerful support to the moral vision Bernardin proposed. The encyclical states:

> *Where life is involved, the service of charity must be profoundly consistent.* It cannot tolerate bias and discrimination, for human life is sacred and inviolable at every stage and in every situation; it is an indivisible good. (John Paul II 1995, n. 87)

The commanding attitude in the consistent ethic of life is *respect*. It is engendered by the co-relative objective *dignity* of life. John Paul II's *Evangelium Vitae* weaves together a wealth of biblical texts to highlight this dignity. Basically such dignity is linked to life's beginning from God and its ultimate destiny with God. The sacredness of life as God's gift gives rise to its inviolability.

I believe it is accurate to say that Catholic consciousness, as it confronts the Genome Project, will be dominantly shaped by this dignity-respect duo.

But respect engendered by dignity cannot be left at that level of generality. Respect, if it is to merit the name, must translate into practical honoring of two aspects of the human person: our basic equality and radical sociality.

First, *sociality*. One does not have to be a Christian to acknowledge this. But Christians do find in the sources of their faith powerful supports against any cultural calousing of this dimension of the person.

In the Judeo-Christian story, God relates to and makes convenants with a people. Both the Old and the New Testament stories yield such abundant evidence of this that it need not be documented. As Christians, we live, move, and have our Christian being as a believing group, an *ecclesia*. Our being in Christ is a shared being. We are vines of the same branch, sheep of the same shepherd. This sociality suggests that our well-being is interdependent. It cannot be conceived of or realistically pursued independently of the good of others. Sociality is part of our being and becoming. As Joseph Sittler put it, "personhood is a social state" (Sittler 1981, 98).

Our radical sociality gives birth in the Catholic tradition to the notion of the common good. *Gaudium et Spes* defines the common good as "the sum of those conditions of social life by which individuals, families, and groups can achieve their own fulfillment in a relatively thorough and ready way" (Flannery 1996, 256 n. 74). As Todd Whitmore has observed, "the concept of the common good . . .forms the basis for both societal claims on the individual and individual claims on society" (Whitmore 1995, 336). While such claims remain unspecified and need to be concretized and adapted in differing times and cultures, they have the distinct virtue of resisting and rejecting a notion of the person as absolutely autonomous and totally independent.

Societal claims on individuals and individual claims on society are specifications of the kind of society we want to create. Such a society, in Catholic tradition, is one characterized by interdependence of persons and countries. This interdependence must be cemented by distributive justice both in the priorities of genetic research (allocation of resources) and the enjoyment of its benefits.

Take allocation of resources as an example. Such allocation reflects the values of a society. As Roger Shinn has put it:

> Most countries of the world can afford little or nothing for elaborate genetic research. In this country, what resources should go into genetic therapies that may some day cure cancer as compared with correcting environmental causes of cancer that are operational right now? What resources should be assigned to research into the unknown as compared with correction of nutritional deficiencies for which answers are available now? (U.S. House 1982, 305)

Too few of our resources go to the most needy. Rather, as Shinn notes, the assignment of resources is proportioned "to the glamor of the project or the interests of groups who influence politics" (ibid.). In other words, it neglects the responsibilities inseparable from our sociality and interdependence.

Something similar must be said about the benefits of genetic innovation. They should be "generally available (not coercively) to all, regardless of geographic location, economic ability or racial lines," in the words of a study group of the National Council of Churches (U. S. Council of Churches 1984, 34).

Second, *equality*. Since our dignity roots in our origin and our destiny (God), it is clear that we are equally dignified. Bringing such a conviction to the Genome Project and the technologies it generates will not be easy. The recent destruction of thousands of preembryos in England is a symbol of my concern. Regardless of one's views on the personhood of preembryos, this happening should evoke deep sadness. It is transparent of our willingness to subordinate virtually anything to our desires and purposes. Referring to "utilitarian lawmaking," *The Tablet* (London) recently editorialized:

> Its primary purpose was not the upholding of absolute principles, nor even the resolving of agonising medical dilemmas when principles conflict, but the advancement of personal happiness by the manipulation of human fertility. Embryos and fetuses have become the means to that end, instruments to be created, kept or disposed of accordingly. Children (born children as well as unborn) are not to be seen as beings in their own right, but as aids to the psychological fulfillment of their parents. (*Tablet* 10 August 1996, 1039)

The Genome Project will very likely confront us with a host of practical moral problems. Some are already with us. For instance, a test is now available for identifying a gene (BRCAI) associated with breast and ovarian cancers. Should a woman with a family history of breast cancer be tested for the mutated gene? Does the prospect of early detection outweigh the burden of knowledge of potential risk?

But the most immediate and obvious problem generated by the Genome Project is genetic privacy. Wendy McGoodwin of the Council for Responsible Genetics, put the problem succinctly: "The same technology that can help doctors diagnose and treat disease is also being misused to deny people jobs or insurance" (McGoodwin in *American Medical News* 1996, 3). In a word, genetic discrimination.

A moral tradition that highlights the equality of persons should provide strong resistance to such discrimination. Needless to say, this tradition sees the most radical discrimination as existence-discrimination (abortion).

In relating moral theology to the Genome Project, I have avoided what I consider extreme positions. I would propose that theology provides the essential context for moral reasoning and therefore affects it deeply. Love of and loyalty to Jesus Christ, the perfect man, sensitizes us to the meaning of persons. The Christian tradition is anchored in faith in the meaning and decisive significance of God's covenant with humanity, especially as manifested in the saving incarnation of Jesus Christ, his eschatological kingdom which is here aborning but will finally only be given. Faith in these events, love of and loyalty to this central figure, yields a decisive way of viewing and intending the world, of interpreting its meaning, of hierarchizing its values. In this sense the Christian tradition only illumines human values, supports them, provides a context for their reading at given points in history. It aids us in staying human by underlining the truly human against all cultural attempts to distort the human. It is by steadying our gaze on the basic human values that are the parents of more concrete norms and rules that faith influences moral judgment and decision-making. That is how I understand "reason informed by faith."

In summary, then, Christian emphases do not immediately yield moral norms and rules for decision-making. But they affect them. The stories and symbols that relate the origin of Christianity and nourish the faith of the individual, affect one's perspectives. They sharpen and intensify our focus on the human goods definitive of our flourishing. It is persons so informed, persons with such "reasons" sunk deep in their being, who face new situations, new dilemmas, and reason together as to what is the best policy, the best protocol for the service of all the values. They do not find concrete answers in their tradition, but they bring a world-view that informs their reasoning—especially by allowing the basic human goods to retain their attractiveness and not be tainted by cultural distortions. This world-view is a continuing check on and challenge to our tendency to make choices in light of cultural enthusiasms which sink into and take possession of our unwitting, pre-ethical selves. Such enthusiasms can reduce the good life to mere adjustment in a triumph of the therapeutic; collapse an individual into his funtionability; exalt his uniqueness into a lonely individualism or crush it into a suffocating collectivism. In this sense I believe it is true to say that the Christian tradition is much more a value-raiser than an answer-giver.

To expect more from moral theology will be, I believe, disappointing. To settle for less will be impoverishing.

16

Afterword: The Geneticization of Western Civilization: Blessing or Bane?*

John M. Opitz

It seems only fair that I precede my formal allocution with some personal remarks to help locate me in the context of this conference, and to expose from the outset the experiences and prejudices I bring to this discussion.

As a native of Germany and long time citizen of the U.S. I have developed the perspective of a *cives mundi* with special regard for the history of "man's inhumanity to man," particularly that founded on arguments of national or racial interest.

With a forty-year interest in developmental biology (Opitz 1995), I have tried to combine my education in zoology and training in pediatrics in order to attain a better perspective on the relationship between evolution and development, knowing and respecting the fact that all living organisms are related and interdependent in a frail and

* Acknowledgments: I thank Mrs. Yvonne Stevenson for expert manuscript preparation and Prof Pierce Mullen (Dept. of History and Philosophy, MSU, Bozeman), Prof. John Hart (Dept. of Theology, Carroll College, Helena), Robert Resta, M.S. (Perinatal Medicine, Swedish Hospital, Seattle, WA), Dr. Susan Lewin (Helena) and many other friends and colleagues who have discussed many of the ideas expressed here and who have scrutinized this manuscript critically. Above all, I am deeply grateful to Prof. Phillip R. Sloan of Notre Dame University for the impetus provided by his invitation to put these thoughts on paper. A version of this paper has appeared previously in German as Opitz 1998. Used here with permission of *Medizinische Genetik*.

vulnerable web of life, earth, air, light and water of which we must be responsible stewards.

As a clinical geneticist who has cared for and about thousands of patients, fetuses and families for over thirty years, I am committed to the primacy of the patient's right to care, respect, autonomy and integrity above any other consideration, patients having been throughout my career my most important teachers and an inexhaustible source of grace in the face of death, miscarriage, malformation, misfortune and life-long suffering.

During the course of my work in this field I was privileged to become a founder of the American Board and of the American College of Medical Genetics, of the Wisconsin Clinical Genetics Center at the University of Wisconsin, the Department of Medical Genetics at Shodair Hospital in Helena and of the statewide Montana Medical Genetics Program (with a permanent funding source secured through a small legislative surcharge on each health insurance policy issued in Montana). In my fifteen-year affiliation at Shodair my coworkers and I tried to combine, at a secondary health care center, primary and tertiary health care approaches in a clinical setting combined with an outstanding genetics laboratory, and information resources center and library, and a strong scholarly and didactic orientation. One of the main sources of strength of that program was the fact that throughout the time I was affiliated with it we had contact with virtually all scientifically and scholarly active workers in medical genetics throughout the world through the editorship of the *American Journal of Medical Genetics*. Hence, I speak on the subject of medical or clinical genetics from the perspective of a physician, administrator, legislative lobbyist, university faculty member and clinical scholar-investigator.

As a humanist (Opitz 1991) I am passionately committed to all of those manifestations of thought and spirit in the musical, artistic and humanistic institutions of western civilization that have over the centuries bestowed meaning on human life and helped to redeem its suffering, especially that incurred innocently.

And as someone who has experienced at first hand the horrors of war, a father who has lost a son, a physician who has sat at the bedside of many a desperately ill or dying child, a fetal pathologist who has autopsied over a thousand fetuses and infants of grieving parents, I am poignantly aware of the evanescence of life, but also know that without hope, commitment to the welfare of humankind, love, and a smile of joy life can be a living death.

Finally, as one whose very survival has depended so frequently and substantially on the kindness of others, I know in the very marrow of my bones that selfless and courageous generosity is one of the greatest of human virtues.

A MATTER OF DEFINITION

After submission of an outline of this talk I was asked to define and to clarify the term "geneticization," a concept that is not original with me. I do not know when and where the term originated, but recently Ruth Hubbard (1995) characterized it as "thinking of ourselves as readouts of our genes"—an *individualization* of wellness/illness concepts. However, I think the term means more than that. I should like to submit that *geneticization* is a *cultural* concept, derived from biology, introduced into medicine and, more recently, into the social consciousness of most of western civilization, and reflecting, in essence, an ever-increasing pre-occupation with abnormal or potentially abnormal *parts* of ourselves, and the fear that these may adversely affect our health or quality of life and that of our children. In Western culture this process had its beginnings in late Medieval/early Renaissance days through the work of the anatomist-physicians and surgeons who showed, that far from constituting a homogeneous substance animated by the spirit, humans, like the animals, consisted of demonstrably different parts. Later anatomist-embryologists demonstrated that each part had a separate prenatal origin and specific developmental integrity; the pioneer pathologists then documented that each part could be the "seat and cause" of illness.

Geneticization represents another manifestation of a long-standing western cultural process whereby alienation from ourselves as *whole* persons and our failure of a serene acceptance of normal life phenomena and processes, including pain, grief, illness, miscarriage and death, has been accompanied by an internalized demonization of ill or potentially ill body *parts* and an obsession of what to do about them. Genes are regarded, in this context, as somewhat mysterious, but nevertheless very concrete parts of ourselves capable of predisposing to or creating a multitude of physical, functional, intellectual, and mental mischiefs in a particularly threatening manner because of the deterministic inevitability of their effects, effects which, for example, sometimes cause close relatives of persons with Huntington chorea to be identified or to identify themselves, consciously or unconsciously, as carriers of the gene and as being "fated" to develop HC without a shred of evidence that they in fact carry it (Kessler 1988). Lay persons, on the whole, lack a sophisticated understanding of the epigenetic nature of pre-and postnatal developmental processes and of the probabilistic, rather than deterministic, relationship between mutant gene and developmental outcome. Having a Huntington disease mutation confers a high probability, but not absolute certainty of developing the disease. The proven presymptomatic carrier of the gene fears its apparently inescapable effect, often feeling a helpless victim beyond human aid and taking little comfort from the fact that the penetrance of Huntington chorea is not 100% even in high old age.

Western populations are particularly susceptible to alarm (or elation) upon reading or hearing about the latest news ("advances") in genetics because of a long preceding cultural tradition of incorporating these frequently preliminary and incompletely understood pronouncements into a specific view or conception of themselves as the sum of potentially defective, and nowadays potentially fixable *parts* and of making concrete lifestyle decisions based on these perceptions. Long before there was the specialty of genetics, persons made decisions on marriage, procreation or continuation of pregnancies based on their perceived risk of developing or transmitting hereditary conditions or ailments, and advised their sons and daughters "not to do to your children as I did to you" and elevating to the status of hereditary evil many conditions either totally or largely environmentally caused.

During the nineteenth and early twentieth century the word "heredity" covered a "multitude of sins" including many social ills such as pauperism, "harlotry," illegitimacy, criminality, "moral depravity," alcoholism, syphilis, deformity, and "degeneracy" in general. In my own family it was tuberculosis "inherited" over at least five generations. It caused me a long stay in a sanitarium as a child, killed my father and affected nearly every relative, including some who married into the family, over many generations. And when, in 1882, Koch demonstrated the infectious cause of tuberculosis, "society" never flinched; what was inherited instead was a "constitutional weakness or predisposition" to the disease (the inflammatory-exudative diathesis) as powerful as the "hereditary" nature of the infectious organism had been before (q.v. Stockel 1996; Burgio 1996). This cultural tradition had a strong influence on medicine early in this century when constitutionally interested physicians as late as 1940 (von Pfaundler 1940) claimed to be able to identify numerous "diatheses," concepts that arose in Greek medicine (Galen, possibly Hippocrates) but that now have vanished from medicine, while living on in the cultural consciousness of society. It was of great interest to hear Jean-Paul Gaudillière speak at this conference of the persistence of the concepts of heredo-alcoholism, -tuberculosis, and -syphilis in France until the middle of this century, a phenomenon I find confirmed while beginning to edit the memoirs of my distinguished colleague and friend, Pierre Maroteaux, arguably the world's greatest authority on skeletal dysplasias, who has worked at the *Hôpital des enfants malades* in Paris for the last half century. He describes vividly how at mid-century his mentor and the founder of pediatric genetics in France, the great Maurice Lamy, was still waging his campaign to eradicate the concept of heredo-syphilis from French constitutional medicine.

In this context an important discussion point may be Ruth Hubbard's contention that "[n]ot only will predictions based on 'knowing our genes' not help most people, but, this *genomania* [emphasis added] may aggravate health problems by overemphasizing the health problems of individuals and deemphasizing the need for adequate social and public health policies" (Hubbard 1995, 10), a point also made by Philip Kitcher in his lovely review of Utopian Eugenics and Social Inequality presented in this volume. In an increasingly self-centered, materialistic and divided world, genomania has the potential of becoming the predominant western obsession, more so than war, famine or pestilence ever were in our consciousness. In a perceptive commentary, Jonsen et al. (1996) speak of the great numbers of future, worried, well "un-patients," proven carriers of a deleterious gene mutation, waiting for "the inevitable" to strike, powerless to do something about the "evil" lurking within. The former perception and acceptance of the inevitability of illness and death as a *normal* attribute of life and of living accordingly and with serenity is increasingly giving way to despair and determination to defer the inevitable as long and as painlessly as possible by any means including quackery, faith healing, organ transplants, prophylactic amputations and massive doses of tranquilizers. The ability to test for genetic predisposition should not lead to the conclusion that therefore it *must* be applied, and physicians should think twice before succumbing to the pressure of commercial enterprises that peddle carrier or predictive tests.

EUGENICIZATION

The term *eugenicization* is a subconcept of geneticization and refers to the political and institutional responses of societies seeking to implement means to deal concretely with real or perceived genetic fears and of making these means available to broad segments of or entire societies, i.e., eugenicization is geneticization institutionalized. The benevolent implication of this formulation being the *voluntary* nature of the specialist-population interaction, the involuntary form being condemned almost universally and rightly so, as "eugenic crimes against humanity."

Thus, another discussion point may be—given the largely cultural context of our view of normal/abnormal, or more to the point—of desirable or undesirable human traits, how are we, as professionals, philosophers, ethicists, to frame the discussion so as to balance the "right" of individuals to the integrity of their decision-making capabilities versus the perceived need of many families to terminate pregnancies of thousands of fetuses known or suspected to be abnormal in their view. Note, that while in every case these involve *individual* decisions, in effect, they are directed at whole segments of the *population*, i.e., most fetuses with anencephaly/spina bifida, Down syndrome or other forms of aneuploidy, homo-, hetero- or hemizygotes for many Mendelian mutations, etc. And while I am a strong supporter of the concept of free and informed reproductive choices made in good conscience and with the best of intentions, I cannot help but view these many dead fetuses as being an internalized population minority who fell victim precisely to the view of fetuses being potentially defective *parts* of the mother, fixable through lethal removal from the mother's body. It is a sad irony that these usually anguished decisions made from the perspective of a "part" are so frequently mourned consciously or unconsciously as the death of "my baby." After a therapeutic abortion in Japan, as well as after a spontaneous loss, this process of mourning and atonement is formalized in a Buddhist ceremony through the dedication of a small statue of a Mizuko-Jizu at temples specially dedicated to the spiritual welfare of the dead babies and their parents.

In speaking of these therapeutically aborted malformed and/or genetically abnormal babies as a "minority," I am not trying to be provocative or confrontational or trying to give aid or comfort to those who bomb abortion clinics or murder their staff members in cold blood. Rather, I feel a need to point out an inexorable analogy with the history of what might be called "external" minorities, e.g. Native Americans in the US in the eighteenth, nineteenth or early twentieth century or the Jews in Germany between 1933 and 1945, and their collective treatment as defective *parts* of, and hence fixable through lethal removal from, the body politic of the dominant population group. Wish that such treatment of (external) minorities were a thing of the past, but the recent catastrophic history of former Yugoslavia and certain parts of Asia and Africa demonstrates that intolerance and sadistic brutality in dealing with human differences may occur in all countries.

In the words of Meg Greenfield:

> . . . .there is no clearer index of savagery and inhumanity and lapsed civilization than the kinds of mass butchery and garbage-like disposal of humankind that we have witnessed in every part of the world in recent years. (Greenfield 1996, 80)

The concept of individuals being disposable parts was put into a eugenic perspective at the beginning of the most violent of all centuries in human history by Galton who concluded his memoirs saying, "Individuals appear to me as partial detachments from the infinite ocean of being. . . ," evolving formerly "principally. . .by means of Natural Selection, which achieved the good of the whole with scant regard to that of the individual." It is "precisely the aim of Eugenics". . . "to replace Natural Selection by other processes that are more merciful and not less effective" (Galton 1909, 323).

The "Law of the People's Republic of China on Maternal and Infant Health Care" of 1995 (Bobrow 1995) stands in direct line with Galton's philosophy in mandating compulsory premarital medical examination. Where this shows "genetic disease of a serious nature which is considered to be inappropriate for child bearing. . . , the two may be married only if both sides agree to take long term contraceptive precaution or to take ligation operation for sterility" (ibid.,409).

According to Bobrow, reference is also made to "relevant mental diseases" such as schizophrenia or manic depressive psychosis. "Termination of pregnancy shall be advised 'in case of fetuses with a defect or genetic disease of a serious nature'" with or without consent as implied in Article 19: "Any termination of pregnancy or application of ligation operation shall be agreed and signed by the concerned person" (ibid.).

The likelihood of abuse of even the most well-intentioned eugenic institutionalization of a genetical perspective of life makes it imperative that all such regulative, legislative or legal actions be subjected to the most intensive ethical scrutiny, preferably by an international panel of experts in the case of national enactments potentially affection millions of individuals. Governments involved, such as the People's Republic of China, may scream infringement of sovereignty; however, where a well considered opinion of an international panel expresses unanimous, grave concern, it is probably wise to maintain relentless international scrutiny and publicity to allow the people of the nation in question to equilibrate, at last, the edicts into a more humane, voluntary and ethically generally accepted norm.

THE PROBLEM OF BIOLOGICAL LITERACY

I mentioned above some problems with public pronouncements on genetic disease research. A major problem, as I see it, is that with exception of the specialists, virtually no one, most physicians included, possess the *biological literacy*, as John Moore calls it (Moore 1995), to discuss specific or general aspects of a given genetic issue with individuals or families in a manner and with the knowledge required to address their questions and most urgent needs and concerns from essential biological, medical, social and cultural perspectives. I concur with Moore that: 1) the lack of cultural and scientific literacy is a grave weakness of contemporary civilization, especially in the U.S.; 2) it has its roots in the K–12 curricula of our school systems; 3) its alleviation will require a major reform of our educational systems; and 4) it is an important factor preventing individuals from understanding "the difference between faith and science—a fundamental point that most schools seek to avoid." Those of us who seek to alleviate the problem by working e.g. on BSCS textbook material for 10th graders, are appalled to know that that will probably be the only biology course U.S. children will take during their secondary school education.

Recently, Dr. Patricia Jacobs of the Wessex Regional Genetics Laboratory, Salisbury, England was quoted as saying:

> People assume that medical schools provide an excellent genetics education for the next generation of doctors. . . : unfortunately this is far from true; few medical schools incorporate genetics teaching throughout the curriculum. . . . As a result, we are going to have another generation of doctors who will be unable to cope with the so-called genetics revolution, and this also needs to be addressed urgently. (Dickson 1995a, 619)

May I suggest as a third point for discussion that the ability of the Banbury *ad hoc* Working Group "to educate primary-care physicians in genetics and to develop models for the delivery of genetic services in primary health-care settings" (Touchette 1995, 501) may be doomed *ab initio* if it presumes a level of general biological, genetic and cultural literacy which, in general, primary-care physicians do not possess. To me, an additional concern is that, according to Dr. James Allen of the AMA, the Working Group,

> . . .will not directly develop (its) own model for either genetics education or the delivery of genetic services. . . , [b]ut . . . will ask whether there are any models in place that would be used for broader application." (Ibid., 501)

Clinical geneticists, individually and collectively, play an extremely important role in information dissemination and popular education on genetic issues. In this process they are aided very effectively by the truly spectacular advances in informatics and genetic library and information services (such as the one at Shodair Hospital in Helena that serves the people and professionals of Montana), and the fact that there is a slowly but ever increasing production of excellent reading material for lay persons on numerous topics and specific conditions.

However, the respectively 780 and 820 board-certified clinical geneticists and genetic counselors in the U.S. are spread extremely thin; they are concentrated mostly in large population centers, and usually have additional academic didactic, administrative and research duties. The kind of community outreach required for effective liaison and consultative didactic support of primary health care providers around a state is expensive, time consuming, exhausting and reaches, in any event, only a small fraction of practicing physicians caring for patients with genetic disorders. All of us teach at many conferences and meetings for lay or professional organizations during the year; however, these address only a minute fraction of the need of the population for accurate genetic information.

Education in genetics itself is a tough challenge. Far too many lecturers I have heard in medical and graduate schools or at postgraduate conferences miss the forest for the trees with a relentless emphasis on technical details rather than the broad principles of genetic biology pertinent to human welfare. Genetic knowledge is advancing so rapidly that no one can keep up with it in all aspects except in one's own field of interest, and much that is presented today is overtaken tomorrow by new developments. The ethical basis of medicine has not been changed substantially by the advances in human genetics, and neither have the imperatives of effective and compassionate communication with patients about their concerns (rather than the physician's obsession with biological detail). Genetic thinking is no different from that of physicians searching for causes and predispositions, cascades of pathogenetic events and their effective management in the patients' best interest and with his or her conscious, voluntary participation. Such a perspective does not dehumanize or medicalize the affected or predisposed individuals into categories of disease, but rather treats them first and foremost as human beings sharing a common heritage of birth, growth, illness, change, need to adapt and to mature and to have faith in the eternal verities of love, gentleness, faithfulness, compassion and enthusiasm about life and its limitless possibilities. *All* human beings carry deleterious genetic alleles or mutations, and there is no such things as a disease without genetic involvement, for better or for worse. It is imperative for physicians to learn that in spite of developmental limitation or common genetic disease predisposition, they must stress the normality and health of the rest of the body, and above all of the mind which will make it possible for individuals to adopt an attitude of wellness and to cultivate health habits that will or may go far to ameliorate or prevent the deleterious environmental effect on genetic predisposition.

A recent, to me rather startling, educational development has been the sudden availability on computer of information and databases on genetic diseases accessible to lay persons, with the consequence, that many of them, frustrated with medicine, genetics or both, are taking the diagnostic initiatives into their own hands and are educating themselves on their own or their child's condition. Having done so, in a recent experience, the mother then called me to discuss prognosis in her two boys, when it became obvious to me that she was in fact talking about syndrome X, not Y as she had persuaded herself. Here then is a truly wonderful opportunity for genetic professionals and educators to disseminate accurate information in lay terms and to take on the challenge to keep it updated. Such a development in mass communication may in fact, constitute a part solution to the problem of genetic education by physicians unable to keep up with progress in all genetic diseases and conditions.

THE FUTURE OF CLINICAL GENETICS

The geneticization of western life is going to place unprecedented demands for service on the medical system, foremost on medical/clinical genetics, but basically on the entire medical establishment from family practitioners interpreting abnormal results of newborn screening to anxious parents to internist/hematologist dealing with unusual thalassemias in the Hmong people of Missoula or Minneapolis.

Is medicine ready for this challenge? I can speak only for medical genetics and pediatrics; but from my expediences in a university setting and in a rural secondary care setting, I rather suspect that medicine at the moment is as ill-prepared for the molecular revolution in biology as it was for the microorganismal era a century ago. I have the impression that on the whole, pediatricians have a better "feel" for constitutional medicine than most other specialists, and there probably is a good reason for their preponderance in the field of clinical genetics.

However, clinical genetics is under siege, and its present condition, at least in the U. S., may be life-threatening. Clinical genetics is an extremely labor-intensive branch of medicine and cannot operate without expensive library and informatics support. It rarely generates in fees and third-party reimbursement even half of the budget of such a unit and clinical geneticists generally are regarded as the fiscal pariahs of any unit, division, department or other organization, e.g., medical school. Clinical geneticists are saddled with enormous clinical duties, consulting on virtually every service of the hospital, but especially pediatrics and the neonatal Intensive Care Unit, and see hundreds of patients throughout the year in the genetics clinic, the prenatal diagnosis and fetal pathology services, outreach or field clinics, in the institution or schools for the mentally retarded, etc. They teach medical students, house officers, counselors, graduate students, postdoctoral fellows and other groups mentioned above. Many have additional clinical duties in "home" departments such as pediatrics and obstetrics where they may have to "round" several weeks or months per year. Thus, as a group they are grossly overworked, frequently underpaid, have an exceptionally difficult time attaining the recognition, promotion and appointments appropriate for their years of service and vast experience, suffer additional ignominy for failing to publish enough or for publishing reviews or case reports rather than "solid science," nowadays meaning mostly molecular biology. Writing grants takes much time and energy and removes the scholar-clinician from the wards; clinical geneticists are at a particular disadvantage in obtaining grants since their work is by far not as attractive to granting agencies as that of their basic science colleagues. Indeed, on the federal level there is virtually no money available for clinical research. Those who try to do both, i.e. clinical and molecular work, frequently live schizophrenic, frantic lives, being additionally vulnerable to the fierce competition in the field of molecular biology and frequently falling prey to secretiveness and paranoia. Maddox has stated:

> But it [competitiveness in the field] has also sacrificed large numbers of postdoctoral fellows on the altar of achievement. It seems a waste of young talent that the academic life should be so intellectually bruising. . . . Civility has been the other casualty of competitiveness, especially in the United States. . . . (Maddox 1995, 522)

In spite of their best efforts, many administrators consider clinical geneticists neither "fish nor fowl" and continue a more or less open policy of discrimination against clinical genetics in general or specifically against its practitioners. One of the most notable exceptions, namely the DD/MR Branch of the NICHHD under the leadership of Felix de la Cruz (who has been called "the greatest friend of clinical genetics in the federal government"), has such an inadequate budget that it can fund only a minute fraction of the potentially meritorious proposals that may come to it. These disciplinary, economic and administrative stresses are causing more and more clinical geneticists to leave the field to enter a primary or secondary care practice or to work exclusively in molecular biology.

I for one (and here I may represent a minority of clinical geneticists) also perceive the establishment of what its practitioners have called "dysmorphology" as an additional threat to the unity and integrity of clinical genetics. No two "dysmorphologists" I have asked have given me the same definition of their specialty and I get the feeling that it is a substitute for the older terms "syndromologist" or "clinical teratologist," namely experts who predominantly study malformed fetuses and children/adults. Since this must therefore also involve the study of the causes of the malformation(s), they are, by definition, practising clinical genetics, and one is led to wonder why the distinction is necessary. Having been involved (some might be tempted to say "embroiled") in this controversy since its inception in Madison in 1964, I strongly suspect that the development of "dysmorphology" represents a political need by some pediatric geneticists for administrative autonomy which would emphasize their apparently unique expertise in contradistinction to that of their more theoretically or laboratory-oriented colleagues in other branches of developmental biology. In its literal sense the term "dysmorphology" means "bad morphology," underlining, in an unintended manner, that this is more than a semantic argument. A schism between clinical genetics and dysmorphology will have most unfortunate consequences for both; a harbinger of this is the fact that with increasing frequency and impatience referring physicians tell me:"I don't need a geneticist, refer me to a (good) dysmorphologist."

Méhes (1996), in a passionate plea that medical genetics not forsake its heritage of classical methods "in the era of molecular genetics," has documented a dramatic effect of a "pleasurably growing number of young research fellows (including physicians) who are very skillful in Southern blot, PCR. . .techniques but less familiar with the traditional methods, including physical examination of the patient" (ibid., 394). Whereas their former efficiency and reliability in finding and documenting birth defects in Czeizel's Hungarian Congenital Malformation Registry was 78%, presently it is only 43%, with the birth prevalence of trisomy 18 detected clinically declining from one in 8,600 to one in 20,570. I should like to submit that it ought to be possible for clinical genetics to retain or to recover these skills in phenotype analysis without making clinical genetics into something less than it is and ought to be—namely an equal partnership of morphological, physiological and genetic methods.

CLINICAL RESEARCH AND MORPHOLOGY

To develop the above points in greater depth it seems appropriate for me to step back and to view the recent precipitous, and I am inclined to say catastrophic, decline in *clinical research* in the field of medical genetics from the perspective of the history of the relationship between morphology and genetics in this century (Sapp 1983; Gilbert 1998; Gilbert et al., 1996), a subject also treated with authority by Evelyn Fox Keller (1995).

In his Herbert Spencer Lecture (1912), William Bateson, the militant protagonist of Mendelism in Britain, began a curious intellectual process which was to find its culmination in the New World in 1932 through another student of W. K. Brooks, namely Thomas Hunt Morgan. Bateson, like Morgan, had approached the problem of the causes of evolutionary change through the study of embryology, and in his magisterial book of 1894, *Materials for the Study of Variation Treated with Especial Regard to Discontinuity in the Origin of Species*, he documented his most important discoveries and insights, including the concept of *homeotic* malformations. After his Mendelian metanoia in 1900, Bateson became not only a vigorous opponent of Galtonian/Darwinian continuity, but began to distance himself critically from embryology, the mother discipline from which genetics had arisen. "One of Bateson's roles would be to destroy the notion that embryology contributed anything to our understanding of the mechanism of evolution"; genetics was to supplant embryology in that role (Gilbert 1998,170). Bateson vigorously championed a Mendelian understanding of human developmental and functional attributes, encouraging on one hand Archibald Garrod's work and insights into the inborn errors of metabolism, but fostering, on the other hand, a eugenic perspective whereby genetics "was to improve the physical, social and moral life of mankind" (ibid.).

Morgan went further. His Mendelian metanoia did not occur until 1910/1911; and being late in taking up the banner of Mendelism he was all the more vigorous in repudiating its embryologic origins. In

1926, Morgan began the process of separating genetics from embryology (Morgan 1926a, 1926b). In his presidential address at the Sixth International Congress of Genetics at Cornell University in 1932 (Morgan 1932a, 1932b), Morgan went so far as to denounce embryologists as invoking "philosophical platitudes" and to conclude that "from this source" (i.e. embryology) "we cannot add to the three contributory lines of research which led to the rise of genetics. . . ." In her recent book, *Refiguring Life* (1995), Evelyn Fox Keller puts it this way: "The very glow of the geneticists' spotlight cast a deep and debilitating shadow on the questions, on the methods, indeed, on the very subject of embryology" (Keller 1995, xv). The rest of her discussion is of fundamental importance in this connection.

The geneticization of "heredity" (the science of the causal analysis of development and evolution during the nineteenth century) led to a severe devaluation of morphology during most of the twentieth century to a point where at mid-century there was serious concern about the very survival of natural history, anatomy, embryology and fetal pathology. When in 1962 I began postdoctoral training in medical genetics at the University of Wisconsin, this anti-morphological bias was as fresh as on the day it was uttered by Morgan thirty years before.

As Gilbert et al. have shown (1996), one of the most lamentable effects of the *geneticization of morphology* was the loss, for several decades, of the immensely fruitful concept of the developmental field from the intellectual patrimony of Western biology. It is one of the great ironies in the history of biology that the very recent "triumphant" vindication of morphology and affirmation of the validity of the field concept occurred through *molecular biology* which, at the present time, comes closer, in many ways, to the epistemological core of nineteenth century "heredity" than to that of twentieth century Mendelism, especially that part subsumed under the concept of "The Modern Synthesis." Professor Keller puts it this way: "after decades of occlusion, the subject of embryogenesis has been restored to biology, the discourse of "gene action" has been supplanted by a more adequate language of "gene activation", and the complexity and agency of the organismic body is finally being accorded its due" (Keller 1995, xvii).

Another fateful consequence of the geneticization of morphology was a de-emphasis of an *evolutionary* perspective on all, but especially human development, to the point where today virtually no clinical geneticist known to me can discuss with any degree of sophistication the topic of the relationship between ontogeny and phylogeny or the clinical relevance of the concepts *atavisms* and *vestigia* which had so exercised the imagination of early nineteenth century morphologists and teratologists. Maddox makes an amusing comment in that connection. At dinner Maddox told another guest, a distinguished lady architect, about his new job as editor of *Nature*. She replied "laconically, 'I suppose that you must be one of those frightful Darwinists then'" (Maddox 1995, 522).

Now, having come to the belated but highly appropriate rescue of morphology (*the science of the form, formation, transformation, malformation and deformation of living organisms*), molecular biology is beginning to "swamp the lifeboat" and to repeat the "struggle for authority in the field of heredity" that characterized genetics in the first third of this century (Sapp 1983).

At this point I must affirm emphatically that I am a strong supporter of the Human Genome Project, that its funding by the federal government is a wise investment of tax dollars, and that the problems related to the Project, as I see them, are not inherent flaws in the science or the scientists but, rather the result of policy decisions which did not take the needs of the *entire* field into account. It is one of the most admirable aspects of the Project to support work on the *ethics* of this form of biological research and its implications for human welfare.

THE CRISIS IN CLINICAL RESEARCH

Having said this, I need to express my concerns about present developments in this field. The first, and to me most urgent "fall-out" of the molecular biology "revolution," is the grave, perhaps *fatal de-emphasis on clinical research* (Ahrens 1992; Feinstein 1995; Weatherall 1995). Until recently, most insights in medical genetics, whether biochemical, developmental or cytological, were based on the study of malformed or ill human beings who came to us for diagnosis, care, counselling and prognostic consultation. One of the happiest times in my career was that brief period some twenty years ago when I participated in an NIGMS (later NICHD) program project establishing clinical genetics research centers at several universities (including the University of Washington-Seattle, the University of Pennsylvania at the Children's Hospital of Philadelphia, and Johns Hopkins University). At that time I also became the principal investigator of a multi- and interdisciplinary Clinical Genetics Research Center at the University of Wisconsin supporting research in clinical, pediatric, obstetrical, anatomical, mental retardation and fetal genetics with many affiliated projects which led to the great flourishing of the Wisconsin Clinical Genetics group (still going strong). After the cessation of support for these centers, clinical research in genetics in the U.S. entered a precipitous decline from which it may not recover unless a drastic remedy becomes available. The need for such research is undeniable. Almost daily I receive calls from molecular biologists for families, pedigrees, blood and skin samples for linkage, mapping, and sequencing purposes in conditions I either described or have worked on. Some 5,000 conditions are listed in McKusick's catalogues of *Mendelian Mutations in Man*. Even if there are only 50,000 (rather than 100,000) coding genes in the human genome, some 90% of potential mutations may be undescribed, never mind the work necessary to delineate the complete spectrum of effects of each mutation, its pre- and postnatal natural history, and prognostic characteristics, knowledge required for a half-way useful counseling session with patients, parents and relatives. Probably hundreds of submicroscopic chromosome abnormalities await discovery with newer molecular methods for clinical correlation. The science of the epigenetic modification of the human genome (e.g., by imprinting) is in its infancy, and almost daily there are new reports of uniparental disomy for one or another pair of human chromosomes requiring clinical delineation. Molecular studies of placental development and its effects on mother and/or fetus are only beginning, and the range of potential mitochondrial pathology in humans can only be guessed at the present time.

I should like to emphasize that this state of affairs *must* be changed before research programs in clinical genetics die altogether. Genotype analysis without phenotype analysis cannot stand on its own and ultimately will fail without it. Transgenic mice are highly informative about mice, but less so, for obvious reasons, about human beings.

I hope, therefore, that it is not inappropriate to raise, as another discussion point, the question whether there is any chance at all to revisit the $282 million budget of the Human Genome Project in order to allocate a reasonable part of it for *clinical* research in medical genetics without depriving the ELSI program of its needed funding. It would seem to me that the speed with which the HGP will be "completed" matters less than the quality of its end result which can only be enhanced by excellent complementary clinical research.

REDUCTIONISM

Goethe may have been a preformationist, but one aspect of his philosophy and natural history that makes it attractive to this very day was his vigorous (philosophical) anti-reductionism. Neel has claimed, with considerable justification, that the "succession of amazing developments in twentieth century genetics resulted primarily from a rigorous application of the reductionist approach, that is, from finding the simplest possible system in which the phenomenon of interest can be studied" (Neel 1976, 39). In this connection it is important to make a distinction between a reductionist *methodology* of the type alluded to in Neel's statement, and a reductionistic *philosophy* which claims that the totality of the life manifestations of an organism is the sum of the parts studied by molecular biologists. Ethically applied there is nothing wrong with the former; however, reductionism of the latter sort is an ever recurrent theme in western biology, and, to my mind, ought to be a concern to all biologists, historians of biology and epistemologists. It is discussed in greater detail by Arthur Peacocke in this volume. Recent comments by such distinguished investigators as Nijhout (1990), Holliday (1994), Weiss (1993), and Strohman (1993; 1994) point to the epistemological obtuseness of much of the present day molecular biological work which is highly reductionistic, representing, as it frequently does, a linear rather than an organicist, that is to say non-linear, dialectical view of development.

The accomplishments of the HGP are staggering and represent one of the most phenomenal of knowledge explosions in biology since the days of the explorers and naturalists of the nineteenth century responsible for the discovery, study and description of tens of thousands of living organisms and their "natural history." However, I think it is only reasonable to point out that once the 100,000 or so genes of the human genome are mapped, the real work will begin of attempting to understand their function under *normal* circumstances of life and development. Kenneth Weiss (1993) ends his wise and extremely important book on genetic variation and human disease with these words:

The inevitable rush of enthusiasm to screen samples, families, or populations for causal alleles for every type of trait will produce many irreproducible results and excessive claims. I think we will be forced to accept that we cannot understand a trait well by enumerating all of its individual 'causes', which will be quixotically ephemeral and environmentally plastic. Instead, we need to identify deeper structures that can reduce the dimensionality of variation and explain it in a simpler way.

* * *

The glamor of finding 'the' gene for a given disease overshadows the greater effort to understand its full causal spectrum. Clearly there are a few rare alleles at major genes that affect a trait. These may be of great biomedical importance, but they are of little population importance. It is from the many more genotypes with individually small effects that the principles of evolution, which mold the pattern of variation, are to be understood. Biomedical research is concerned mostly with abnormal variation, but most variation, in most traits, in most people, for most of their lives is within the *normal* range, and deserves more attention. Diseases are just part of the natural phenotype distribution that we happen to choose to study. (Weiss 1993, 306, 313)

COMMUNICATION AND JARGON

If the problem of genetic literacy at all levels is to be addressed effectively, then human/medical genetics will need to develop means of clear, concise, consistent and correct communication with a minimum of jargon.

From the perspective of an editor it is evident to me and to other editors that the "new biology" creates *jargon* at such an alarming rate that many manuscripts are virtually inaccessible to the intelligent, well-educated lay person, a criterion of good writing valued above all by Buffon (. . .*le style c'est l'homme même*). This aids and abets a new kind of illiteracy which helps to obscure concepts and principles previously accessible to a large readership steeped in the classical idiom of Western biology and medicine. This point is also made by John Maddox in his peroration upon retirement as editor of *Nature* (1995). He goes so far as to state that this trend at obfuscation may be deliberate:

> The English language is another casualty. It used to seem that *Nature's* contributors wrote clearly, but no longer. . . . The obscurity of

the literature is now so marked that one can only believe it to be deliberate. Do people hide their meaning from insecurity, for fear of being found out or, in the belief that what they have to say is important, to hide the meaning from other people? (Maddox 1995, 522)

I am *not* objecting to the creation of new technical terms (if correctly and unambiguously defined in non-circular, non-tautological manner). However, what makes communication and understanding difficult is the graceless and ill-considered use of these terms as substitutes for equivalent lay terms at the expense of style, clarity and grammatical correctness. One of the side effects of becoming addicted to the use of jargon is a terrible impoverishment of vocabulary with loss of synonyms which handicaps correct description of newly observed phenomena. The incredible richness of the English language ought to be a guarantee of being able to keep up with the equal richness of biological phenomena that will need to be described in the future.

Biology is the most historical of the *Naturwissenschaften*; all living phenomena have a developmental historical and an evolutionary historical basis, the latter extending perhaps over 3.8 billion years. Any protein or nucleic acid sequence compared to that of another species is an historical document whereby the degree of similarity reflects homology by virtue of descent from a common ancestor with prototypic sequence. No discovery in this field is made *ex nihilo* but represents a consequence of prior thought, work and publication by others. As an editor and historian it is sadly obvious to me that many workers in the field seem overcome by a historical laziness or indifference in failing to quote (by name) those who came before them and who did the initial (frequently clinical) "spade work" in the field; in this way they are contributing to a devaluation of that (clinical) contribution and science and to obscuring much of classical Western pathography. In part this also reflects the loss of civility mentioned by Maddox above.

Authors are developing a "touchiness" (on bad days I am tempted to speak of paranoia) about the review process, fearing theft of intellectual property by competitors asked to function as referees of manuscripts and requesting that their paper not be sent to X, Y, or Z for that reason. A careful scrutiny of the bibliography may show that the list of potential reviewers submitted with or without my request may contain former collaborators of the authors who may have difficulties reviewing the MS without a "positive" bias. Sorting collabora-tors into co-authors or persons thanked in the acknowledgments is be-coming a nightmare, more for authors than for editors (who must nevertheless consider the reasonableness of a title page with many, at times a dozen or two of co-authors). Merited co-authorship is an ethi-cal problem, and I have yet to see a reasonable consensus on the sub-ject. The push for rapid publication to secure priority is a daily headache and additional severe responsibility for already overbur-dened reviewers. Yet, when all is said and done, I am the first to sup-port the peer-review system as the best system we have, given the fact that almost inevitably authors will also end up as reviewers and re-viewers as authors.

The *pedigree* has been a time-honored way for clinical geneticists to illustrate the relationships between individuals in a family to aid in an understanding of the segregation of alleles or haplotypes. Literally thousands have been published, almost never with the consent of ev-ery person depicted; indeed, until recently, the pedigree, unlike clini-cal photos, was considered exempt from confidentiality/release con-siderations. This is no longer so; and unless there is written release by all living adults portrayed in the pedigree in the future it will proba-bly not be possible to publish them anymore.

CONCLUSIONS

The "blessings" of the Human Genome Project are evident and too numerous to mention; at a minimum they include a vast increase in knowledge of human and comparative genome structure and organi-zation, more accurate diagnostic, carrier and prenatal tests, a deeper understanding of molecular evolution and of disease heterogeneity, clarification of the inheritance of many presently inadequately under-stood conditions, and perhaps a beginning understanding of allelic, epistatic and epigenetic sources of phenotypic variability.

But, these advances may become a decidedly mixed blessing if they lead to genetic discrimination (Culliton 1995a; 1995b), an irrational eugenicization or genetic fatalism of the population, a further weak-ening of the specialty of clinical genetics and of clinical research in medical genetics (Dickson 1995; Culliton 1995c), de-emphasis or outright denigration of phenotypic/morphological and functional analyses of pre- and postnatal phenotypes at the expense of genotype analysis, and to reductionistic/linear rather than organicist/dialectical thinking about development. It is my fervent hope that the dialogues established at this conference may avert these potentially baneful effects of the molecular revolution and restore some balance in the future development of medicine and developmental biology.

References

Published Materials

Abir-Am, Pnina. 1982. "The Discourse of Physical Power and
 Biological Knowledge in the 1930's: A Reappraisal of the
 Rockefeller Foundation's 'Policy' in Molecular Biology." *Social
 Studies of Science* 12: 341–82.

Adam, Charles and P. Tannery (eds.). 1897–1910. *Oeuvres de
 Descartes.* 12 vols. Paris: Cerf.

Adams, Mark, (ed.). 1990. *The Wellborn Science: Eugenics in
 Germany, France, Brazil, and Russia.* Oxford: Oxford
 University Press.

Ahrens, E. H. Jr. 1992. *The Crisis in Clinical Research: Overcoming
 Institutional Obstacles.* New York: Oxford University Press.

Allen, Garland. 1975. "Genetics, Eugenics, and Class Struggle."
 Genetics 79 (supplement, June): 29–45.

American Film Institute. 1989. *Catalog of Feature Films, 1911–1920.*
 Berkeley: University of California Press.

Anderson, Christopher and Peter Alhous. 1991. "Secrecy and the
 Bottom Line." *Nature* 354: 96.

Anderson, W. French. 1989. "Human Gene Therapy: Why Draw a
 Line?" *Journal of Medicine and Philosophy* 14: 681–93.

———. 1992. "The First Signs of Danger." *Human
 Gene Therapy* 3: 359–60.

Andrews, Lori et al. 1994. *Assessing Genetic Risks: Implications for
 Health and Social Policy.* Report of the Institute of Medicine;
 Washington: National Academy Press.

Angier, Natalie. 1990. "Great 15-Year Project to Decipher Genes
 Stirs Opposition." *The New York Times* 139 (5 June): B5, B12.

Angier, Natalie. 1992. "Blueprint for a Human." *The New York Times* 142 (6 October): C6.

Anon. 1915. "Jackie Swims to Increase Beauty." *Call* (24 November): 5.

Anon. 1960. "A Proposed Standardized System of Nomenclature of Human Mitotic Chromosomes." Unsigned article. *Lancet* 278: 1063–65.

Anon. 1989. "Diplomacy Please; The U.S. Human Genome Project Must Be Mounted Internationally, and Others Must Help Pay For It." Unsigned editorial. *Nature* 342: 1–2.

Anon. 1990. "Spinal Muscular Atrophies." Unsigned Editorial. *Lancet* 336: 280–81.

Aquinas, Thomas. 1984. *Quaestiones de Anima (Questions on the Soul)*. Trans. James Robb. Milwaukee: Marquette University Press.

———. 1945. *Summa Theologica*. In: *The Basic Writings of Thomas Aquinas*. Ed. A. C. Pegis. New York: Random House.

———. 1957. *Summa contra Gentiles (On the Truth of the Catholic Faith)*. Trans. Charles J. O'Neil. New York: Doubleday.

Arbib, Michael and Mary B. Hesse. 1986. *The Construction of Reality*. Cambridge: Cambridge University Press.

Arcus, Doreen and J. Kagan. 1995. "Temperament and Craniofacial Variation in the First Two Years." *Child Development* 66: 1529–40.

Ariew, Roger. 1992. "Descartes and the Tree of Knowledge." *Synthese* 92: 101–16.

Aronson, L., E. Tobach, D. Lehrman, and J. Rosenblat (eds.). 1970. *Development and Evolution of Behavior*. San Francisco: Freeman.

Aspen Institute. 1993. *Harness the Rising Sun: American Strategies for Managing Japan's Rise as a Global Power*. Lanham, MD: University Press of America.

Aspray, William. 1990. *John von Neumann and the Origins of Modern Computing*. Cambridge, MA: MIT Press.

Atkinson, A. B. 1970. "On the Measurement of Inequality." *Journal of Economic Theory* 2: 244–63.

Atlan, Henri and Moshe Koppel. 1990. "The Cellular Computer DNA: Program or Data." *Bulletin of Mathematical Biology* 52: 335–48.

Atlan, Henri et al. 1987. "The Self Creation of Meaning." *Physica Scripta* 36: 563–76.

Augustine, Leroy, Herman Branson, and Eleanore B. Carver. 1953. "A Search for Intersymbol Influence in Protein Structure." In: Quastler 1953a, 105–118.

Avery, L., C. Bargmann, and H. R. Horvitz. 1993. "The *Caenorhabditis elegans unc–31* Gene Affects Multiple Nervous System-Controlled Functions." *Genetics* 134: 455–64.

————, D. Raizen, and S. R. Lockery. 1995. "Electrophysiological Methods." In: Epstein and Shakes 1995, 251–69.

Ayala, F. J. and T. Dobzhansky (eds.). 1974. *Studies in the Philosophy of Biology: Reduction and Related Problems*. London: Macmillan.

Bacon, Francis. 1937. *Essays, Advancement of Learning, New Atlantis, and Other Pieces*. Ed. Richard Foster Jones. New York: Odyssey Press.

Bailey, J. M. and R. C. Pillard. 1991. "A Genetic Study of Male Sexual Orientation." *Archives of General Psychiatry* 48: 1089–96.

————, M. C. Neale, and Y. Agyei. 1993. "Heritable Factors Influence Sexual Orientation in Women." *Archives of General Psychiatry* 50: 217–23.

Baker, Lynn Rudder. 1995. "Need a Christian Be a Mind-Body Dualist?" *Faith and Philosophy* 12: 489–504.

Baker, R. and F. Elliston (eds.). 1984. *Philosophy and Sex*. Buffalo: Prometheus.

Baldwin, Simeon. 1899. "The Natural Right to a Natural Death." *Journal of Social Science* 37: 1–17

Baltimore, David. 1984. "The Brain of a Cell." *Science* 84: 150.

Bamford, James. 1982. *The Puzzle Palace: A Report on America's Most Secret Agency*. Boston: Houghton Mifflin Co.

Banner, Lois. 1983. *American Beauty*. New York: Knopf.

Banner, M. 1990. *The Justification of Science and the Rationality of Belief*. Oxford: Clarendon Press.

Banta, Martha. 1987. *Imaging American Women*. New York: Columbia University Press.

Barataud, Bernard. 1992. *Au Nom de nos enfants*. Paris: Le Grand Livre du Mois.

Barbour, Ian. 1990. *Religion in an Age of Science*. San Francisco: Harper and Row.

Bargmann, C. 1993. "Genetic and Cellular Analysis of Behavior in *C. elegans*." *Annual Review of Neuroscience* 16: 47–51.

Bartels, D. M., B. LeRoy, A. L. Caplan, and P. McCarthy. 1998. "Nondirectiveness in Genetic Counseling: A Survey of Practicioners." *American Journal of Medical Genetics* 72:172–79.

————, B. LeRoy, and A. L. Caplan (eds.). 1993. *Prescribing Our Future: Ethical Challenges in Genetic Counseling*. New York: Aldine de Gruyter.

Barth, Karl. 1938. *The Knowledge of God and the Service of God According to the Teaching of the Reformation*. London: Hodder and Stoughton.

Bateson, William. 1894. *Materials for the Study of Variation, Treated with Especial Regard to Discontinuity in the Origin of Species*. New York: Macmillan.

————. 1912. *Biological Fact and the Structure of Society*. Oxford: Clarendon.

Bauer K. H., E. Hanhart, J. Lange, and G. Just (eds.). 1940. *Handbuch der Erbbiologie den Menschen*. Vol. 2. Berlin: Springer.

Beardsley, Tim. 1991. "Smart Genes." *Scientific American* 271 (August): 87–95.

Beatty, John. 1991. "Genetics in the Atomic Age: The Atomic Bomb Casualty Commission, 1947–1956." In: Benson, Rainger, and Maienschein 1991, 284–324.

————. 1993. "Scientific Collaboration, Internationalism, and Diplomacy: The Case of the Atomic Bomb Casualty Commission," *Journal of the History of Biology* 26: 205–31.

Bechtel, W. 1992. "Emergent Phenomena and Complex Systems." In: Beckermann, Flohr, and Kim 1992a, 257–88.

————, and A. Abrahamsen. 1991. *Connectionism and the Mind: An Introduction to Parallel Processing in Networks*. Cambridge, MA: Blackwells.

————, and R. Richardson. 1993. *Discovering Complexity: Decomposition and Localization As Strategies in Scientific Research*. Princeton: Princeton University Press.

Beckermann, Ansgar. 1992b. "Supervenience, Emergence, and Reduction." In: Beckermann, Flohr, and Kim 1992a, 94–118.

Beckerman, Asgar, H. Flohr, and J. Kim (eds.). 1992a. *Emergence or Reduction?: Essays on the Prospects of Nonreductive Physicalism.* Berlin; New York: W. de Gruyter.

Bell, A. P., and M. S. Weinberg. 1978. *Homosexualities: A Study of Diversity Among Men and Women.* New York: Simon and Schuster.

Belmaker, Robert. 1994. "Genetic Markers, Temperament, and Psychopathology." *Biological Psychiatry* 36: 71–72.

Bem, Daryl J. 1996. "Exotic Becomes Erotic: A Developmental Theory of Sexual Orientation." *Psychological Review* 103: 320–35.

Benda, C. 1969. *Down's Syndrome, Mongolism and its Management.* New York: Grune and Stratton.

Benjamin, J., L. Li, C. Patterson, B. Greenberg, D. Murphy, and D. Hamer. 1996. "Population and Familial Association Between the D4 Dopamine Receptor Gene and Measures of Novelty Seeking." *Nature* 12: 81–84.

Bennett, Charles H. 1986. "On the Nature and Origin of Complexity in Discrete, Homogeneous, Locally-Interacting Systems." *Foundations of Physics* 16: 585–92.

Benson, K., R. Rainger, and J. Maienschein (eds.). 1991. *The Expansion of American Biology.* New Brunswick, N J: Rutgers University Press.

Berkner, Lloyd V. 1950. *Science and Foreign Relations: International Flow of Scientific and Technological Information.* Washington, D.C.: Department of State, Division of Publications.

Bernard, Claude. 1957 [1865]. *Introduction to the Study of Experimental Medicine.* Trans. H. C. Green. New York: Dover
———. 1867. *Rapport sur les progrès et la marche de la physiologie générale en France.* Paris: Imprimerie Impériale.

Bernardin, Joseph Cardinal. 1988. *The Consistent Ethic of Life.* Kansas City: Sheed and Ward.

Berntson, G, J. T. Cacioppo, and K. S. Quigley. 1991. "Autonomic Determinism: The Modes of Autonomic Control, The Doctrine of Autonomic Space, and the Laws of Autonomic Constraint." *Psychological Review* 98: 459–87.

Bertalanffy, Ludwig von. 1927. "Eine mnemonische Lebenstheorie als Mittelweg zwischen Mechanismus und Vitalismus." *Biologia Generalis* 3: 405–10.

Bertalanffy, Ludwig von.. 1928. "L'Etat actuel du problème de l'évolution." *Scientia* 46: 37–47; 79–88.

Beurton, Peter. 1995. "Genes: A Unified View From the Perspective of Population Genetics." In: Buerton, Lefevre, and Rheinberger 1995, 15–20.

———, W. Lefevre, and H. J. Rheinberger (eds.). 1995. *Gene Concepts and Evolution*. Berlin: Max-Planck Institut für Wissenschaftsgeschichte, Preprint No. 18.

Biederman, Joseph. 1995. "High Risk for Attention Deficit Hyperactivity Disorder Among Children of Parents With Childhood Onset of the Disorder: A Pilot Study." *American Journal of Psychiatry* 152: 431–35.

Birken, Lawrence. 1989. *Consuming Desire: Sexual Science and the Emergence of a Culture of Abundance 1871–1914*. Ithaca: Cornell University Press, 1989.

Blumenberg, Hans. 1986. *Die Lesbarkeit der Welt*. Frankfurt am Main: Suhrkamp.

Bobrow, M. 1995. "Redrafted Chinese Law Remains Eugenic." *Journal of Medical Genetics* 32: 409.

Bock, G. R. and J. A. Goode (eds.). 1996. *Genetics of Criminal and Antisocial Behavior*. Ciba Foundation Symposium 194. Chichester and New York: John Wiley.

Bogdan, Robert. 1988. *Freak Show*. Chicago: University of Chicago Press.

Bowers. J. Z. and E. F. Purcell (eds.). 1976. *Advances in American Medicine: Essays at the Bicentennial*. Vol. 1. New York: Josiah Macy Foundation.

Bowler, Peter J. 1983. *The Eclipse of Darwinism: Anti-Darwinian Evolution Theories in the Decade Around 1900*. Baltimore: Johns Hopkins University Press.

Brahe, C. et al. 1993. "Presymptomatic Diagnosis of SMA III by Genotype Analysis." *American Journal of Human Genetics* 45: 409–11.

Brandt, Allan. 1987. *No Magic Bullet*. New York: Oxford University Press.

Branson, Herman R. 1953. "Information Theory and the Structure of Proteins." In: Quastler 1953a, 84–104.

Braun, Marta. 1992. *Picturing Time: The Work of Etienne-Jules Maret*. Chicago: University of Chicago Press.

Brenner, Sydney. 1957. "On the Impossibility of All Overlapping
    Triplet Codes in Information Transfer from Nucleic Acid to
    Proteins." *Proceedings of the National Academy of Sciences* 43:
    687–94.
Brenner, Sydney. 1988. "Foreword." In: Wood 1988, ix–xiii.
Brinster R. L. and J. W. Zimmermann. 1994. "Spermatogenesis
    Following Male Germ-cell Transplantation." *Proceedings of the
    National Academy of Sciences* 91 (24): 298–302.
Brock, D. 1992. "The Human Genome Project and Human Identity."
    Houston Law Review 29: 8–22.
Brown, R. S. and K. Marshall (eds.). 1993. *Advances in Genetic
    Information: A Guide for State Policy Makers.* 2nd ed.
    Lexington, KY: Council of State Governments.
Browne, Malcolm. 1994. "What Is Intelligence, and Who Has It?" *New
    York Times Book Review* (16 October): 3, 41, 45.
Brunner, H. G. 1996. "MAOA Deficiency and Abnormal Behavior:
    Perspectives on an Association." In: Bock and Goode 1996,
    155–67.
——, M. Nelen, P. van Zandvoort, et al. 1993a. "X-linked
    Borderline Mental Retardation With Prominent Behavioral
    Disturbance: Phenotype, Genetic Localization, and Evidence for
    Disturbed Monoamine Metabolism." *American Journal of
    Human Genetics* 52: 1032.
——, M. Nelen, X. Breakefield, H. Ropers, and B. van Oost. 1993b.
    "Abnormal Behavior Associated with a Point Mutation in the
    Structural Gene for Monoamine Oxidase A." *Science* 262: 578–
    80.
Brzustowicz, L. M. et al. 1990. "Genetic Mapping of Chronic
    Childhood-Onset Spinal Muscular Atrophy to Chromosome
    5q11.2–13.3." *Nature* 344: 540–41.
Burgio, G. R. 1996. "Diathesis and Predispositions: the Evolution of a
    Concept." *European Journal of Pediatrics* 155: 163–64.
Buchler, Justus, (ed.). 1955. *Philosophical Writings of Peirce.* New
    York: Dover.
Burian, R. M. 1985. "On Conceptual Change in Biology: the Case of
    the Gene." In: Depew and Weber 1985, 21–42.
——. 1995. "Too Many Kinds of Genes? Some Problems Posed by
    Discontinuities in Gene Concepts and the Continuity of Gene

Materials." In: Beurton, Lefevre, and Rheinberger 1995, 43–51.

Burian, R. M. 1996. "Underappreciated Pathways Toward Molecular Genetics as Illustrated by Jean Brachet's Cytological Embryology." In: Sarkar 1996b, 67–85.

———, J. Gayon, and D. Zallen. 1988. "The Singular Fate of French Genetics in the History of French Biology." *Journal of the History of Biology* 21: 357–402.

——— and J. Gayon. 1990. "Genetics after World War II: The Laboratories at Gif." *Cahiers pour l'histoire du CNRS* 7: 25–48.

——— and Jean Gayon. 2000. *See* Unpublished.

Burr, C. 1995. "The Destiny of You." *The Advocate* 697 (Dec. 26): 36–42.

Byne, W. 1994. "The Biological Evidence Challenged." *Scientific American* 270: 50–55.

Bynum. W. F. and R. Porter (eds.). 1985. *William Hunter and the Eighteenth Century Medical World.* Cambridge: Cambridge University Press.

———. 1993. *Encyclopaedia of the History of Medicine.* London: Routledge.

Callen, David F. 1992. "The Human Genome Project—Australian Science Must Be Involved!" *Search* 23 (9): 264–66.

Campbell, Donald T. 1974. "'Downward Causation' in Hierarchically Organised Systems." In: Ayala and Dobzhansky 1974, 179–86.

———. 1976. "On Conflicts Between Biological and Social Evolution and Between Psychology and Moral Tradition." *Zygon* 11: 176–208.

Caneva, Kenneth. 1993. *Robert Mayer and the Conservation of Energy.* Princeton: Princeton University Press.

Canguilhem, Georges. 1994. *A Vital Rationalist: Selected Writings From Georges Canguilhem.* Ed. François Delaporte; Trans. Arthur Goldhammer. New York: Zone.

Cantor, Charles. 1992, "The Challenges to Technology and Informatics." In: Kevles and Hood 1992, 98–111.

Caplan, Arthur L. (ed.). 1992. *When Medicine Went Mad: Bioethics and the Holocaust.* Totowa, NJ: Humana Press.

———. 1993a. "The Concepts of Health, Illness and Disease." In: Bynum and Porter 1993, 233–48.

Caplan, Arthur L. 1993b. "Neutrality is Not Morality: the Ethics of Genetic Counseling." In: Bartels, LeRoy, and Caplan 1993, 149–65.

———. 1994. "The Relevance of the Holocaust to Current Biomedical Issues." In: Michalczyk 1994, 3–13.

———. 1995. *Moral Matters: Ethical Issues in Medicine and the Life Sciences.* New York: John Wiley & Sons.

———. 1998a. *Am I My Brother's Keeper?: The Ethical Frontiers of Biomedicine.* Bloomington: Indiana University Press.

———. 1998b. *Due Consideration: Controversy in the Age of Medical Miracles.* New York: John Wiley and Sons.

Carey, John. 1992. "This Genetic Map Will Lead to a Pot of Gold." *Business Week* 3254 (2 March): 74.

Carlson, Elof A. 1991. "Defining the Gene: an Evolving Concept." *American Journal of Human Genetics* 49: 475–87.

———. 1981. *Genes, Radiation, and Society: The Life and Work of H. J. Muller.* Ithaca: Cornell University Press.

Carol, Anne. 1995. *Histoire de l'eugénisme en France: Les médecins et la procréation.* Paris: Seuil.

Carter, C. O. et al. 1960. "Chromosome Translocation as a Cause of Familial Mongolism." *Lancet* 2: 678–80.

Carter, Richard B. 1983. *Descartes' Medical Philosophy: the Organic Solution to the Mind-Body Problem.* Baltimore: Johns Hopkins University Press.

Chadwick, Ruth. 1993. "What Counts as Success in Genetic Counseling?" *Journal of Medical Ethics* 19: 43–46.

Chaitin, Gregory J. 1979. "Algorithmic Information Theory." *I.B.M. Journal of Research Development* 21: 350–59.

Chalfie, M. and J. White. 1988. "The Nervous System." In: Wood 1988, 337–91.

Child, C. M. 1915. *Senescence and Rejuvenation.* Chicago: University of Chicago Press.

Churchland, Paul M. 1984. *Matter and Consciousness.* Cambridge, MA: MIT Press.

———. 1989. *A Neurocomputational Perspective: The Nature of Mind and the Structure of Science.* Cambridge, MA: MIT Press.

——— and P. Churchland. 1986. *Neurophysiology.* Cambridge, MA: MIT Press.

——— and T. J. Sejnowski. 1988. "Perspectives on Cognitive Neuroscience." *Science* 242: 741–45.

Cimino, Guido and François Duchesneau (eds.). 1997. *Vitalism from Haller to the Cell Theory*. Bibliotheca di Physis No. 5. Florence: Olschki.

Clarke, C. M. et al. 1961. "21 Trisomy/Normal Mosaicism in an Intelligent Child With Some Mongoloid Characters." *Lancet* 1: 1028–30.

Cloninger, C. Robert, et al. 1997. "Integrative Psychobiological Approach to Psychiatric Assessment and Treatment." *Psychiatry Interpersonal and Biological Process* 60: 120–41.

―――. 1996. "A General Quantitative Theory of Personality Development, Fundamentals of a Self-Organizing Psychobiological Complex." *Development and Psychopathology* 8: 247–72.

―――, R. Adolfsson, and N. M. Svrakic. 1996. "Mapping Genes for Human Personality." *Nature Genetics* 12: 3–4.

Cockburn, David (ed.). 1991. *Human Beings*. Cambridge: Cambridge University Press.

Cohen, Daniel. 1993. *Les Gènes de l'espoir. A la découverte du génome humain*. Paris: Robert Laffont.

Cohen, Jon. 1997. "Corn Genome Pops Out of the Pack." *Science* 276: 1960–62.

Cohen, R. S. and T. Schnelle (eds.). 1986. *Cognition and Fact: Materials on Ludwik Fleck*. Boston: D. Reidel.

Cole, Michael and Sheila Cole. 1993. *The Development of Children*. 2nd ed. San Francisco: Freeman.

Coleman, W. and F. L. Holmes (eds.). 1988. *The Investigative Enterprise: Experimental Physiology in 19th C. Medicine*. Berkeley: University of California Press.

Coles, Peter. 1988. "The Pros and Cons of Freedom of Access to Human Genome Data." *Nature* 333 (6175): 692.

Collins, Harry. 1985. *Changing Order*. London: Sage.

Connelly, Robert. 1986. *The Motion Picture Guide: Silent Film*. Chicago: Cinebooks Inc., 1986.

Cook-Deegan, Robert M. 1988. "Testimony on Human Genome Project." In: U.S. Congress Office of Technology Assessment 1988.

―――. 1989. "The Alta Summit, December 1984." *Genomics* 5 (1989): 661–63.

Cook-Deegan, Robert M. 1994. *The Gene Wars: Science, Politics and the Human Genome*. New York: Norton.

Coutagne, G. 1903. "Recherches expérimentales sur l'hérédité chez les vers à soie." *Bulletin scientifique de la France et de la Belgique* 37:1–194.

Couvares, Francis. 1992. "Hollywood, Main Street, and the Church: Trying to Censor the Movies Before the Production Code." *American Quarterly* 44 (December): 584–615.

Cowan, W. M. (ed.). 1981. *Studies in Developmental Neurobiology*. New York: Oxford University Press.

Crabbe, J., J. Belknap, and K. Buck. 1994. "Genetic Animal Models of Alcohol and Drug Abuse." *Science* 264: 1715–23.

Craig, Gordon. 1986. "The German Mystery Case." *New York Review of Books* (30 January): 20–23.

Cranefield, Paul. 1957. "The Organic Physics of 1847 and the Biophysics of Today." *Journal of the History of Medicine and Allied Science* 12: 407-23.

Crick, Francis. 1957. "Discussion note." In: Crook 1957, 25–26.

———. 1959. "The Present Position of the Coding Problem." *Brookhaven National Laboratory Symposia* June: 35–39.

———. 1966. *Of Molecules and Men*. Seattle: University of Washington Press.

———. 1988. *What Mad Pursuit: A Personal View of Scientific Discovery*. New York: Basic Books.

———, L. Orgel, and J. Griffith. 1957. "Codes Without Commas." *Proceedings of the National Academy of Sciences* 43: 416–21.

Crocker, L. 1979. "Meddling with the Sexual Orientation of Children." In: O'Neil 1979, 145–54.

Cronin, Helena. 1992. *The Ant and the Peacock: Altruism and Sexual Selection from Darwin to Today*. New York: Cambridge University Press.

Crook, E. M. (ed.). 1957. *The Structure of Nucleic Acids and Their Role in Protein Synthesis*. Biochemical Society Symposium, 14 (18 February 1956). London: Cambridge University Press.

Crowe, L. 1985. "Alcohol and Heredity: Theories About the Effects of Alcohol Use on Offspring." *Social Biology* 32: 146–61

Culliton, B. 1995a. "Editorial: Genes and Discrimination." *Nature Medicine* 1: 385.

———. 1995b. " Politics and Genes." *Nature Medicine* 1: 181.

Culliton, B. 1995c. "Clinical Investigation: An Endangered Science." *Nature Medicine* 1: 281.

Culotta, Charles. 1968. *See Dissertations*

———.1972. *Respiration and the Lavoisier Tradition: Theory and Modification*. Philadelphia: American Philosophical Society.

Cunningham, A. and P. Williams (eds.). 1992. *The Laboratory Revolution in Medicine*. Cambridge: Cambridge University Press.

Cunningham, Susan. 1997. "Two Faces of Intentionality." *Philosophy of Science* 64: 445–60.

Dagognet, François. 1992. *Etienne-Jules Maret*. Trans. R. Galeta. New York: Zone.

Dancoff, S. and H. Quastler. 1953. "Information Content and Error Rate in Living Things." In: Quastler 1953a, 263–73.

Daniels, R. J. et al. 1992. "Prenatal Prediction of Spinal Muscular Atrophy." *Journal of Medical Genetics* 29: 165–70.

Danks, D. M. 1994. "Germ-line Gene Therapy: No Place in Treatment of Genetic Disease." *Human Gene Therapy* 5 (2): 151–52

Darrow, Clarence. 1926. "The Eugenics Cult." *American Mercury* 8 (June): 129–37.

Davenport, Charles. 1911. *Heredity in Relation to Eugenics*. New York: Henry Holt.

Davidson, Donald. 1970. "Mental Events." In: Foster and Swanson 1970, 79–101.

Davis, M. 1992. "The Role of the Amygdala in Conditioned Fear." In: J. P. Aggleton (ed.). *The Amygdala*. New York: Wiley.

Davis, Bob. 1990. "Watson Doesn't Use Gentle Persuasion to Enlist Japanese and German Support for Genome Effort." *Wall Street Journal* (18 June): A12.

Dawkins, Richard. 1976. *The Selfish Gene*. Oxford: Oxford University Press.

De Grazia, Edward and Roger K. Newman. 1982. *Banned Films: Movies, Censors and the First Amendment*. New York: R. R. Bowker.

De Grouchy, J. and M. Lamy. 1961. "Etude des chromosomes humains à partir de leucocytes sanguins." *Revue française d'études cliniques et biologiques* 6: 825–27.

Deacon, Terrence. 1997. *The Symbolic Species: The Co-evolution of Language and the Brain*. New York: Norton.

Debré, R. 1974. *L'Honneur de vivre*. Paris: Stock.

Debré, R.. 1976. "Maurice Lamy." *Bulletin de l'académie de médecine* 160: 111–25.

———, M. Lamy, G. Sée, and Mme St. Schramek. 1936. "La Maladie hémolytique familiale. Etude de vingt-cinq cas personnels." *Société médicale des hôpitaux de Paris* (15 May): 797–809.

Degler, Carl. 1991. *In Search of Human Nature*. New York: Oxford University Press.

Delbrück, M. and G. Stent. 1957. "On Mechanisms of DNA Replication." In: McElroy and Glass 1957, 699–736.

Delehanty, J., R. L. White, and M. L. Mendelsohn. 1986. "Approaches to Determining Mutation Rates in Human DNA." *Mutation Research* 167: 215–32.

DeLisi, Charles. 1988. "The Human Genome Project." *American Scientist* 76: 488–92.

Denzinger, Heinrich and Adolf Schönmetzer. 1962. *Enchiridion symbolorum definitionum et declarationum de rebus fidei et morum*. Friburg: Herder.

Depew, David and Bruce Weber (eds.). 1985. *Evolution at a Crossroads: the New Biology and the New Philosophy of Science*. Cambridge, MA: MIT Press.

Derrida, Jacques. 1976. *Of Grammatology*. Trans. Gayatri Chakravorty Spivak. Baltimore: Johns Hopkins University Press.

Descartes, R. [1641] 1984. *The Philosophical Writings of Descartes*. Trans. J. Cottingham, R. Stoothoff, and D. Murdoch. 2 vols. Cambridge: Cambridge University Press.

———. [1647] 1983. *Principles of Philosophy*. Trans. V. R. and R. P. Miller. Dordrecht: Reidel.

———. [1664] 1972. *Treatise of Man*. Translated and edited by T. S. Hall. Cambridge, MA: Harvard University Press.

DeVries, Hugo. [1889] 1910. *Intracellular Pangenesis*. Trans. C. Stuart Gager. Chicago: Open Court.

Dewey, Mrs. Melvil. 1914. "College Courses in Euthenics." *National Conference on Race Betterment Proceedings* 1: 349.

Dickman, Steven. 1988. "West Germany Voices Objections to European Genome Project." *Nature* 336: 416.

Dickson, David. 1989. "Watson Floats a Plan to Carve Up the Genome." *Science* 244: 521–22.

———. 1995a. "Britain Must Tackle 'Genetic Illiteracy' Among New Doctors: News." *Nature* 377: 466.

Dickson, David. 1995b. "UK Panel Warns of Crisis in Clinical Research Careers." *Nature* 375: 619.

DiLalla, L. et al. 1994. "Genetic Etiology of Behavioral Inhibition Among 2 Year-Old Children." *Infant Behavior and Development.* 17: 405–12.

Dobzhansky, Theodosius. 1962. *Mankind Evolving: the Evolution of the Human Species.* New Haven: Yale University Press.

Dörner, G. 1976. *Hormones and Brain Differentiation.* Amsterdam: Elsevier.

DSM–IV. 1994. (= *Diagnostic and Statistical Manual of Mental Disorders*). 4th ed. Washington, D.C.: American Psychiatric Association.

Dubos, René and Jean Dubos. 1953. *The White Plague: Tuberculosis, Man and Society.* London: Victor Gollancz.

Dubowitz, V. 1964. "Infantile Muscular Atrophy: A Prospective Study With Particular Reference to a Slowly Progressive Variety." *Brain* 87: 707–18.

Duchesneau, François. 1982. *La Physiologie des lumières: empirisme, modèles et théories.* The Hague: Martinus Nijhoff.

———. 1985. "Vitalism in Late Eighteenth-Century Physiology: The Cases of Barthez, Blumenbach and John Hunter." In: Bynum and Porter 1985, 259–95.

Dudley, William. 1990. *Genetic Engineering: Opposing Viewpoints.* San Diego: Greenhaven Press.

Duster, Troy. 1990. *Backdoor to Eugenics.* New York: Routledge.

Dworkin, Ronald. 1993. *Life's Dominion.* New York: Knopf.

Eaves, L. J. and L. M. Gross 1990. "Theological Reflections on the Cultural Impact of Human Genetics." *Insights* (Chicago Center for Religion and Science) 2 (2): 15–18.

Ebstein, R., O. Novick, Umansky, et al. 1996. "Dopamine D4 Receptor *DRD4.* Exon III Polymorphism Associated With the Human Personality Trait of Novelty Seeking." *Nature* 12: 78–80.

Eccles, John. 1993. *Rethinking Neural Networks: Quantum Fields and Biological Data.* Hillsdale, NJ: Erlbaum.

Edelman, Gerald M. 1988. *Topobiology: An Introduction to Molecular Embryology.* New York: Basic Books.

———. 1989. *The Remembered Mind: A Biological Theory of Consciousness.* New York: Basic Books.

Edelman, Gerald M. 1992. *Bright Air, Brilliant Fire: On the Matter of Mind.* New York: Basic.

Eden, Paul. 1917. "Eugenics, Birth Control, and Socialism." In: Eden Paul (ed.). 1917. *Population and Birth Control.* New York: Critic & Guide.

Edge, David. 1974. "Technological Metaphor and Social Control." *New Literary History* 6: 135–48.

Edwards, Paul. 1995. *The Closed World: Computers and the Politics of Discourse in Cold War America.* Cambridge MA: MIT Press.

———. (ed.). 1967. *Encyclopedia of Philosophy.* 8 vols. New York: Colliers.

Egan, S. E. and R. A. Weinberg. 1993. "The Pathway to Signal Achievement." *Nature* 365: 781–83.

Eichrodt, W. 1961. *Theology of the Old Testament.* Trans. A. S. Peake. Philadelphia: Westminster.

Elias, Sherman and George J. Annas 1987. *Reproductive Genetics and the Law.* New York: Yearbook Medical Publishers.

Elkana, Yehuda. 1974. *The Discovery of the Conservation of Energy.* Cambridge, MA: Harvard University Press.

Emde, Robert et al. 1992. "Temperament, Emotion, and Cognition at Fourteen Months: The MacArthur Longitudinal Twin Study." *Child Development* 63: 1437–55.

Engel, Sigmund. 1912. *The Elements of Child-Protection.* Trans. Paul Eden. New York: Macmillan.

Engelhardt, H. Tristam. 1975. "The Concepts of Health and Disease." In: Engelhardt, H. Tristam (ed.). 1975. *Evaluation and Explanation in the Biomedical Sciences.* Pp. 125–41. Dordrecht: D. Reidel.

———. 1984. "Persons and Humans: Refashioning Ourselves in a Better Image and Likeness." *Zygon* 19: 281–95.

Ephrussi, B., U. Leopold, J. D. Watson, and J. J. Weigle. 1953. "Terminology in Bacterial Genetics." *Nature* 171: 701.

Epstein, H. F. and D. C. Shakes (eds.). 1995. Caenorhabditis elegans: *Modern Biological Analysis of an Organism.* (Methods in Cell Biology, 48) New York: Academic Press.

Everson, William K. 1978. *American Silent Film.* New York: Oxford University Press.

Faden, Ruth R. et al. 1985. "What Participants Understand About a Maternal Serum Alpha-Fetoprotein Screening Program." *American Journal of Public Health* 75: 1381–84.

Falconer, D. S. and T. Mackay. 1996. *Introduction to Quantitative Genetics*. 4th ed. Essex, UK: Longman.

Falk, Raphael. 1986. "What Is a Gene?" *Studies in the History and Philosophy of Science* 17: 133–73.

————. 1990. "Between Beanbag Genetics and Natural Selection." *Biology and Philosophy* 5: 313–25.

Fallows, James. 1989. "Containing Japan." *The Atlantic* (May): 40–54.

Farber, Paul Lawrence. 1994. *The Temptations of Evolutionary Ethics*. Berkeley: University of California Press.

Farmer, Linda L. 1997. "Subsistence and Substantiality According to Thomas Aquinas." *Memini* 1: 133–48.

Feinstein, A. R. 1995. "Essay Review: The Crisis in Clinical Research." *Bulletin of the History of Medicine* 69: 288–91.

Ferguson-Smith, M. A. 1991. "European Approach to the Human Gene Project." *FASAB Journal* 5 (1): 61–65.

Fiedler, Leslie. 1978. *Freaks*. New York: Simon and Schuster.

Finkel, D. et al. 1996. "Cross-Sequential Analysis of Genetic Influence on Cognitive Ability in the Swedish Adoption/Twin Study of Aging." *Aging, Neuropsychology, and Cognition* 3: 84–89.

Fish, Stanley. 1980. *Is There a Text in This Class? The Authority of Interpretive Communities*. Cambridge, MA: Harvard University Press.

Fisher, Irving. 1915. "Eugenics—Foremost Plan of Human Redemption." *National Conference on Race Betterment Proceedings* 2: 63–66.

———— and E. L. Fisk. 1917. *How to Live*. 12th ed. New York: Funk & Wagnalls.

Fisher, Ronald A. 1923. "The Evolution of the Conscience in Civilized Communities in Special Relation to Sexual Vices." In: *Second International Congress of Eugenics* 1923, 313–18; and "Discussion," 464–65.

Flannery, Austin, O.P. (ed.). 1996. *Vatican II Constitutions, Decrees, Declarations*. Northport, NY: Costello Publishing Co.

Fleck, Ludwig. 1986. "Scientific Observation and Perception in General." In: Cohen and Schnelle 1986, 59–78.

Fleming, Donald. 1968. "Emigre Physicists and the Biological Revolution." *Perspectives in American History* 2: 152–89.

Fletcher, R., S. Fletcher, and E. Wagner. 1982. *Clinical Epidemiology: The Essentials*. Baltimore: Williams and Wilkins.

Fogel, T. 1990. "Are Genes the Units of Inheritance?" *Biology and Philosophy* 5: 349–71.

Foster, John. 1991. *The Immaterial Self: A Defence of the Cartesian Conception of the Self.* London: Routledge.

Foster, L. and J. W. Swanson (eds.). 1970. *Experience and Theory*. Amherst: University of Massachusetts Press.

Foucault, Michel. 1970. *The Order of Things*. Trans. Anon. New York: Vintage.

Frank, Robert L. 1987. "American Physiologists in German Laboratories, 1865–1914." In: Geison 1987, 11–46.

Frézal, J. 1959a. "La Consultation de génétique." *La Médecine infantile* (April): 21–25.

———."L'Etude des maladies héréditaires chez l'homme." *La Médecine infantile* (April): 11–19.

———. 1967. Editorial. "La Responsabilité eugénique du pédiatre." *La Presse médicale* (26 May): 75.

——— and J. Rey. 1964. "La Détection des hétérozygotes."*La Presse médicale* 72: 831–34.

Fried, Michael. 1987. *Realism, Writing, Disfiguration*. Chicago: University of Chicago Press.

Friedman, George and Meredith Lebord. 1991. *The Coming War with Japan*. New York: St. Martin's Press.

Frost, W. N. and P. S. Katz. 1996. "Single Neuron Control Over a Complex Motor Program." *Proceedings of the National Academy of Sciences* 93: 422–26.

Fuchs, Joseph, S. J. 1987. *Christian Morality: The Word Becomes Flesh*. Washington: Georgetown University Press.

Fujimura, Joan. 1988. "The Molecular Bandwagon in Cancer Research: Where Social Worlds Meet." *Social Problems* 35: 261–83.

Fulker, D. and L. Cardon. 1994. "A Sib-Pair Approach to Interval Mapping of Quantitative Trait Loci." *American Journal of Human Genetics* 54: 1092–1103.

Galas, David. J. 1990. "Testimony to Hearing Before Subcommittee on Energy Research and Development." See U.S. Senate, 11 July 1990: 18–19, 28–29.

Galilei, Galileo. [1615] 1957. "Letter to the Grand Duchess." In: S. Drake (ed.). *Discoveries and Opinions of Galileo.* New York: Doubleday.

Galison, Peter. 1994. "The Ontology of the Enemy: Norbert Wiener and the Cybernetic Vision." *Critical Inquiry* 21: 228–66.

Galton, Francis. 1875. "A Theory of Heredity." *Contemporary Review* 27: 80–95.

———. 1909. *Memories of My Life.* New York, Dutton.

Gamow, George. 1954a. "Possible Relation between Deoxyribonucleic Acid and Protein Structure." *Nature,* 173: 318.

———. 1954b. "Possible Mathematical Relations between Deoxyribonucleic Acid and Proteins." *Det Kongelige Danske Videnskabernes Selkab, Biologiske Meddelelser* 22: 1–13.

——— and N. Metropolis. 1954. "Numerology of Polypeptide Chains." *Science* 120: 779–80.

——— and M. Ycăs. 1955. "Statistical Correlation of Protein and Ribonucleic Acid Composition." *Proceedings of the National Academy of Sciences* 41: 1011–19.

———, A. Rich, and M. Ycăs. 1956. "The Problem of Information Transfer from Nucleic Acids to Proteins." *Advances in Biological and Medical Physics* 4: 41–51.

Gannon, T. and C. Rankin. 1995. "Methods of Studying Behavioral Plasticity in *Caenorhabditis elegans.*" In: Epstein and Shakes 1995, 205–33.

Gardner, D. 1993. *The Neurobiology of Neural Networks.* Cambridge, MA: MIT Press.

Garver, K. L. 1991. "Eugenics: Past, Present, and the Future." *American Journal of Human Genetics* 49 (5): 1109–18.

Gasman, Daniel. 1971. *The Scientific Origins of National Socialism.* London: MacDonald.

Gaudillière, Jean-Paul. 1992. "Biochimistes et biomédecine dans l'après-guerre. Deux itinéraires entre laboratoire et hôpital." *Sciences Sociales et Santé* 10: 107–47.

———. 1995. "How Weak Bonds Stick? Genetic Diagnosis Between the Laboratory and the Clinic." *The Genetic Engineer and Biotechnologist* 15: 99–112.

Gayon, Jean. 1991. "Un Objet singulier dans la philosophie biologique bernardienne: l'hérédité." In: Michel 1991, 170–82.

Gayon, Jean. 1992. "Animalité et végétalité dans les représentations de l'hérédité." *Revue de synthèse* (4th series) 3/4: 423–38.

————. 1995a. "De La Mesure à l'ordre; histoire philosophique du concept d'hérédité." In: Port 1995, 629–45.

————. 1995b. "From Measure to Order: A Philosophical Scheme for the History of the Concept of 'Heredity'." In: Beurton, Lefevre, and Rheinberger 1995, 53–60.

Gazzaniga, Michael S. (ed.). 1995. *The Cognitive Neurosciences.* Cambridge, MA: MIT Press.

Geison, Gerald. 1987. *Physiology in the American Context, 1850–1940.* Bethesda: American Physiological Society

Gell-Mann, Murray. 1995. "What Is Complexity?" *Complexity* 1:1.

Gelman, D., D. Foote, T. Barrett, and M. Talbot. 1992. "Born or Bred: The Origins of Homosexuality." *Newsweek* 119 (24 February): 46–53.

Genome Sequencing Consortium. 1998. "Genome Sequence of the Nematode *C. elegans*: A Platform for Investigating Biology." *Science* 282 (5396): 2012-18.

Gerhart, J. and M. Kirschner. 1997. *Cells, Embryos and Evolution.* Oxford: Blackwell.

Gershon, E. S. 1990. "Genetics [of Manic-Depressive Illness]." In: Goodwin and Jamison 1900, 373–401.

———— and C. R. Cloninger (eds.). 1994. *Genetic Approaches to Mental Disorders.* Washington, D.C.: American Psychiatric Press.

Gibson, D. et al. 1965. "Morphological and Behavioral Consequences of Chromosome Subtype in Mongolism." *American Journal of Mental Deficiencies* 69: 801–04.

Gilbert, Scott F. 1998. "Bearing Crosses: The Historiography of Genetics and Embryology." *American Journal of Medical Genetics* 76: 168–82.

————, J. M. Opitz, and R. A. Raff. 1996. "Review: Resynthesizing Evolutionary and Developmental Biology." *Developmental Biology* 173: 357–72.

Gilbert, Walter. 1992. "A Vision of the Grail." In: Kevles and Hood 1992, 83–97.

Gilliam, T. C. et al. 1990. "Genetic Homogeneity Between Acute and Chronic Forms of Spinal Muscular Atrophy." *Nature* 345: 823–25.

Gilman, Sander L. 1985. *Difference and Pathology*. Ithaca: Cornell University Press.

————. 1995. *Picturing Health and Illness*. Baltimore: Johns Hopkins University Press.

Globus, G. M. and I. Savodnik (eds.). 1976. *Consciousness and the Brain*. New York: Plenum Press.

Golomb, Solomon W. 1962. "Efficient Coding for the Desoxyribonucleic Channel," *Proceedings of Symposia in Applied Mathematics* 14: 87–100.

————, B. Gordon, and L. Welch. 1958."Comma-free Codes." *Canadian Journal of Mathematics* 10: 202–09.

————, L. Welch, and M. Delbrück. 1958b. "Construction and Properties of Comma-Free Codes." *Det Kongelige Danske Videnskabernes Selkab, Biologiske Meddelelser* 23: 1–34.

Goodfellow, P.N. 1992. "Variation is Now the Theme." *Nature* 359: 777–78.

Goodwin, F. and K. R. Jamison (eds.). 1990. *Manic-Depressive Illness*. Oxford: Oxford University Press.

Gorman, Christine. 1992. "Sizing Up the Sexes." *Time* 140 (20 January): 42–51.

————. 1993. "The Race to Map Our Genes." *Time* 141 (8 February): 57.

Gortmaker, S. L., J. Kagan, et al. 1997. "Daylength During Pregnancy and Shyness in Children: Results from Northern and Southern Hemispheres." *Developmental Psychobiology* 31: 107–14.

Gottlieb, G. 1992. *Individual Development and Evolution: The Genesis of Novel Behavior*. New York: Oxford University Press.

Gould, Stephen J. 1977. *Ontogeny and Phylogeny*. Cambridge, MA: Harvard University Press.

————. 1983. *Hen's Teeth and Horse's Toes*. New York: W. W. Norton.

————. 1989. *Wonderful Life: the Burgess Shale and the Nature of History*. London: Penguin.

————. 1993. "The Great Seal Principle." In: S. Gould. 1993. *Eight Little Piggies*. Pp. 371–81. New York: W. W. Norton.

Gray, R. "Death of the Gene: Developmental Systems Strike Back." In: Griffiths 1992, 165–209.

Greenfield, M. 1996. "Respecting the Dead." *Newsweek* (11 April): 80.

Greenspan, Patricia S. 1993. "Free Will and the Genome Project." *Philosophy and Public Affairs* 22 :31–43.

Greenspan, R. J. 1995a. "Understanding the Genetic Construction of Behavior." *Scientific American* 272: 72–78.

Greenspan, R. J. 1995b. "Flies, Genes, Learning, and Memory." *Neuron* 15: 747–50

———, E. Kandel, and T. Jessel. 1995. "Genes and Behavior." In: Kandel, Schwartz, and Jessel 1995, 555–77.

Greenwald, John. 1989. "Friend or Foe?" *Time* (24 April): 44–45.

Greenwood, J. W. 1971a. "The Scientist-Diplomat: A New Hybrid Role in Foreign Affairs." *Science Forum* 19: 15.

———. 1971b. "The Science Attaché: Who He Is and What He Does." *Science Forum* 20: 21–25.

Grene, Marjorie. 1967. "Biology and the Problem of Levels of Reality." *New Scholasticism* 41: 427–29.

———. 1971. "Reducibility: Another Side Issue." In: M. Grene. 1971. *Interpretations of Life and Mind.* Pp. 14–37. New York: Humanities Press.

———. 1988. "Hierarchies and Behavior." In: Tobach and Greenberg 1987, 3–17.

Gregory, Frederick. 1977 *Scientific Materialism in Nineteenth-Century Germany.* Dordrecht: Reidel.

Griesemer, James R. and William C. Wimsatt. 1989. "Picturing Weismannism: A Case Study of Conceptual Evolution." In: Ruse 1989, 75–137.

Griffin, James. 1985. *Well-Being.* Oxford: Oxford University Press.

Griffiths, Paul (ed.). 1992. *Trees of Life: Essays in Philosophy of Biology.* Boston: Kluwer.

Grob, G. 1994. *The Mad Among Us: A History of the Care of America's Mentally Ill.* New York: Free Press.

Gros, F. 1986. *Les Secrets du gène.* Paris: Odile Jacob.

Guéroult, Martial. [1952] 1985. *Descartes' Philosophy Interpreted According to the Order of Reasons.* 2 vols. Trans. Roger Ariew. Minnesota: University of Minnesota Press.

Gustafson, James M. 1974. *Theology and Christian Ethics.* Philadelphia: Pilgrim Press.

———. 1984. *Ethics From a Theocentric Perspective.* Vol. 2: *Ethics and Theology.* Chicago: University of Chicago Press.

Gustafson, James M. 1993. "Where Theologians and Genetics Meet." *Center for Theology and Natural Science Bulletin* 13 : 4.

Guyer, Michael. 1927. *Being Well-Born: An Introduction to Heredity and Eugenics*. Indianapolis: Bobbs-Merrill.

Hacking, Ian. 1982. *Representing and Intervening*. Cambridge: Cambridge University Press.

Haeckel, Ernst. 1876. *Die Perigenesis des Plastidule: oder Die Wellenzeugnung der Lebenstheilchen*. Berlin: Reimer.

———. 1905. *The Wonders of Life*. Trans. Joseph McCabe. New York: Harper & Brothers.

———. [1868] 1876. *The History of Creation*. New York: D. Appleton.

Hagner, M., H. J. Rheinberger, and B. Wharig-Schmidt (eds.). 1994. *Objekte, Differenzen und Konjunkturen: Experimentalsysteme im historischen Kontext*. Berlin: Akademie Verlag.

Haiselden, Harry. 1915a. "Baby Bollinger." *Chicago American* (23 November through 10 December 1915).

———. 1915b. "Baby Bollinger." *Boston American* (10 December): 16.

Haldane, J. B. S. 1932. *The Inequality of Man*. London: Chatto and Windus.

———. [1928] 1932. "Origin of Life." Reprinted in: Haldane 1932.

Haldane, John Scott. 1931. *The Philosophical Basis of Biology*. New York: Doubleday, Doran and Co.

Hall, B. 1962. "Down's Syndrome With Normal Chromosomes." *Lancet* 2: 1026–27.

Hall, G. Stanley. 1910. "What is to Become of Your Baby?" *Cosmopolitan* 47 (April): 661–68.

Hall, J. A.Y. and D. Kimura. 1994. "Dermatoglyphic Asymmetry and Sexual Orientation in Men. " *Behavioral Neuroscience* 108: 1203–06.

Hall, J. C. 1994. "The Mating of a Fly." *Science* 164: 1702–14.

Haller, Mark H. 1963. *Eugenics: Hereditarian Attitudes in American Thought*. New Brunswick, N.J.: Rutgers University Press.

Hamer, Dean H., S. Hu, V. Magnuson, N. Hu, and A. M. L. Pattatucci. 1993a. "A Linkage Between DNA Markers on the X Chromosome and Male Sexual Orientation." *Science* 261: 321–27.

———. 1993b. "Response." *Science* 261: 1259.

Hamer, Dean H., and P. Copeland. 1994. *The Science of Desire: The Search for the Gay Gene and the Biology of Behavior.* New York: Simon and Schuster.

Hamilton, Alice. 1915. "The Bollinger Case." *Survey* 35 (4 December): 265–66.

Haraway, Donna. 1979. "The Biological Enterprise: Sex, Mind, and Profit from Human Engineering to Sociobiology." *Radical History Review* 29: 206–37.

Harding, Sandra (ed.). 1993. *The Racial Economy of Science.* Bloomington, IN: Indiana University Press.

Harris, John. 1993. *Wonderwoman and Superman.* Oxford University Press.

Harwood, Jonathan. 1993. *Styles of Scientific Thought: The German Genetics Community, 1900-1933.* Chicago: University of Chicago Press.

Haseman, J. and R. Elston. 1972. "The Investigation of Linkage Between a Quantitative Trait and a Marker Locus." *Behavior Genetics* 2: 3–19.

Haskell, Thomas. 1977. *The Emergence of Professional Social Science.* Urbana: University of Illinois Press.

Hasker, William. 1974. "The Souls of Beasts and Men." *Religious Studies* 10: 265–77.

Hay, H.A. 1985. *Essentials of Behaviour Genetics.* Oxford: Blackwells.

Heims, Steven J. 1980. *John von Neumann and Norbert Wiener: From Mathematics to the Technologies of Life and Death.* Cambridge, MA: MIT Press.

Heldke, Lisa M. and Stephen H. Kellert. 1995. "Objectivity as Responsibility," *Metaphilosophy* 26 (4): 360–78.

Heller, Jan. 1996. *Human Genome Research, and the Challenge of Contingent Future Persons.* Omaha: Creighton University Press.

Hempel, Carl G. and Paul Oppenheim. [1948] 1965. "Studies in the Logic of Explanation." In: Carl G. Hempel (ed.). 1965. *Aspects of Scientific Explanation.* New York: Free Press.

Henshaw, Paul S. and Austin Brues. 1947. *General Report, Atomic Bomb Casualty Commission, January 1947.* Washington, D.C.: National Research Council.

Hering, Ewald. [1870] 1920. "Memory as Universal Function of Organized Matter." In: *Unconscious Memory.* Translated and edited by S. Butler. Pp. 63–86. London: A. C. Fifield.

Herrnstein, Richard J. and Charles Murray. 1994. *The Bell Curve: Intelligence and Class Structure in American Life.* New York: Free Press.

Hesse, Mary. 1985. "Reductionism in the Sciences: Some Reflections on Part I." In: Peacocke 1985, 105–13.

Himes, Michael J. and Stephen J. Pope (eds.). 1996. *Finding God in All Things*. New York: Crossroad, 1996.

Hinsley, F. H. and Alan Stripp (eds.). 1993. *Codebreakers: The Inside Story of Bletchley Park*. New York: Oxford University Press.

Holden, C. 1995. "More on Genes and Homosexuality." *Science* 268: 1571.

Hollander, Russell. 1989. "Euthanasia and Mental Retardation." *Mental Retardation* 27 (April): 53–61.

Holliday, R. 1994. "Epigenetics: An Overview." *Developmental Genetics* 15: 453–57.

Holmes, Frederic L. 1985. *Lavoisier and the Chemistry of Life: An Exploration of Scientific Creativity*. Madison: University of Wisconsin Press.

Holmes, S. J. 1915. "Misconceptions of Eugenics." *Atlantic* 115 (February): 222–27.

Holtzman, Neil A. 1989. *Proceed With Caution*. Baltimore: Johns Hopkins University Press.

Hood, L. 1992. "Biology and Medicine in the Twenty-First Century." In: Kevles and Hood 1992, 136–63.

Hu, S., A. Pattatucci, C. Patterson, et al. 1995. "Linkage Between Sexual Orientation and Chromosome Xq28 in Males But Not in Females." *Nature Genetics* 11: 248–56.

Hubbard, Ruth. 1995. "Genomania and Health." *American Scientist* 83: 8–10.

———, and Elijah Wald. 1993. *Exploding the Gene Myth*. Boston: Beacon.

———, and Elijah Wald, with Nicholas Hildyard. 1993. "The Eugenics of Normalcy: The Politics of Gene Research." *The Ecologist* 23 (5): 185–91.

Humphreys, Paul. 1997a. "How Properties Emerge." *Philosophy of Science* 64: 1–17.

———. 1997b. "Emergence, not Supervenience." *PSA 1996*, Part 2. 337–45.

Huntington, Ellsworth. 1935. *Tomorrow's Children*. New York: John Wiley.

Huntington, Samuel P. 1991. "America's Changing Strategic Interests." *Survival* 33: 3–17.

Huysteen, W. van 1989. *Theology and the Justification of Faith.* Grand
    Rapids, MI: Eerdmans.
Hyman, Steven E. and E. J. Nestler. 1996. "Initiation and Adaptation:
    A Paradigm for Understanding Psychotropic Drug Action."
    American Journal of Psychiatry 153: 151–62.
Illinois State Charities Commission. 1914. *Fourth Annual Report,
    Institution Quarterly* 5 (June).
Inman, B. R. and Daniel F. Burton. 1990. "Technology and
    Competitiveness: The New Policy Frontier." *Foreign Affairs* 69
    (Spring): 116–34.
International SMA Collaboration. 1991. "Workshop Report."
    Neuromuscular Disorders 1: 81.
Jacob, François. 1974a. "Le Modèle linguistique en biologie." *Critique*
    322: 197–205.
———.1974b. *The Logic of Life.* Trans. Betty E. Spillmann. New
    York: Pantheon.
Jacobus, Mary. 1982. "Is There a Woman in This Text?" *New Literary
    History* 14: 117–42.
Jaggar, A. M. and K. J. Struhl. 1995. "Human Nature." In: Reich
    1995, 1172–73.
Janssens, Louis. 1980. "Artifical Insemination: Ethical Consider-
    ations." *Louvain Studies* 8: 3–29.
Jasny, B. 1995. "The Genome Maps 1995." *Science* 270: 415.
Jayne, E. T. 1959. "Note on Unique Decipherability." *IRE
    Transactions on Information Theory, IT–5*: 98–102.
Jeeves, M. A. 1991. "Minds and Brains: Then and Now."
    Interdisciplinary Reviews 16: 69–81.
Jenkins, Trefor. 1992. "The Human Genome Project—Does South
    Africa Have a Role to Play in It?" *South African Medical
    Journal* 80: 536–38.
Johannsen, W. 1911. "The Genotype Conception of Heredity."
    American Naturalist 45: 129–59.
Johnsson, E. G. et al. 1997. "Lack of Evidence for the Allelic
    Association Between Personality Traits and the Dopamine D-
    sub-4 Receptor Gene Polymorphisms." *American Journal of
    Psychiatry* 154: 697–99.
John Paul II, Pope. 1981. "The Path of Scientific Discovery." *Origins*
    11: 178–80.

John Paul II, Pope. 1990. "Letter to Rev. George Coyne, S. J." In:
Russell, Stoeger and Coyne 1990, M1–M14.

———. 1992. "Lessons of the Galileo Case."*Origins* 22: 370–74.

———. 1995. *Evangelium Vitae. Origins* 24: 71–79.

———. 1996. "Message to the Pontifical Academy of Sciences on
Evolution." *Origins* 26: 350–52.

Johnston, T. 1987. "The Persistence of Dichotomies in the Study of
Behavioral Development." *Developmental Review* 7: 149–82.

———. 1988. "Developmental Explanation and Ontogeny of
Birdsong: Nature/Nurture Redux." *Behavioral and Brain
Science* 11: 617–63.

Jonsen A. R., S. J. Durfy, W. Burke, and Arno G. Motulsky. 1996.
"The Advent of the Unpatients." *Nature Medicine* 2: 622–24.

Josso, N., J. de Grouchy, J. Frézal, M. Lamy. 1963. "Le Syndrome de
Turner familial, étude de deux familles avec caryotypes XO et
XX." *Annales de pédiatrie* (2 April): 775–78.

Jowett, Garth. 1976. *Film: The Democratic Art*. Boston: Little,
Brown.

Judson, Horace Freeland. 1992. "A History of the Science and
Technology Behind Gene Mapping and Sequencing." In: Kevles
and Hood 1992, 37–80.

———. 1996. *The Eighth Day of Creation*. Revised edition. New
York: Cold Spring Harbor Press.

Kagan, J. 1962. *Birth to Maturity: a Study in Psychological
Development*. New York: Wiley.

———. 1989. *Unstable Ideas: Temperament, Cognition, and Self*.
Cambridge, MA: Harvard University Press.

———. 1994. *Galen's Prophecy: Temperament in Human Nature*.
New York: Basic Books.

———. 1997a. "Temperament and the Reactions to Unfamiliarity."
Child Development 68: 139–43.

———. 1997b. "Conceptualizing Psychopathology: The Importance of
Developmental Profiles." *Development and Psychopathology* 9:
321–34.

Kahn, A. (ed.). 1993. *Thérapie génique—L'ADN médicament*.
Montrouge, France: Editions John Libbey Eurotext.

Kahn, David. 1967. *The Codebreakers: A Story of Secret Writing*.
New York: Macmillan Co.

Kandel Eric (ed.). 1987. *Molecular Neurobiology in Neurology and Psychiatry.* New York: Raven Press.

Kandel, Eric, J. Schwartz, and T. Jessel. 1995. *Principles of Neural Science.* New York: Elsevier.

Kant, Immanuel. [1790] 1963. *Critique of Judgement.* Trans. J. C. Meredith. Oxford: Clarendon, 1963.

Kass, Leon R. 1985. *Toward a More Natural Science: Biology and Human Affairs.* New York: Free Press.

Kater, M. 1992. "Unresolved Questions of German Medicine and Medical History in the Past and Present." *Central European History* 25: 407–23.

Kausch, K. et al. 1991. "No Evidence For Linkage of Autosomal Dominant Proximal Spinal Muscular Atrophies to Chromosome 5q Markers." *Human Genetics* 86: 317–18.

Kay, Lily E. 1985. "Conceptual Models and Analytical Tools: The Biology of Physicist Max Delbrück." *Journal of the History of Biology* 18: 207–46.

———. 1993. *The Molecular Vision of Life: Caltech, The Rockefeller Foundation, and the Rise of the New Biology.* Oxford and New York: Oxford University Press.

———. 1994. "Who Wrote the Book of Life: Information and the Transformation of Molecular Biology." In: Hagner, Rheinberger, and Wharig-Schmidt 1994, 151–79.

———. 1995. "Who Wrote the Book of Life: Information and the Transformation of Molecular Biology." *Science in Context* 8: 609–34.

———. 1996. "Life as Technology: Representing, Intervening, and Molecularizing." In: Sarkar 1996b, 87–99.

———. 1997. "Cybernetics, Information, Life: Emergence of Scriptural Representations of Heredity." *Configurations* 5: 23–91.

Keller, Evelyn Fox. 1990. "Physics and the Emergence of Molecular Biology: A History of Cognitive and Political Synergy." *Journal of the History of Biology* 23: 389–409.

———. 1992. "Nature, Nurture and the Human Genome Project." In: Kevles and Hood 1992, 264–99.

———. 1994. "Language and Science: Genetics, Embryology, and the Discourse of Gene Action." In: *Great Ideas Today.* Chicago:

Encyclopaedia Brittanica. Pp. 2–29. Reprinted in: Keller 1995b: 3–42.

Keller, Evelyn Fox. 1995a. "The Body of a New Machine: Situating the Organism Between Telegraphs and Computers." In: Keller 1995b, 79–118.

————. 1995b. *Refiguring Life: Metaphors of Twentieth-Century Biology.* New York: Columbia University Press.

———— and E. Lloyd (eds.). 1992. *Keywords in Evolutionary Biology.* Cambridge, MA: Harvard University Press.

———— and Helen Longino. 1996. *Feminism and Science.* New York: Oxford University Press.

Keller, Helen. 1915a. "Physician's Juries for Defective Babies." *The New Republic* (18 December): 173–74.

————. 1915b. "Helen Keller, Blind, Deaf and Dumb Genius, Writes for Daily Call on Defective Baby Case." *New York Call* (26 November): 5.

Kelsoe, J., E. Ginns, J. Egeland, et al. 1989. "Reevaluation of the Linkage Relationship Between Chromosome 11p Loci and the Gene for Bipolar Affective Disorder in the Old Order Amish." *Nature* 342: 238–43.

Kennan, George F. 1947a. "The Sources of Soviet Conduct." *Foreign Affairs* 25: 566–82.

————. 1947b. "Results of Planning Staff Study of Questions Involved in the Japanese Peace Settlement." In: U.S. Government 1972: 537–43.

Kennedy, J., L. Giuffra, H. Moises, et al. 1988. "Evidence Against Linkage of Schizophrenia to Markers On Chromosome 5 in a Normal Swedish Pedigree." *Nature* 336: 167–69.

Kessler, S. 1988. "Invited Essay on the Psychological Aspects of Genetic Counseling V. Preselection: A Family Coping Strategy in Huntington Disease." *American Journal of Medical Genetics* 31: 617–21.

Kevles, Daniel J. 1992. "Out of Eugenics: The Historical Politics of the Human Genome." In: Kevles and Hood 1992, 3–36.

————. [1985] 1995. *In the Name of Eugenics.* Cambridge, MA: Harvard University Press. First published 1985.

————. 2000. *See in press.*

Kevles, Daniel, and L. Hood (eds.). 1992. *The Code of Codes— Scientific and Social Issues in the Human Genome Project.* Cambridge, MA: Harvard University Press.

Kieswetter, S., M. Macek, C. Davis, et al. 1993. "A Mutation in CFTR Produces Different Phenotypes Depending on Chromosomal Background." *Nature Genetics* 5: 274–78.

Kim, Jaegwon. 1992. "'Downward Causation' in Emergentism and Non-reductive Physicalism." In: Beckermann et al. 1992a, 119–38.

——— and E. Sosa (eds.). *The Companion to Metaphysics.* Oxford: Blackwells.

Kinsey, Alfred. 1948. *Sexual Behavior in the Human Male.* Philadelphia: W. B. Saunders.

———. 1953. *Sexual Behavior in the Human Female.* Philadelphia: W. B. Saunders.

Kitcher, Philip. 1984. "1953 and All That. A Tale of Two Sciences." *Philosophical Review* 18: 335–73. Reprinted in: Sober 1994, 379–99.

———. 1989. "Explanatory Unification and the Causal Structure of the World." In: Kitcher and Salmon 1989, 410–505.

———. 1992. "Gene: Current Usages." In: Keller and Lloyd 1992, 128–31.

———. 1995. "Who's Afraid of the Human Genome Project?" In: *PSA 1994: Proceedings of the 1994 Biennial Meeting of the Philosophy of Science Association.* Vol. 2. Pp. 313–21. East Lansing, MI: Philosophy of Science Association.

———. 1996. *The Lives to Come.* New York: Simon & Schuster.

——— and W. S. Salmon (eds.). 1989. *Scientific Explanation.* Minneapolis: University of Minnesota Press.

Klass, Perri. 1989. "The Perfect Baby?" *New York Times Magazine* (29 January).

Kockelmans, Joseph. 1966. *Phenomenology and Physical Science: An Introduction to the Philosophy of Physical Science.* Pittsburgh: Duquesne University Press.

Kodani, M. 1960. "Three Diploid Chromosome Numbers of Man." *Proceedings of the National Academy of Sciences* 43: 285–92.

Kohler, Robert. 1976. "Management of Science: The Experience of Warren Weaver and the Rockefeller Foundation Programme in Molecular Biology." *Minerva* 14: 279–306.

Kolata, Gina. 1992. "Biologist's Speedy Gene Method Scares Peers But Gains Backers." *The New York Times* (28 July): B5, B8.

———. 1993. "Unlocking the Secrets of the Genome." *New York Times* (30 October): B5–6.

Kolker Aliza and B. M. Burke. 1994. *Prenatal Testing: A Sociological Perspective.* Westport, CT: Bergin and Garvey.

Koshland, Daniel E., Jr. 1989. "Sequences and Consequences of the Human Genome." *Science* 246: 189.

Kotlowitz, Alex. 1991. *There Are No Children Here.* New York: Doubleday.

Kottler, M. J. 1974. "From 48 to 46: Cytological Technique, Preconception, and the Counting of Human Chromosomes." *Bulletin of the History of Medicine* 48: 467–71.

Kozol, Jonathan. 1992. *Savage Inequalities.* New York: Harper.

Kuhn, Annette. 1988. *Cinema, Censorship and Sexuality, 1909–1925.* London: Routledge.

Kuhn, Thomas S. 1962. *The Structure of Scientific Revolutions.* Chicago: University of Chicago Press.

Kupferman I. 1992. "Genetic Determinants of Behavior." In: Kandel, Schwartz, and Jessell 1992, 987–96.

Kuschnick, T. 1961. "Mongolism: Recent Studies and Developments." *Quarterly Review of Pediatrics* 16: 150–58.

Lamy, M. 1943. *La Génétique et ses applications à la médecine.* Paris: Doin.

———. 1951. "La Génétique médicale: ses méthodes et des perspectives." *La Semaine des hôpitaux de Paris* 27: 1675–81.

———. 1956. "Les Applications de la génétique à l'eugénisme." *Archiv für Klaus Schift Verorforschung* 31: 251–60.

———. 1970. "Le Conseil génétique." *Bulletin de l'académie nationale de médecine* 154: 175–83.

——— and J. de Grouchy. 1954. "L'Hérédité de la myopathie (formes basses)." *Journal de génétique humaine* 3: 219–61.

———, P. Royer, and J. Frézal. 1959. *Maladies héréditaires du métabolisme chez l'enfant.* Paris: Masson.

Lander, E. and D. Botstein. 1989. "Mapping Mendelian Factors Underlying Quantitative Traits Using RFLP Linkage Maps." *Genetics* 121: 185–99.

——— and N. J. Schork. 1994. "Genetic Dissection of Complex Traits." *Science* 265: 2037–48.

Lavoisier, Antoine. [1777] 1780. "Expériences sur la respiration des animaux, et sur les changements qui arrivent à l'air en passant par leur poumon." *Mémoires sur l'académie royale des sciences.* 1777: 185–94.

———, and Pierre Simon LaPlace. [1780] 1784. "Mémoire sur la chaleur." *Mémoires sur l'académie royale des sciences.* 1780: 355–408.

Lawrence, J. H. 1938. *See unpublished.*

Lecuyer, R. 1958. *Le Mongolisme. Principaux problèmes médicaux, psychologiques et sociaux.* Paris: Doin.

Ledley, Robert S. 1955. "Digital Computational Methods in Symbolic Logic, With Examples in Biochemistry." *Proceedings of the National Academy of Sciences* 41: 498–511.

Lee, C. H. et al. 1961. "Definitive Diagnosis of Mongolism in Newborn Infants by Chromosome Studies." *Journal of the American Medical Association* 178: 1030–32.

Lehrman, D. 1970. "Semantic and Conceptual Issues in the Nature-nurture Problem." In: Aronson, Tobach, Lehrman, and Rosenblat 1970, 17–50.

Lejeune, J. 1960a. "Les Maladies humaines par aberrations chromosomiques." *Revue française d'études cliniques et biologiques* 5: 341–47.

———. 1960b. "Le Mongolisme, trisomie dégressive." *Annales de génétique* 2: 1–34.

———. 1970. "Le Pronostic vital." *Annales de biologie clinique* 28: 1–2.

——— and R. Turpin. 1960. "Etude de l'excretion urinaire de certains métabolites du tryptophane chez les enfants mongoliens." *Comptes-Rendus de l'académie des sciences* 251: 474–76.

———, M. Gauthier, R. Turpin. 1959a. "Etude des chromosomes somatiques de neuf enfants mongoliens," *Comptes-Rendus de l'académie des sciences* 248: 1721–22.

———, R. Turpin, and M. Gauthier. 1959b. "Le Mongolisme, premier exemple d'aberration autosomique humaine." *Annales de génétique* 1: 41–49.

———, M. Gauthier, and R. Turpin. 1959c. "Les chromosomes humains en culture de tissus." *Comptes-Rendus de l'académie des sciences* 248: 602–03.

Lejune, J., R. Turpin, and M. Gauthier. 1960. "Etude des chromosomes somatiques humains: Technique de culture de fibroblastes *in vitro.*" *Revue française des études cliniques et biologiques* 5: 406–08.

Lenoir, T. 1982. *The Strategy of Life: Teleology and Mechanics In Nineteenth Century German Biology.* Dordrecht: Reidel.

———. 1988. "Science for the Clinic: Science Policy and the Formation of Carl Ludwig's Institiute in Leipzig." In: Coleman and Holmes 1988, 139–78.

———. 1992. "Laboratories, Medicine and Public Life in Germany 1830–1849." In: Cunningham and Williams 1992, 14–71.

Leplin, J. 1984. *Scientific Realism.* Berkeley: University of California Press.

Leslie, Stuart W. 1993. *The Cold War and American Science: The Military-Industrial Complex.* New York: Columbia University Press.

LeVay, S. 1991. "A Difference in Hypothalamic Structure Between Heterosexual and Homosexual Men." *Science* 253: 1034–37.

———. 1995. *The Sexual Brain.* Cambridge, MA: MIT Press.

——— and D. Hamer. 1994. "Evidence for a Biological Influence in Male Homosexuality." *Scientific American* 270: 44–49.

Levay-Lehad E., A. Lahad, E. Wijaman, et. al. 1995. "Apolipoprotein E Genotypes and the Age of Onset in Early-onset Familial Alzheimer's Disease." *Annals of Neurology* 38: 678–80.

Lewin, Benjamin. 1983. *Genes.* New York: Wiley.

Lewin, Roger. 1989. "Genome Planners Fear Avalanche of Red Tape." *Science* 244 (4912): 1543.

Lewontin, R. C. 1992a. "The Dream of the Human Genome." *New York Review of Books* 39: 31–40. Reprinted in: Lewontin 1992b, 61–83.

———. 1992b. *Biology as Ideology: The Doctrine of DNA.* New York: Harper-Collins.

———. 1992c. "Doubts about the Human Genome Project." *The New York Review of Books* 34: 19–24.

———. 1995. *Human Diversity.* New York: Scientific American Library.

Lhoff, Brigitte. 1997. "The Concept of Vital Forces as a Research Program (From Mid-XVIIIth Century to Johannes Müller)." In: Cimino and Duchesneau 1997, 127–42.

Lindee, Susan M. 1994. *Suffering Made Real: American Science and the Survivors at Hiroshima*. Chicago: University of Chicago Press.

Liu, K. S. and P. W. Sternberg. 1995. "Sensory Regulation of Male Mating Behavior in *Caenorhabditis elegans*." *Neuron* 14: 79–89.

Locke, John. [1690] 1959. *An Essay Concerning Human Understanding*. Ed. A. C. Alexander. 2 vols. New York: Dover.

Loeb, Jacques. 1912. *The Mechanistic Conception of Life*. Chicago: University of Chicago Press.

Lockery, S. 1995. "Signal Propagation in the Nerve Ring of the Nematode *C. elegans*." *Society for Neuroscience Abstracts* 569.7, Pt. 2: 1454.

Longino, Helen. 1990. *Science as Social Knowledge: Values and Objectivity in Scientific Inquiry*. Princeton: Princeton University Press.

Longmore, Paul K. 1985. "Screening Stereotypes: Images of Disabled People." *Social Policy* 16 (Summer 1985): 31–37.

Los Alamos National Laboratory. 1992. *The Human Genome Project*. *Los Alamos Science* 20.

Lucretius. 1995. *On the Nature of Things*. Trans. Anthony M. Esolen. Baltimore: Johns Hopkins University Press.

Ludmerer, Kenneth M. 1972. *Genetics and American Society: A Historical Appraisal*. Baltimore: Johns Hopkins.

Lwoff, André. 1965. *Biological Order*. Cambridge, MA: M.I.T. Press.

Lynch, W. E. 1967. "Soul in the Bible." *New Catholic Encyclopedia* Vol. 13. New York: McGraw Hill, 449–50.

Macer, Darryl. 1992. "The 'Far East' of Biological Ethics." *Nature* 359: 770.

Macke, J. P., N. Hu, S. Hu, et al. 1993. "Sequence Variation in the Androgen Receptor Gene Is Not a Common Determinant of Male Sexual Orientation." *American Journal of Human Genetics* 53: 844–52.

MacNamara, Vincent. 1985. *Faith and Ethics*. Washington: Georgetown University Press.

Maddox, J. 1995. "Valediction From an Old Hand." *Nature* 378: 521–23.

Mahowald, Mary B. "Baby Doe Committees: A Critical Evaluation." *Clinics in Perinatology* 15 (December): 789–800.

Mallet, R. 1964. "Aspect clinique et diagnostic du mongolisme." *Revue du praticien* 14: 7–18.

March of Dimes Birth Defects Foundation. 1992. *Genetic Testing and Gene Therapy: National Survey Findings.* White Plains, NY: Louis Harris and Associates.

Marshall, Eliot. 1995. "A Strategy for Sequencing the Genome 5 Years Early." *Science* 267 (10 February): 783–84.

McBrien, Richard (ed.). 1995. *The Harper-Collins Encyclopedia of Catholicism.* San Francisco: Harper Collins.

McConaughy, Walter P. 1961. "A Pacific Partnership." *Department of State Bulletin* 45: 634–37.

McCormick, Richard A and Charles E. Curran. 1980. *Readings in Moral Theology II: The Distinctiveness of Christian Ethics.* New York: Paulist Press.

McElroy, William and Bentley Glass (eds.). 1957. *A Symposium on the Chemical Basis of Heredity.* Baltimore: Johns Hopkins University Press.

McEwen, B. S. 1995. "Stressful Experience, Brain, and Emotions: Developmental, Genetic, and Hormonal Influences." In: Gazzaninga 1995, 1117–36.

McGourty, Christine. 1989. "Speak Softly or Carry a Big Stick." *Nature* 341: 679.

McKim, William. 1900. *Heredity and Human Progress.* New York: G. P. Putnam and Sons.

McLaughlin, Brian P. 1992. "British Emergentism." In: Beckerman, Flohr and Kim 1992, 49–93.

McMillan, Brockway. 1956. "Two Inequalities Implied by Unique Decipherability." *IRE Transactions on Information Theory* 2: 115–16.

McMullin, E. 1972. "The Dialectics of Reduction." *Idealistic Studies* 2: 95–115.

———. (ed.). 1978. *The Concept of Matter in Modern Philosophy.* Notre Dame: University of Notre Dame Press.

———. 1984. "The Case for Scientific Realism." In: Leplin 1984, 8–40.

McMullin, E. 1985. (ed.). *Evolution and Creation*. Notre Dame: University of Notre Dame Press.

———. 1993. "Evolution and Special Creation." *Zygon* 28: 299–335.

———. 1995. "Matter." In: Kim and Sosa 1995, 299–302.

———. 1999. "Materialist Categories." *Science and Education* 8:37–44.

Mechanic, David. 1994. *Inescapable Decisions: The Imperatives of Health Reform*. New Brunswick, NJ: Transaction Publishers.

Méhes, K. 1996. "Invited Editorial Comment: Classical Clinical Genetics in the Era of Molecular Genetics." *American Journal of Medical Genetics* 61: 394–95.

Mehler, Barry. 1987. "Eliminating the Inferior." *Science for the People* (November/December): 14–18.

Melki, J. et al. 1990. "Mapping of Acute (type I) Spinal Muscular Atrophy to Chromosome 5q12-q14." *Lancet* 336: 271–73.

Melki, J. et al. 1992. "Prenatal Prediction of Werdnig-Hoffmann Disease Using Linked Polymorphic DNA Probes." *Journal of Medical Genetics* 29: 171–74.

Mendelsohn, Everett. 1964. *Heat and Life*. Cambridge, MA: Harvard University Press.

Mendelsohn, Mortimer. 1986. "Prospects for DNA Methods to Measure Human Heritable Mutation Rates." In: Ramel, Lambert, and Magnusson 1986, 337–44.

Michalczyk, J. J. (ed.). 1994. *Medicine, Ethics and the Third Reich*. Kansas City: Sheed &Ward.

Michel, J. (ed.). 1991. *La Nécessité de Claude Bernard*. Paris: Klincksieck.

Midgley, Mary. 1994. *The Ethical Primate*. New York: Routledge.

Mitchell, David and Sharon Snyder (eds.). 1997. *The Body and Physical Difference: Discourses of Disability in the Humanities*. Ann Arbor: University of Michigan Press.

Mitchell, W. J. Thomas. 1986. *Iconology: Image, Text, Ideology*. Chicago: University of Chicago Press.

Monmonier, Mark S. 1996. *How to Lie With Maps*. 2nd. ed. Chicago: University of Chicago Press.

Moore John A. 1995. "Cultural and Scientific Literacy." *Molecular Biology of the Cell* 6: 1–6.

Morea, Peter. 1990. *Personality: an Introduction to the Theories of Psychology*. London: Penguin Books.

Morgan, Thomas Hunt. 1910. "Chromosomes and Heredity."
American Naturalist 44: 449–96.

———. 1926a. *The Theory of the Gene.* New York: Yale University
Press.

———. 1926b. "The Genetics and Physiology of Development."
American Naturalist 60: 489–515.

———. 1932a. "The Rise of Genetics." *Science* 76: 261–88.

Morgan, Thomas Hunt. 1932b. *The Scientific Basis of Evolution.* New
York: W.W. Norton.

———. [1934]1965. "The Relation of Genetics to Physiology and
Medicine." *The Scientific Monthly.* 41: 5–18. Reprinted in:
Anon. (ed.). *Nobel Lectures in Physiology or Medicine: 1922–
1941.* Pp. 313–28. Amsterdam: Elsevier.

Mosse, George. 1978. *Toward the Final Solution: A History of
European Racism.* London: J. M. Dent.

———. 1985. *Nationalism and Sexuality.* New York: H. Fertig, 1985

Muller, H. J. 1922. "Variation Due to Change in the Individual Gene."
American Naturalist 56: 32–50.

———. [1926]1929. "The Gene as the Basis of Life." *Proceedings of
the International Congress of Plant Science* 1: 897–921.

Müller-Hill, Benno. 1988. *Murderous Science: Elimination By
Scientific Selection of Jews, Gypsies, and Others, Germany
1933–1945.* Trans. George Fraser. Oxford; New York: Oxford
University Press.

Munson, R. and L. H. Davis, 1992. "Germ-line Gene Therapy and the
Medical Imperative." *Kennedy Institute of Ethics Journal* 2 (2):
137–58.

Murphy, Timothy F. and Marc A. Lappe (eds.). 1994. *Justice and the
Human Genome Project.* Berkeley: University of California
Press.

Musser, Charles. 1990. *The Emergence of Cinema.* New York:
Scribners.

Nagel, E. 1961. *The Structure of Science.* New York: Harcourt,
Brace, and World.

Nanney, D. L. 1957. "The Role of Cytoplasm in Heredity." In:
McElroy and Glass 1957, 134–64.

National Academy of Sciences. 1984. *Science and Creationism.*
Washington, D.C.: National Academy Press.

National Bioethics Advisory Commission. 1997. *Cloning Human Beings*. Rockville: Government Printing Office.

National Conference on Race Betterment (NCRB). 1914. *Proceedings* 1.

———. 1915. *Proceedings* 2.

National Security Council. 1949. "NSC 49/1." Comments on NSC 49 (15 June). In: U.S. Govt. 1972, 858–62.

National Security Council. 1948. "NSC 13/2." Report by the National Security Council with Respect to United States Policy Toward Japan (7 October). In: U.S. Govt. 1972: 858–62.

Neel, James V. 1976. "Human Genetics." In: Bower and Purcell 1976, 39–99.

———. 1990. "Unfolding Perspectives on the Genetic Effects of Human Exposure to Radiation." In: Sutherland and Woodward 1990, 337–49.

———. 1994. *Physician to the Gene Pool: Genetic Lessons and Other Stories*. New York: Wiley.

——— and W. J. Shull. 1956. *The Effect of Exposure to the Atomic Bombs on Pregnancy Termination in Hiroshima and Nagasaki*. Washington, D.C.: National Academy of Sciences / National Research Council.

——— et al. 1988. "Search for Mutations Altering Protein Charge and/or Function in Children of Atomic Bomb Survivors: Final Report." *American Journal of Human Genetics* 42: 663–76.

———, W. J. Schull, et al. 1989. "Implications of the Hiroshima-Nagasaki Genetic Studies for the Estimation of the Human 'Doubling Dose' of Radiation." *Genome* 31: 853–59.

———, W. Schull, A. A. Awa, C. Satoh, C. Kaho, M. Otake, and Y. Yashimoto. 1990. "The Children of Parents Exposed to Atomic Bombs: Estimates of the Genetic Doubling Dose of Radiation for Humans." *American Journal of Human Genetics* 46: 1053–72.

Nelkin, Dorothy and Laurence Tancredi. 1989. *Dangerous Diagnostics*. New York: Basic Books.

——— and M. Susan Lindee. 1995. *The DNA Mystique*. New York: W. H. Freeman.

New York Times. 1946. "Muller, Biologist Wins Nobel Prize." (1 November): A7; C21.

———. 1991. "Gene Experts Tell of Possible Abuse." (18 October): A18

New York Times 1998. "Interview With Eric Lander." *New York Times* (6 September): B6.

Nijhout, H. F. 1990. "Metaphors and the Role of Genes in Development." *Biological Essays* 12 (9): 441–46.

Norden, Martin F. 1994. *The Cinema of Isolation: A History of Physical Disability in the Movies.* New Brunswick, NJ: Rutgers University Press.

Normile, Dennis. 1998. "Bid for Better Beef Gives Japan a Leg Up on Cattle." *Science* 282: 1975–76.

O'Toole, G. J. A. 1991. *Honorable Treachery: Intelligence, Espionage, and Covert Action from the American Revolution to the CIA.* New York: Atlantic Monthly Press.

Okimoto, Daniel I. and James H. Raphael. 1993. "Ambivalence, Continuity, and Change: American Attitudes Toward Japan and U.S.-Japanese Relations." In: Aspen Institute 1993, 117–63.

Olby, Robert C. 1974. *The Path to the Double Helix.* Seattle: University of Washington Press.

———. 1986. "Biochemical Origins of Molecular Biology: A Discussion." *Trends in Biochemical Sciences* 11: 303–05.

O'Neil, O. (ed.). 1979. *Having Children.* New York: Oxford University Press.

Opitz, John M. 1991. *Honors Lecture: The Wolf of Agobio: Prolegomena to the Medical Humanities.* Bozeman, MT: Montana State University Press.

———. 1996. "Forty-four Years of Work in Morphology." *Proceedings of the Greenwood Genetics Center* 15: 97–99.

Osmundsen, John A. 1961. "New Way to Read Life's Code Found." *New York Times* (7 April): 12.

Oyama, S. 1985. *The Ontogeny of Information: Developmental Systems and Evolution.* Cambridge: Cambridge University Press.

Packard, George R. 1966. *Protest in Tokyo: The Security Crisis of 1960.* Princeton: Princeton University Press.

Pappé, H. O. 1967. "Philosophical Anthropology." In: Edwards 1967, Vol. 6, 159–66.

Parfit, Derek. 1984. *Reasons and Persons.* Oxford: Oxford University Press.

Parsons, Louella. 1917. "Seen on the Screen." *Chicago Herald* (2 April): 11.

Paterson, R. and C. Barral. 1994. "L'Association française contre les myopathies: Trajectoire d'une association d'usagers et construction associative d'une maladie." *Sciences sociales et santé* 12: 79–111.

Pattee, Howard. 1969. "How Does a Molecule Become a Message?" *Developmental Biology.* Supplement 3: 1–16.

Paul, Diane. 1992. "Eugenic Anxieties, Social Realities, and Political Choices." *Social Research* 59: 663–83. Reprinted in: Paul 1998, 95–115.

Paul, Diane. 1994. "Toward a Realistic Assessment of PKU Screening." In: *PSA 1994: Proceedings of the 1994 Biennial Meeting of the Philosophy of Science Association.* Vol. 2. Pp. 322–38. East Lansing, MI: Philosophy of Science Association.

———. 1995. *Controlling Human Heredity: 1865 to the Present.* New York: Humanities Press.

———. 1998. *The Politics of Heredity: Essay on Eugenics, Biomedicine, and the Nature-Nuture Debate.* Albany: State University of New York Press.

Pauly, Philip. 1987. *Controlling Life: Jacques Loeb and the Engineering Ideal in Biology.* New York: Oxford University Press.

Peacocke, Arthur. 1971. *Science and the Christian Experiment.* London: Oxford University Press.

———. 1984. *Intimations of Reality: Critical Realism in Science and Religion.* Notre Dame: University of Notre Dame Press.

——— (ed.). 1985. *Reductionism in Academic Disciplines.* Worcester: Billin and Sons.

———. [1986] 1994. *God and the New Biology.* Reprint, Gloucester, MA: Peter Smith.

———. 1993. *Theology for a Scientific Age: Being and Becoming— Natural, Divine and Human.* Oxford: Blackwells. 2nd expanded edition. London: S. C. M. Press; Minneapolis: Fortress Press.

Pear, Robert. 1992. "U.S. Will Tighten Health-Lab Goals: Research Funds To Be Steered with an Eye toward Profit." *New York Times* (24 August): A-1, 16.

———. 1996. "Health Research, Once Facing Big Budget Cut, Gets an Increase," *New York Times* (16 January): A-13

Pearn, J. 1982. "Infantile Motor Neuron Disease." *Advances in Neurology* 36: 121–30.

Peirce, Charles Sanders. [1893/1910] 1955. "Logic as Semiotic: The Theory of Signs." In: Buchler 1955, 98–119.

Pennsylvania State Board of Censors. 1918. *Rules and Standards.* Harrisburg: J. L. L. Kuhn.

Penrose, Lionel S. and G. F. Smith. 1966. *Down's Anomaly.* London: J. A. Churchill.

Pernick, Martin S. 1993. "Sex Education Films, U.S. Government." *Isis* 84: 766–68.

———. *The Black Stork: Eugenics and the Death of "Defective" Babies in American Medicine and Motion Pictures Since 1915.* New York: Oxford University Press.

Peters, Ted. 1997. *Playing God? Genetic Determinism and Human Freedom.* New York and London: Routledge.

———. (ed.). 1998. *Science and Theology: The New Consonance.* Boulder: Westview Press.

Pfaundler, M. von. 1940. "Erbpathologie der Diathesen. Betrachtet vom pädiatrischen Standpunkte." In: Bauer, Hanhart, Lange, and Just 1940. Vol. 2, 640–84.

Picard, J. F. 1992. "Poussée scientifique ou demande de médecins ? La recherche médicale en France de l'Institut National d'Hygiène à l'INSERM." *Sciences sociales et santé* 10: 47–107.

Pinell, P. 1992. *La Naissance d'un fléau.* Paris: A.M. Metailié.

——— and M. Zafiropoulos. 1983. *Un Siècle d'échec scolaire (1882-1982).* Paris: Editions Ouvrières.

Plomin, Robert. 1990. "The Role of Inheritance in Behavior." *Science* 248: 183–88.

———. 1995. "Genetics and Children's Experiences in the Family." *Journal of Child Psychology and Psychiatry and Allied Disciplines* 36: 33–68.

———, et al. 1993. "Genetic Change and Continuity From Fourteen to Twenty Months: The MacArthur Longitudinal Twin Study." *Child Development* 64: 1354–76.

———, M. J. Owen, and P. McGuffin. 1994. "The Genetic Basis of Complex Human Behaviors." *Science* 264: 1733–39.

——— et al. 1997. "No Association Between General Cognitive Ability and the A1 Allele of the D2 Dopoamine Receptor Gene." *Behavior Genetics* 27: 29–31.

Polani, P. E. et al. 1960. "A Mongol Girl With 46 Chromosomes." *Lancet* 278: 721.

Polanyi, Michael. 1964. *Personal Knowledge*. Reprint of 1958 edition with new preface. New York: Harper.

———. 1968. "Life's Irreducible Structure." *Science* 160:1308–12.

Pool, R. 1993. "Evidence for Homosexuality Gene." *Science* 261: 291–92.

Popper, Karl R. 1968. "Epistemology Without a Knowing Subject." In: Van Rootselaar and Staal 1968, 333–73.

Port, M. (ed.). 1995. *Passion des formes (hommage à René Thom)*. Paris: ENS Editions Fontenay St. Cloud.

Portin, Petter. 1993. "The Concept of the Gene: Short History and Present Status." *Quarterly Review of Biology* 68: 173–223.

Pratt, Fletcher. 1942. *Secret and Urgent: The Story of Codes and Ciphers*. New York: Blue Ribbon Books.

Press, N. A. and C. H. Browner. 1994. "Collective Silences, Collective Fictions: How Prenatal Testing Became Part of Routine Prenatal Care." In: Rothenberg and Thomson 1994, 201–18.

Prestowitz, Clyde. 1988. *Trading Places: How We Allowed Japan to Take the Lead*. New York: Basic Books.

Proctor, Robert N. 1988. *Racial Hygiene: Medicine Under the Nazis*. Cambridge, MA: Harvard University Press.

Quastler, Henry (ed.). 1953a. *Essays on the Use of Information Theory in Biology*. Urbana: University of Illinois Press.

———. 1953b. "The Measure of Specificity." In: Quastler 1953a, 41–74.

Rafferty, C. 1995. "Homosexuality or 'Disorder'." *Chicago Tribune* (1 August): 1–2.

Rahner, Karl. 1961–74. *Theological Investigations*. 23 vols. New York: Crossroad.

———. 1965a. *Hominisation: The Evolutionary Origin of Man as a Theological Problem*. New York: Herder and Herder.

———. 1965b. "The Official Teaching of the Church on Man in Relations to the Scientific Theory of Evolution." In: Rahner 1965a, 11–31.

———. 1966. "Christology Within an Evolutionary View." In: Rahner 1961–74. vol. 5, 157–92.

———. 1969. *Hearers of the Word*. Trans. M. Richards. New York: Herder and Herder

———. 1969. "The Unity of Spirit and Matter in the Christian Understanding of Faith." In: Rahner 1961–74, vol. 6, 153–77.

Rahner, Karl. 1972a. "The Problem of Genetic Manipulation." In: Rahner 1961–74, vol. 9, 225–52.

———. 1972b. "The Experiment With Man." In: Rahner 1961–74, vol. 9, 205–24.

———. 1985. *Foundations of Christian Faith*. New York: The Crossroad Publishing Company.

Raizen, D. and L. Avery. 1994. "Electrical Activity and Behavior in the Pharynx of *Caenorhabditis elegans*." *Neuron* 12: 483–95.

Ramel, C., B. Lambert, and J. Magnusson (eds.). 1986. *Genetic Toxicology of Environmental Chemicals: Proceedings of the Fourth International Conference on Environmental Mutagens, held in Stockholm, Sweden, June 24-28, 1985*. New York: A. R. Liss.

Rapp, Rayna. 1988. "Moral Pioneers: Women, Men and Fetuses on a Frontier of Reproductive Technology." *Women and Health*. 13: 101–16.

Ravin, Arnold. 1977. "The Gene as Catalyst; The Gene as Organism." *Studies in History of Biology* 1: 1–45.

Rawls, John. 1971. *A Theory of Justice*. Cambridge, MA: Harvard University Press.

Rechsteiner, Martin. 1991. "The Human Genome Project: Misguided Science Policy." *Trends in Biochemical Sciences* 16: 453–59.

Riech, William Thomas (ed.). 1995. *The Encyclopedia of Bioethics*. Revised edition. 5 vols. New York: MacMillan.

Reilly, Philip. 1977. *Genetics, Law, and Social Policy*. Cambridge, MA: Harvard University Press.

Reilly, Philip. 1983. "The Surgical Solution: The Writings of Activist Physicians in the Early Days of Eugenical Sterilization." *Perspectives in Biology and Medicine* 26: 637–56.

———. 1991. *The Surgical Solution: A History of Involuntary Sterilization in the United States*. Baltimore: Johns Hopkins University Press.

Rensberger, Boyce. 1990. "Survey: Firms Back Human Genome Project." *The Washington Post* 113 (22 January): A2.

Rheinberger, H. J. 1995. "Genes: A Disunified View from the Perspective of Molecular Biology." In: Beurton, Lefevre, and Rheinberger 1995, 7–13.

Rhodes, Richard. 1986. *The Making of the Atomic Bomb*. New York: Simon and Schuster.

Richelson, Jeffrey. 1989. *The U.S. Intelligence Community*. 2nd edition. Cambridge, MA: Ballinger Publishing Co.

Rignano, E. 1923. *La Mémoire biologique*. Paris: Alcan.

Risch, N., E. Squires-Wheeler, and B. J. Keats. 1993. "Male Sexual Orientation and Genetic Evidence." *Science* 262: 2063–65.

———, and K. Merikangas. 1996. "The Future of Genetic Studies of Complex Human Diseases." *Science* 273: 1516–17.

Roberts, Leslie. 1987. "Who Owns the Human Genome?" *Science* 237: 358–61.

———. 1989a. "Genome Project Under Way, at Last." *Science* 243: 167–68.

———. 1989b. "Watson Versus Japan." *Science* 246: 576–77.

Roberts, Leslie. 1992. "NIH Gene Patents, Round Two." *Science* 255: 912–13.

——— and Peter Aldhous. 1992. "Two Chromosomes Down, 22 to Go." *Science* 258: 28–30.

Robie, Theodore. 1934. "Selective Sterilization and Mental Retardation." In: *Third International Congress* , 201–9.

Robinson, J. L. et al. 1992. "The Heritability of Inhibited and Uninhibited Behavior: A Twin Study." *Developmental Psychology* 28: 1030–37.

Robinson, William J. 1917. *Eugenics and Marriage*. New York: Critic & Guide.

———. 1922. *Eugenics, Marriage and Birth Control*. 2nd ed. New York: Critic & Guide.

Roger, Jacques. 1997. *The Life Sciences in Eighteenth-Century French Thought*. Trans. R. Ellrich; Ed. K. R. Benson. Stanford: Stanford University Press.

Roll-Hansen, Nils. 1969. "On the Reduction of Biology to Physical Science." *Synthese* 20: 277–89.

———. 1976. "Critical Teleology: Immanuel Kant and Claude Bernard on the Limitations of Experimental Biology." *Journal of the History of Biology* 9: 59–91.

Romm, Joseph J. 1993. *Defining National Security: The Nonmilitary Aspects*. New York: Council on Foreign Relations Press.

Rose, S. 1995. "The Rise of Neurogenetic Determinism." *Nature* 373: 380–82.

Rose, Steven. 1985. "The Roots and Social Functions of Biological Reductionism." In: Peacocke 1985, 24–42.

Rosenberg, Alex. 1994. "Subversive Reflections on the Human Genome Project." In: *PSA 1994: Proceedings of the 1994 Biennial Meeting of the Philosophy of Science Association.* Vol. 2. Pp. 329–35. East Lansing, MI: Philosophy of Science Association.

———. 1985. *The Structure of Biological Science.* Cambridge: Cambridge University Press.

Rosenberg, Charles. 1976. *No Other Gods: On Science and American Social Thought.* Baltimore: Johns Hopkins University Press.

———. 1992. "Framing Disease." In: C. Rosenberg and J. Golden, (eds.). 1992. *Framing Disease.* Pp. xiii–xxvi. New Brunswick, NJ: Rutgers University Press.

Rosenblueth, A., N. Wiener, and J. Bigelow. 1943. "Behavior, Purpose, and Teleology." *Philosophy of Science* 10: 18–24.

Ross, Dorothy. 1972. *G. Stanley Hall.* Chicago: University of Chicago Press.

Rothenberg, K. H. and E. J. Thomson (eds.). 1994. *Women and Prenatal Testing: Facing the Challenge of Genetic Technology.* Columbus, OH: Ohio State University Press.

Rowen, L., G. Mahairas, and L. Hood. 1997. "Sequencing the Human Genome." *Science* 278: 605–07.

Rubinsztein, D., J. Leggo, R. Coles, et al. 1996. "Phenotypic Characterization of Individuals with 30–40 CAG Repeats in the Huntington Disease (HD) Gene Reveals HD Cases With 36 Repeats and Apparently Normal Elderly Individuals with 36–39 Repeats." *American Journal of Human Genetics* 59: 16–22.

Ruse, Michael (ed.). 1989. *What Philosophy of Biology Is.* Dordrecht: Kluwer.

Russell, E. S. 1930. *The Interpretation of Development and Heredity.* Oxford: Clarendon.

Russell, O. Ruth. 1975. *Freedom to Die: Moral and Legal Aspects of Euthanasia.* New York: Human Sciences Press.

Russell, R. J., W. R. Stoeger, and G. Coyne (eds.). 1988. *Physics, Philosophy, and Theology: A Common Quest for Understanding.* Notre Dame: University of Notre Dame Press.

Russell, R. J., Stoeger, W. R. and G. Coyne (eds.). 1990. *John Paul II on Science and Religion: Reflections on the New View From Rome*. Notre Dame: University of Notre Dame Press.

Russett, Cynthia Eagle. 1989. *Sexual Science*. Cambridge, MA: Harvard University Press.

Sadley, William. 1923. "Endocrines, Defective Germ-Plasm, and Hereditary Defectiveness." In: *Second International Congress*: 341–50.

Saleeby, C. W. 1923. "Preventive Eugenics: The Protection of Parenthood from Racial Poisons." In: *Second International Congress* : 309–12.

Sanger, Frederick, S. Nicklen, and A. R. Coulson. 1977. "DNA Sequencing With Chain Terminating Inhibitors." *Proceedings of the National Academy of Sciences* 74: 5463–67.

Sapp, Jan. 1983. "The Struggle for Authority in the Field of Heredity 1900–1932: New Perspectives on the Rise of Genetics." *Journal of the History of Biology* 16: 311–42.

Sapp, Jan. 1987. *Beyond the Gene: Cytoplasmic Inheritance and the Struggle for Authority in Genetics*. New York: Oxford University Press.

Sarkar, Sahotra. 1992. "Models of Reduction and Categories of Reductionism." *Synthese* 91:167–94.

———. 1996a. "Biological Information: a Skeptical Look at Some Central Dogmas of Molecular Biology." In Sarkar 1996b, 187–231.

———. (ed.). 1996b. *The Philosophy and History of Molecular Biology*. Dordrecht: Kluwer.

———. 1998. *Genetics and Reductionism*. Cambridge: Cambridge University Press.

Saürländer, Willibrald. 1994. "The Nazis' Theater of Seduction." *New York Review of Books* (21 April): 16–19.

Schaffner, K. F. 1993. *Discovery and Explanation in Biology and Medicine*. Chicago: University of Chicago Press.

———. 1996. "Theory Structure and Knowledge Representation in Molecular Biology." In: Sarkar 1996b. 27–45.

———. 1998 "Genes, Behavior, and Developmental Emergentism: One Process, Indivisible." *Philosophy of Science* 65: 209–52.

———. 1999. *See in press*.

———. 2000. *See in press*.

Schaffner, K. F., and R. Wachbroit. 1994. "Il Cancero come malitta genetics: problemi sociali ed ethici." *L'Arco di Giano: rivista di medical humanities 6 settembre-dicembre 1994*: 13–29.

Schneider, W. 1990. *Quality and Quantity*. Cambridge: Cambridge University Press.

Schofield, Robert. 1970. *Mechanism and Materialism: British Natural Philosophy in an Age of Reason*. Princeton: Princeton University Press.

Schrödinger, Erwin. 1944. *What is Life?* Cambridge: Cambridge University Press.

Schull, William J. 1990. *Song Among the Ruins*. Cambridge, MA: Harvard University Press.

———. 1995. *Effects of Atomic Radiation: A Half-Century of Studies from Hiroshima and Nagasaki*. New York: Wiley, 1995.

Second International Congress of Eugenics. 1923. *Eugenics in Race and State: Scientific Papers of the Second International Congress of Eugenics . . . New York 1921*. Baltimore: Williams & Wilkins.

Semon, R. 1911. *Die Mneme als Erhaltendes Prinzip im Wechsel des Organischen Geschehens*. 3rd ed. Leipzig: Engelmann.

Sen, Amartya. 1992. *Inequality Reexamined*. Cambridge, MA: Harvard University Press.

Sengupta, P., H. Colbert, and C. Bargmann. 1994. "The *C. Elegans* Gene *odr-7* Encodes an Olfactory-Specific Member of the Nuclear Receptor Superfamily." *Cell* 79: 971–80.

Serres, K. et al. 1990. "Heterogeneity in Proximal Spinal Muscular Atrophy." *Lancet* 336: 749–50.

Shannon, Claude. 1948. "The Mathematical Theory of Communication." *Bell System Technical Journal* 27: 379–423; 623–56.

———. 1949. "Communication Theory of Secrecy Systems." *Bell System Technical Journal* 28: 656–715.

——— and W. Weaver. 1949. *The Mathematical Theory of Communication*. Urbana: University of Illinois Press.

Shapiro, Harold T. 1977. "Ethical and Policy Issues of Human Cloning." *Science* 277: 195–96.

Shapiro, Robert. 1991. *The Human Blueprint: The Race to Unlock the Secrets of Our Genetic Script*. New York: St. Martin's Press.

Shimony, Abner. 1992. *Search for a Naturalistic World View*. 2 vols. Cambridge: Cambridge University Press.

Simon, Herbert A. 1981. *The Sciences of the Artificial*. Enlarged edition. Cambridge, MA: M.I.T. Press.

Singer, Eleanor. 1991. "Public Attitudes Toward Genetic Testing." *Population Research and Policy Review* 10: 235–55.

Singer, M. and P. Berg. 1991. *Genes and Genomes: A Changing Perspective*. Mill Valley, CA: University Science Books.

Sinsheimer, Robert L. 1959. "Is the Nucleic Acid Message in a Two-Symbol Code?" *Journal of Molecular Biology* 1: 218–20.

———. 1967. *The Book of Life*. Reading, MA: Addison-Wesley Publishing, Co.

———. 1989. "The Santa Cruz Workshop—May 1985." *Genomics* 5: 954–56.

———. 1991. "The Human Genome Initiative." *FASAB Journal* 5: 2885.

———. 1994. *The Strands of a Life: The Science of DNA and the Art of Education*. Berkeley: University of California Press.

Sittler, Joseph. 1958. *The Structure of Christian Ethics*. New Orleans: Louisiana State University Press.

Sittler, Joseph. 1981. *Grace Notes and Other Fragments*. Philadelphia: Fortress Press.

Skolnikoff, Eugene B. 1967. *Science, Technology, and American Foreign Policy*. Cambridge, MA: MIT Press.

Sloan, Phillip R. 1977. "Descartes, the Skeptics and the Rejection of Vitalism in Seventeenth Century Physiology." *Studies in History and Philosophy of Science* 8: 1–28.

———. 2000. *See in press*.

Smart, J. J. and B. Williams. 1973. *Utilitarianism: For and Against*. Cambridge: Cambridge University Press.

Sober, E. (ed.). 1994. *Conceptual Issues in Evolutionary Biology*. 2nd ed. Cambridge, MA: MIT Press.

Sorenson, Theodore C. 1990. "Rethinking National Security." *Foreign Affairs* 69: 1–18.

Sournia, Jean-Charles. 1990. *A History of Alcoholism*. Oxford: Basil Blackwell.

Southern, Edward. 1975. "Detection of Specific Sequences Among DNA Fragments Separated by Gel Electrophoresis." *Journal of Molecular Biology* 98: 503–17.

Sperry, Roger W. 1983. *Science and Moral Priority*. Oxford: Blackwell.

Sperry, Roger W. 1988. "Psychology's Mentalist Paradigm and the Religion/Science Tension." *American Psychologist* 43: 607–13.

Star, Susan Leigh. 1986. "Triangulating Clinical and Basic Research: British Localizationists, 1870–1906." *History of Science* 24: 29–48.

———— and James R. Griesemer. 1989. "Institutional Ecology, 'Translations,' and Boundary Objects: Amateurs and Professionals in Berkeley's Museum of Vertebrate Zoology." *Social Studies of Science* 19: 387–420.

Steiner, George. 1989. *Real Presences.* London and Boston: Faber and Faber; Chicago: University of Chicago Press.

Steinfels, Peter. 1973. "Introduction" to"The Concept of Health." *Hastings Center Studies* 1: 3–88.

Stengers, Isabelle. 1987. "La Propagation des concepts." In: Isabelle Stengers (ed.). *D'une science à l'autre. Des Concepts nomades.* Pp. 9–26. Paris: Seuil.

Stent, G. 1981. "Strength and Weakness of the Genetic Approach to the Development of the Nervous System." In: Cowan 1981, 288–321.

Stent, G., and R. Calendar. 1978. *Molecular Genetics: An Introductory Narrative.* San Francisco: W. H. Freeman.

Stevens, William K. 1993. "Want a Room With A View? Idea May Be in the Genes." *The New York Times* (30 November): B5, B9.

Stock, Brian. 1983. *The Implications of Literacy: Written Language and Models of Interpretation in the Eleventh and Twelfth Century.* Princeton: Princeton University Press.

Stockel, S. 1995. "Frühsterblichkeit, Minderwertigkeit, Konstitution. Unterschiedliche Konzepte in der Pädiatrie in den 20er Jahren." *Monatschrift für Kinderheilkunde* 143: 1192–96.

Strohman, R. 1993. "Ancient Genomes, Wise Bodies, Unhealthy People: Limits of a Genetic Paradigm in Biology and Medicine." *Perspectives in Biology and Medicine* 37: 112–45.

————. 1994. "Epigenesis: The Missing Beat in Biotechnology?" *Biology and Technology* 12: 156–64; and 329.

Stump, Eleanore. 1995. "Non-Cartesian Substance Dualism and Materialism Without Reductionism." *Faith and Philosophy* 12: 505–31.

Sullivan, Francis A. 1996. *Creative Fidelity: Weighing and Interpreting Documents of the Magisterium.* New York: Paulist Press.

Sulloway, Frank. 1979. *Freud, Biologist of the Mind.* New York: Basic Books.

Sulston, J. E. and H. R. Horvitz. 1977. "Post-embryonic Cell Lineages of the Nematode *Caenorhabditis elegans.*" *Developmental Biology* 56: 110–56.

————, E. Schierenberg, J. G. White, and J. N. Thomson. 1983. "The Embryonic Cell Lineage of the Nematode *Caenorhabditis elegans.*" *Developmental Biology* 100: 64–119.

Suppe, F. 1984. "Curing Homosexuality." In: Baker and Elliston 1984, 391–420.

Sutherland, B. M. and A. D. Woodward (eds.). 1990. *DNA Damage and Repair in Human Tissues.* New York: Plenum.

Swinburne, Richard. 1986. *The Evolution of the Soul.* Oxford: Clarendon.

T. R. B. 1987. "The Tyranny of Beauty." *The New Republic* (12 October): 4.

Talbot, Eugene S. 1898. *Degeneracy.* New York: Charles Scribner's Sons.

Tauber, A. I. and S. Sarkar. 1992. "The Human Genome Project: Has Blind Reductionism Gone Too Far?" *Perspectives in Biology and Medicine* 35: 220–35.

Temkin, Larry. 1993. *Inequality.* New York: Oxford University Press.

Tesh, Sylvia. 1988. *Hidden Arguments: Political Ideology and Disease Prevention Policy.* New Brunswick, NJ: Rutgers University Press.

The Economist. 1993. "Multimedia: The Tangled Webs They Weave." (16 October): 21–24.

The Tablet. 1996. "Moment of Truth." Editorial. (10 August): 1039.

Third International Congress of Eugenics. 1934. *A Decade of Progress in Eugenics: Scientific Papers of the Third International Congress of Eugenics.* Baltimore: Williams & Wilkins.

Thomas, J. H. 1994. "The Mind of a Worm." *Science* 264: 1698–99.

Thompson, D'Arcy Wentworth. 1942. *On Growth and Form.* New edition. Cambridge: Cambridge University Press.

Tobach, E. and G. Greenberg (eds.). 1987. *Social Behavior and the Concept of Integrative Levels.* Hillside, NJ: Erlbaum.

Tolchin, Martin and Susan J. Tolchin. 1992. *Selling our Security: The Erosion of America's Assets*. New York: Knopf.

Touchette, N. 1995. "Genetics Tests and Services Taken to Task." *Nature Medicine* 1: 501.

Troland, Leonard Thompson. 1914. "The Chemical Origin and Regulation of Life." *The Monist* 24: 92–133.

———. 1917. "Biological Enigmas and the Theory of Enzyme Action." *American Naturalist* 51: 321–50.

Trombley, Stephen. 1988. *The Right to Reproduce: A History of Coercive Sterilization*. London: Weidenfeld and Nicolson.

Tryster, Hillel. 1991. "The Art and Science of Pure Racism." *Jerusalem Post International* (17 August): 13.

Turner, W. J. 1995. "Homosexuality, Type 1: An Xg28 Phenomenon." *Archives of Sexual Behavior* 24:109–34.

Turney, Jon. 1998. *Frankenstein's Footsteps: Science, Genetics and Popular Culture*. New Haven: Yale University Press.

Turpin, R. 1932. "La Stérilisation des inadaptés sociaux." *Revue anthropologique* 7: 14–15.

———. 1937. "Etude étiologique de 104 cas de mongolisme." In: R. Turpin (ed.). 1937. *Premier congrès de la fédération internationale latine des sociétés d'eugénique*. Paris: Masson.

Turpin, R. 1939. "Le Péril pour l'espèce des mutations germinales." *Semaine des hôpitaux de Paris* 13: 339–47.

———. 1941a. *Applications familiales de l'eugénisme*. Rapport au Comité National de l'Enfance, Paris.

———. 1941b. "De L'Importance médicale des phénomènes héréditaires." *Le Progrès médical* (21 June): 458–68.

———. 1951. *L'Hérédité des prédispositions morbides*. Paris: Presses Universitaires de France.

———. 1955. *La Progenèse*. Cours du Centre International de l'Enfance. Paris: Masson.

——— and A. Caratzali. 1934. "Remarque sur les ascendants et les collatéraux des sujets atteints de mongolisme." *La Presse médicale* (25 July): 1186–89.

U. S. Congress, Office of Technology Assessment [OTA]. 1988. *Mapping Our Genes. Genome Projects: How Big, How Fast?* Baltimore: Johns Hopkins University Press.

U. S. Council of Churches of Christ. Panel on Bioethical Concerns. 1984. *Genetic Engineering: Social and Ethical Consequences.* New York: Pilgrim Press.

U. S. Department of Energy. 1990. *Human Genome Program Report (The Human Genome Initiative of the U.S. Department of Energy).* Office of Energy Research; Office of Health and Environmental Research. SuDoc #E1.19:0382. Washington, D.C.: U.S. Government Printing Office.

U. S. Department of State. 1972. "Comments on National Security Council 49/1" (15 June 1949.) *Foreign Relations of the United States, 1947.* Vol. 6: *The Near East.* Pp. 871–73. Washington, D.C.: U.S. Government Printing Office.

U. S. Government. 1972. *Foreign Relations of the United States, 1947.* Vol. 6, *The Far East.* Washington, D.C.: U.S. Government Printing Office.

U. S. House of Representatives. 1983. "Hearings Before the Subcommittee on Investigations and Oversight of the Committee on Science and Technology." # 170, (16–18 November, 1982.) Washington, D.C.: U.S. Government Printing Office.

———. 1987. *Departments of Labor, Health and Human Services, Education, and Related Agencies Appropriation for 1988.* Hearings Before a Subcommittee of the Committee on Appropriations. Washington, D.C.: U.S. Government Printing Office.

———. 1988. *OTA Report on the Human Genome Project.* "Hearing before the Subcommittee on Oversight and Invstigations, of the Committee on Energy and Commerce." (27 April). Washington, D.C.: U.S. Government Printing Office.

———. 1989. *The Role of International Cooperation in Mapping the Human Genome.* "Hearing Before the Subcommittee on International Scientific Cooperation, of the Committee on Science, Space, and Technology." (19 October). Pp. 30–38. Washington, D.C.: U.S. Government Printing Office.

———. 1992. Committee on Government Operations. *Designing Genetic Information Policy: The Need for an Independent Policy Review of the Ethical, Legal, and Social Implications of the Human Genome Project.* 102nd Cong., 2nd sess. SuDoc #Y 1.1/8:102–478. Washington, D.C.: U. S. Government Printing Office.

U. S. National Security Council. 1948. "NSC 13/2," Report by the National Security Council with Respect to United States Policy Toward Japan (7 October 1948). In: U. S. Government 1972. Vol. 6, 858–62.

U. S. Senate. 1987. *U.S.-Japan Science and Technology Agreement.* Hearing before the Subcommittee on Science, Technology, and Space, of the Committee on Commerce, Science, and Transportation. Washington, D.C.: U.S. Government Printing Office.

———. 1988. *Biotechnology Competitiveness Act of 1987,* Senate Report 100–359, 25 May 1988. Washington, D.C.: U.S. Government Printing Office.

———. 1989. *Human Genome Initiative and the Future of Biotechnology.* Hearing before the Subcommittee on Science, Technology, and Space, of the Committee on Commerce, Science, and Transportation, 9 November 1988. 101st Cong., 1st sess.: SuDoc # Y4.C 73/7: S.hrg. 101–528.Washington, D.C.: U.S. Government Printing Office.

———. 1990. *Human Genome Project.* Hearing before the Subcommittee on Energy Research and Development of the Committee on Energy and Natural Resources, United States Senate, 101st Cong., 2nd sess.: SuDoc #Y4.En 2: S.hrg. 101–894. Washington, D.C.: U.S. Government Printing Office.

Vale, Jack R. 1973. "Role of Behavior Genetics in Psychology." *American Psychologist* 28: 871–82.

Van der Sluis, I. 1979. "The Movement for Euthanasia, 1875–1975. *Janus*s66: 131–72

Van Fraassen, Bas C. and Jill Sigman. 1993. "Interpretation in Science and in the Arts." In: George Levine (ed.): *Realism and Respresentation.* Pp. 73–99. Madison: University of Wisconsin Press.

Van Inwagen, Peter. 1993. *Metaphysics.* Boulder, CO: Westview Press.

———. 1995. "Dualism and Materialism: Athens and Jerusalem." *Faith and Philosophy* 12: 475–88.

Van Rootselaar, B. and J. F. Staal (eds.) 1968. *Logic, Methodology and Philosophy of Science III.* Proceedings of The Third International Congress for Logic, Methodology and Philosophy of Science. Amsterdam: North-Holland Pub. Co.

Van Till, Howard. 1996. "Basil, Thomas, and the Doctrine of Creation's Functional Integrity." *Science and Christian Belief* 8: 21–38.

Vatican. 1965. *Schema constitutionis pastoralis de ecclesia in mundo huius temporis: Expension modorum partis secundae.* Vatican City: Vatican Press.

———. 1980. *Declaration on Euthanasia.* Vatican City: Polyglot Press.

Vaughn, Stephen. 1990."Morality and Entertainment: The Origins of the Motion Picture Production Code." *Journal of American History* 77: 39–65.

Veatch, Robert. 1976. *Death, Dying, and the Biological Revolution.* New Haven: Yale University Press.

Vicedo, Marga. 1992. "The Human Genome Project: Towards an Analysis of the Empirical, Ethical, and Conceptual Issues Involved." *Biology and Philosophy* 7: 255–78.

Vogel, Gretchen. 1997. "Genetic Enhancement: From Science Fiction to Ethics Quandary." *Science* 277: 1753–54.

Von Ehrenstein, G. and E. Schierenberg. 1980. "Cell Lineages and Development of *Caenorhabditis elegans* and Other Nematodes." In Zuckerman 1980, 1–71.

Von Engelhardt, Dietrich. 1997. "Vitalism Between Science and Philosophy in Germany Around 1800." In: Cimino and Duchesneau, 1997. Pp. 157–74.

Von Neumann, John. 1951a. "The General and Logical Theory of Automata." In: von Neumann 1951b, 1–32.

———. 1951b. *Cerebral Mechanism in Behavior.* New York: Hafner Publishing Co.

Wachbroit, Robert. 1995. "Genetic Engineering: Human Genetic Engineering." In: Reich 1995, 937.

Waddington, Conrad. 1957. *The Strategy of the Genes: a Discussion of Some Aspects of Theoretical Biology.* With an appendix by H. Kacser. London and New York: Macmillan.

Wade, Nicholas. 1998. "It's a Three-Legged Race to Decipher the Human Genome." *New York Times* (23 June): B11.

———. 1999a. "The Genome's Combative Entrepreneur." *New York Times* (18 May): D1–2.

———. 1999b. "Scientists are Told of Gains in Effort to Decode Genome." *New York Times* (22 May): A11.

Waldholz, Michael, and Hilary Stout. 1992. "Rights to Life: A New Debate Rages Over the Patenting of Gene Discoveries." *The Wall Street Journal* (17 April): A1.

Walton, J. N. 1956. "Amyotonia Congenita." *Lancet* 270: 1023–27.

Warkany, J. 1960. "Etiology of Mongolism."*Journal of Pediatrics* 54: 412–19.

Warner, R. H. and H. L. Rosett. 1975. "Effects of Drinking on Offspring: An Historical Survey of the American and British Literature," *Journal of Studies of Alcohol* 36: 1395–1420.

Warren, Stafford L. 1948. *See unpublished.*

Warthin, Aldred Scott. 1930. *Creed of a Biologist.* New York: P.B. Hoeber.

Watson, James D. 1990a. "First Word." *Omni* 12 (9): 6.

———. 1990b. "The Human Genome Project: Past, Present, and Future." *Science* 248 (4951): 44–49.

———. 1992. "A Personal View of the Project." In: Kevles and Hood 1992, 164–73

——— and Francis Crick. 1953a. "Molecular Structure of Nucleic Acids: A Structure for Deoxyribose Nucleic Acid," *Nature* 171: 737–38.

———. 1953b. "Genetical Implications of the Structure of Deoxyribonucleic Acid," *Nature* 171: 966.

——— et al. 1987. *Molecular Biology of the Gene.* Menlo Park, CA Benjamin/Cummings.

———, and Norton Zinder. 1990. "Genome Project Maps Paths of Diseases and Drugs." *The New York Times* 140 (13 October): 14.

Weatherall, D. 1995. *Science and the Quiet Art.* New York: W.W. Norton.

Weaver, Warren. 1970. "Molecular Biology: Origins of the Term." *Science* 170: 582.

Weindling, Paul. 1989. *Health, Race and German Politics Between National Unification and Nazism 1870–1945.* Cambridge: Cambridge University Press.

Weismann, A. 1883. *Uber die Vererbung.* Jena: Gustav Fischer.

———. [1885] 1889. "Continuity of the Germ Plasm." In: A. Weismann. 1889. *Essays Upon Heredity and Kindred Biological Problems.* Translated and edited by E. B. Poulton. Oxford: Clarendon.

Weiss, K. M. 1993. *Genetic Variation and Human Disease: Principles and Evolutionary Approaches.* Cambridge: Cambridge University Press.

Weiss, Sheila F. 1987. *Race Hygiene and National Efficiency: The Eugenics of Wilhelm Schallmayer.* Berkeley: University of California Press.

———. 1990. "The Race Hygiene Movement in Germany, 1904-1945." In: Adams 1990, 8–68.

Welchman, Gordon. 1982. *The Hut Six Story: Breaking of the Enigma Codes.* New York: McGraw Hill.

Wells, H. G. 1905. *Modern Utopia.* London: Chapman & Hall.

Wertz, Dorothy. 1997. "Society and the Not-So-New Genetics: What Are We Afraid Of? Some Future Predictions From a Social Scientist." *Journal of Contemporary Health Law and Policy* 13: 299–346.

———, J. H. Fanos, and P. R. Reilly, 1994. "Genetic Testing for Children and Adolescents." *Journal of the American Medical Association* 272: 875–81.

Wexler, Nancy. 1992. "Clairvoyance and Caution: Repercussions From the Human Genome Project." In: Kevles and Hood 1992, 211–43.

White, J. G., E. Southgate, J. N. Thomson, and S. Brenner. 1986. "The Structure of the Nervous System of the Nematode *Caenorhabditis elegans.*" *Philosophical Transactions of the Royal Society* (Series B) 314: 1–340.

Whitmore, Todd. 1995. "The Common Good." In: McBrien 1995, 336.

Wiener, Norbert. 1950. *The Human Use of Human Beings: Cybernetics and Society.* Boston: Houghton Mifflin Co.

———. 1965. *Cybernetics: or Control and Communication in the Animal and the Machine.* 2nd edition. Cambridge, MA: MIT Press.

Wiggam, Albert E. 1924. *The Fruit of the Family Tree.* Garden City, NJ: Garden City Publishing.

Wigner, Eugene. 1967. *Symmetries and Reflections.* Bloomington: Indiana University Press.

Williams, Bernard. 1973. "A Critique of Utilitarianism" In: Smart and Williams 1973, 77–150.

Wilson, E. O. 1975. *Sociobiology: The New Synthesis.* Cambridge, MA: Belknap Press of Harvard University Press.

————. 1977. "Biology and the Social Sciences." *Daedalus* 106 (4): 127–40.

————. 1978 *On Human Nature.* Cambridge, MA: Harvard University Press.

————. 1998. *Consilience: The Unity of Knowledge.* New York: Knopf.

Wimsatt, W. 1976. "Reductionism, Levels of Organization, and the Mind-Body Problem." In: Globus and Savodnik 1976, 205–67.

————. 1981 "Robustness, Reliability and Multiple-Determination in Science." In: *Knowing and Validating in the Social Sciences: a Tribute to Donald Campbell.* Pp. 124–63. San Francisco: Jossey Bass.

Winner, Langdon. 1980. "Do Artifacts Have Politics?" *Daedalus* (Winter): 121–36.

Woese, Carl E. 1967. *The Genetic Code: The Molecular Basis for Genetic Expression.* New York: Harper & Row.

Wood, Robert W. 1989. Hearing before the Subcommittee on International Scientific Cooperation, of the Committee on Science, Space, and Technology, 19 October 1989. In: U.S. House of Representatives 1989.

Wood, W. (ed.). 1988. *The Nematode* Caenorhabditis elegans. Cold Spring Harbor: Cold Spring Harbor Press.

Woodcock, George. 1962. *Anarchism.* Cleveland: World Publishing.

Wright, John H. 1996. "Spirit and Matter." In: Himes and Pope 1996 127–39.

Wright, Robert. 1994. "Infidelity: It May Be in Our Genes." *Time* 144: 44–52.

Ycǎs, Martynas. 1969. *The Biological Code.* New York: John Wiley & Sons.

Yesley, Michael S. (ed.). 1993. *Bibliography: Ethical, Legal & Social Implications of the Human Genome Project.* Washington, D.C. U.S. Department of Energy Research.

———— and P. Ossorio (eds.). 1994. *Bibliography: Ethical, Legal & Social Implications of the Human Genome Project, Supplement* Washington, D.C.: U.S. Department of Energy Research.

Yoxen, Edward L. 1979. "Where Does Schrödinger's 'What is Life?' Belong in the History of Molecular Biology." *History of Science* 17: 17–52.

Zuckerman, B. (ed.). 1980. *Nematodes as Biological Models: Volume 1: Behavioral and Developmental Models.* New York: Academic Press.

Unpublished Materials

a. Collections of Archival Papers

Atomic Bomb Casualty Commission Papers. National Academy of Sciences Archives, Washington, D.C.

American Institute of Physics Oral History Archives. New York City.

Robert Cook-Deegan Collection. National Reference Center for Bioethics Literature, Georgetown University, Washington, D.C.

Max Delbrück Papers. California Institute of Technology.

Erwin Chargaff Papers. American Philosophical Society, Philadelphia, PA.

George Gamow Papers. Library of Congress, Washington, D.C.

Herman J. Muller Papers. Lilly Library, Indiana University, Bloomington, IN.

Linus Pauling Papers. Oregon State University Archives, Corvallis, OR.

Lionel Penrose Papers. University College London.

Donald Poulsen and Heinz Hermann Papers. Yale University, Department of Biology Archives.

Leo Szilard Papers. University of California, San Diego.

John Von Neumann Papers. Library of Congress, Washington, D.C.

Norbert Wiener Papers. Massachusetts Institute of Technology Archives, Cambridge, MA.

b. Dissertations

Bogin, Mary. 1990. "The Meaning of Heredity in American Medicine and Popular Health Advice, 1771–1860." Ph.D. diss., Cornell University.

Culotta, Charles. 1968. "A History of Respiration Theory, Lavoisier to Paul Bert." Ph.D. diss., University of Wisconsin.

Curtis, Patrick Almond. 1983. "Eugenic Reformers, Cultural Perceptions of Dependent Populations, and the Care of the Feebleminded in Illinois, 1909–1920." Ph.D. diss., University of Illinois, Chicago Circle.

Moss, Lenny. 1998. "What Genes Can't Do." Ph.D. diss., Northwestern University.

Peterson, James C. 1992. "An Ethical Analysis and Proposal for the Direction of Human Genetic Intervention." Ph.D. diss., University of Virginia.

Risse, Guenther. 1971. "The History of John Brown's Medical System in German During the Years 1790-1806." Ph.D. diss., University of Chicago.

Doyle, Richard. 1993. "On Beyond Living: Rhetorics of Vitality and Post Vitality in Molecular Biology." Ph.D. diss., University of California, Berkeley.

Durbin, R. M. 1987. "Studies on the Development and Organization of the Nervous System of *Caenorhabditis elegans.*" Ph.D. diss., Cambridge University.

Elks, Martin. 1992. "Visual Rhetoric: Photographs of the Feeble-Minded during the Eugenics Era, 1900–1930." Ph.D. diss., Syracuse University.

Gariepy, Thomas. 1990. "Mechanism Without Metaphysics: Henricius Regius and the Establishment of Cartesian Medicine." Ph.D. diss., Yale University.Kuepper, Stephen Louis. 1981. "Euthanasia in America, 1890–1960." Ph.D. diss., Rutgers University.

Mehler, Barry Alan. 1988. "A History of the American Eugenics Society, 1921–1940." Ph.D. diss., University of Illinois, Urbana.

c. Unpublished Papers, Lectures, and Works in Press

Anderson, Norman G. and N. Leigh Anderson. 1986. "A Program for Large-Scale Analysis of Nucleotide Sequences (P.L.A.N.S.), *Genome Sequencing Workshop*, 3-4 March 1986, Santa Fe, NM, Folder "Santa Fe I." Robert Cook-Deegan Collection.

Anon. 1945. "Non-Technical Record of the History of the Medical Program of the Manhattan Project." Typed Manuscript, Oak Ridge National Laboratory Public Reading Room, ORF-00752.

Anon. 1984. "Unpublished agenda for DOE/ICPEMC Meeting, Alta, Utah 10–13 December, 1984." (Personal Communication, Dr. Mortimer Mendelsohn, LLNL).

Anderson, Norman G. and N. Leigh Anderson. 1986. "A Program for Large-Scale Analysis of Nucleotide Sequences (P.L.A.N.S.), *Genome Sequencing Workshop*, 3-4 March 1986, Santa Fe, NM, Folder "Santa Fe I." Robert Cook-Deegan Collection.

Anon. 1945. "Non-Technical Record of the History of the Medical Program of the Manhattan Project." Typed Manuscript, Oak Ridge National Laboratory Public Reading Room, ORF-00752.

Anon. 1984. "Unpublished agenda for DOE/ICPEMC Meeting, Alta, Utah 10–13 December, 1984." (Personal Communication, Dr. Mortimer Mendelsohn, LLNL).

Ariew, Roger. 1998. "Descartes and the Late Scholastics on the Order of the Sciences." Unpublished paper, Second Biennial History of Philosophy of Science (HOPOS) Meeting, University of Notre Dame, March 1998.

Atlan, Henri. 1994. "ADN: programme ou données (ou: Le Génétique n'est pas dans le gène)." In: *1er Congrès mondial médecine et philosophie*. (30 May).

Atlan, Henri. 1998. "DNA: Program or Data?" Unpublished MS, submitted to *Biology and Philosophy*.

Brenner, Sydney. 1956. "On the Impossibility of All Overlapping Codes." Note to the RNA Tie Club, September 1956, Delbrück Papers, Box 30.5, Caltech.

Burian, Richard, and Jean Gayon (eds.). *Conceptions de la science, hier, aujourd'hui et demain. Colloque d'hommage à Marjorie Grene*. Brussels: Ousia (in press).

California Institute of Technology, Jet Propulsion Laboratory. 1961. "Recent Results in Comma-Free Codes." Research Summary, 15 February.

Clarke, W. Norris. "St. Thomas on the Immediate Creation of the Human Soul by God." Unpublished manuscript.

Cohen, Cynthia B. 1985. "The Treatment of Impaired Newborns in American History: Implications for Public Policy." Unpublished MS, Department of Philosophy, Villanova University.

Crick, Francis. 1955. "On Degenerate Templates and the Adapter Hypothesis." Undated note to the RNA Tie Club.

Darling, George. 1958. "Atomic Bomb Casualty Commission Annual Report, 1 July 1957–30 June 1958," ABCC Records NAS.

Falk, Raphael. 1995. "The Gene: From an Abstract to a Material Entity and Back." Unpublished MS.

Fortun, Michael. 1992. "Money Talks: Negotiating the Human Genome Project in the U.S. Congress." Unpublished paper, History of Science Society.

Frézal, J. 1959a. "Titres et travaux." Unpublished MS, Bibliothèque Interuniversitaire de Médecine, Paris.

Gaudillière, Jean-Paul. 1990. "French Strategies in Human Genome Research." Paper presented at the Conference on the Human Genome Project, Harvard University, June.

Kay, Lily. 2000. *Who Wrote the Book of Life: A History of the Genetic Code*. In Press, Stanford: Stanford University Press.

Kevles, Daniel. 2000. "The Ghost of Galton." In: Signer and Wegs 2000.

Lamy, M. 1972. "Titres et travaux." Unpublished MS, Bibliothèque Interuniversitaire de Médecine, Paris.

Lawrence, J. H. "The Effects of Neutrons on Tissues." Two page memo-circular, Lawrence Papers, Bancroft Library, Berkeley CA. Container 1.

Mendelsohn, Mortimer. 1985. "Informal Report of Meeting on DNA Methods for Measuring the Human Heritable Mutation Rate." Lawrence Livermore National Laboratory Internal Report UCID-20315.

Muller, H. J. "Message to University Students Studying Science." Translated from *Kagaku Asaki* 11, no. 6 (1951): 28–29. Muller Papers. Lilly Library, Indiana University, Bloomington, IN.

Penrose, Lionel. 1964. "The Causes of Mongolism." Unpublished Lecture, Queen Square Hospital, April 1964. Penrose Collection, file 62/4, University College, London.

Rimoin, David L. 1994. "Statement From the Joint Committee on Professional Practice, July 1994." Memorandum Circulated to American College of Medical Genetics by the President, 1 August 1994.

Rydell, Robert W. 1984. "Eugenics Hits the Road: The Popularization of Eugenics at American Fairs and Museums Between the World Wars." Unpublished paper, History of Science Society.

Schaffner, K. F. 1999. "Complexity and Research Strategies in Behavioral and Psychiatric Genetics" prepared for UTMB-Galveston/ MacArthur Culture and Biology Project. To be published in R. Carson and M. Rothstein (eds.). *Culture and Biology*. Baltimore: Johns Hopkins University Press.

————. 2000. "Genetic Explanations of Behavior: Of Worms, Flies and Men." To appear in: D. Wasserman and R. Wachbroit (eds.). *Genetics and Criminal Behavior: Methods, Meaning, and Morals*. Cambridge: Cambridge University Press.

Signer, M. (ed.). 2000. *Humanity at the Limit: The Impact of the Holocaust Experience on Christians and Jews.* Indiana University Press (in press).

Sloan, Phillip R. 2000. "Teleology and Form Revisited." In: Burian and Gayon (in press).

Terry, Jennifer. 1994. "The Seductive Power of Science in the Making of Deviant Subjectivity." Unpublished paper, History of Science Society.

Warren, Stafford L. 1948. "Address to American Pharmaceutical Association, 12 August 1948." Warren Papers, UCLA Special Collections, 987, Box 57, Folder 10.

f. Electronic Sources

Avery, L. 1996. Webpage http://eatworms.-swmed.edu/.

CRG (=Council for a Responsible Genetics). 1996. "Do Genes Determine Whether We Are Lesbian, Gay, Bisexual, or Straight?" Website http://essential.org/crg

Lockery, S. 1996." Statement of Research Program." Web Homepage, http://chinook.uoregon.edu.

Palfreman, Jon. 1989. *Decoding the Book of Life*. Written and produced for the *Nova* Series. 58 min. Coronet Film and Video. Videocassette.

Index

Abbott Laboratories, 48
abortion: and economic
  coercion, 239; as existence
  discrimination, 427; eugenic,
  434; and Human Genome
  Project, 235; Japanese
  attitudes toward, 435
Adams, William, 40
adult-onset disorders 266
Aebersold, Paul, 43, 50; as
  director of isotope program,
  44
AEC (Atomic Energy
  Commission): as emulating
  Manhattan Project, 41;
  Atomic Medicine Division,
  55; founding of, 36; funding
  for medical research, 41;
  integrated with UCLA
  medical school, 41– 42;
  Isotope Division, goals of, 47;
  isotope program milestones,
  45; university contracts for
  research, 39
AFM (*Association française
  contre les myopathies*), 83,
  91; and genetic data collection
  in France, 85

African-Americans: depicted as
  defectives, 198
aggression: genetic causes of,
  318
Allen, Herbert: and radioactivity
  scan, 62
Alta Summit (1984), 145; and
  Human Genome Project, 11–
  15, 146; list of participants,
  14; meeting agenda of: 13
Alvarez, Luis, 110
Alzheimer's disease: genetic
  causes of, 307, 318
American Board of Nuclear
  Medicine, 63
American values: and the
  Human Genome Project, 158
amniocentesis, 82, 93; and
  Down's Syndrome, 78; and
  genetic screening 266
androgen: fetal, 316
anencephaly, 434
Anger scintillation camera, 62
Anlagen: and inheritance, 280
anti-codons, 128
anti-egalitarianism, 245
anti-reductionism: and DNA,
  346

experimenter's regress,
65, 71, 78; and clinician's
regress, 94; Harry Collins on,
64
expertise: and eugenics, 206
explanation: and reduction, 16–
17, 302, 373; causal, 303;
network, 312; oligogenetic,
312
explanatory extension: Philip
Kitcher on, 304
faith: and reason, 423; as
grounded on narrative, 421
familial mongolism, 69
FAP (fixed-action pattern), 305
fascinoma, 80
fear circuit: neurobiological
modeling of, 337
Federal Technology Transfer
Act, 145
Fels Institute: behavioral studies,
333
Feynman, Richard, 112
Finnis, John: on human goods,
418
Fisher, Irving: and eugenics,
189, 191, 207
Fisher, Ronald A.: and eugenics,
204
Fisher, Irving, 205
focal levels: and emergence, 347
Ford, Charles, 69
Foucault, Michel, 105
fourfold way: of Lander and
Schork, 321
Frankenstein: myth of, 1
freedom: and moral action, 365
French Human genetics:
scientific style of, 93
Friedell, Hymer, 55
frontier metaphors: and the
Human Genome Project, 162

Fuchs, Josef: on bioethics and
theology, 422
functional genes: and
chromosomes, 276
galactosemia, 83
Galilei, Galileo, 4; on the book
of nature, 102; and scriptural
interpretation, 367
Galton, Francis, 223, 293; and
"beauty map" of Britain, 190;
and eugenics, 190, 435; and
stirp theory, 292
Gamow, George: on genetic
code, 103, 110–111
Garrod, Archibald: on inborn
errors of metabolism, 443
Gassendi, Pierre, 3
Geiger, Hans, 30
gender: and genetics, 397
gene activation, 292
gene concept 17–18, 291; and
abnormality, 432; and
causation, 277, 291; and
chromosomes, 274; defective,
158; developmental concept
of, 18; difficulties in defining
273; epistasis, 329; functional
definition of, 274; hardening
of gene concept, 17; Hermann
J. Muller on, 283; and
hierarchy relations, 364;
historical meaning of, 275; in
transmission and molecular
genetics, 275; instability of,
275; inutility of, 274; as law-
code, 285; as mechanism of
control, 107; as moral
concept, 169; as organism,
277; and organismic
dynamics, 275; and
organisms, 278; operational
definition of, 295; and